The Conquest of Viruses

Van G. Wilson

The Conquest of Viruses

A History of Viral Vaccines

Van G. Wilson
Department of Microbial Pathogenesis & Immunology, College of Medicine
Texas A&M University
Bryan, TX, USA

ISBN 978-3-031-87561-8   ISBN 978-3-031-87562-5   (eBook)
https://doi.org/10.1007/978-3-031-87562-5

© The Editor(s) (if applicable) and The Author(s), under exclusive license to Springer Nature Switzerland AG 2025

This work is subject to copyright. All rights are solely and exclusively licensed by the Publisher, whether the whole or part of the material is concerned, specifically the rights of translation, reprinting, reuse of illustrations, recitation, broadcasting, reproduction on microfilms or in any other physical way, and transmission or information storage and retrieval, electronic adaptation, computer software, or by similar or dissimilar methodology now known or hereafter developed.
The use of general descriptive names, registered names, trademarks, service marks, etc. in this publication does not imply, even in the absence of a specific statement, that such names are exempt from the relevant protective laws and regulations and therefore free for general use.
The publisher, the authors and the editors are safe to assume that the advice and information in this book are believed to be true and accurate at the date of publication. Neither the publisher nor the authors or the editors give a warranty, expressed or implied, with respect to the material contained herein or for any errors or omissions that may have been made. The publisher remains neutral with regard to jurisdictional claims in published maps and institutional affiliations.

Cover illustration: Copyright information: © yurik art / stock.adobe.com / Generated with AI

This Springer imprint is published by the registered company Springer Nature Switzerland AG
The registered company address is: Gewerbestrasse 11, 6330 Cham, Switzerland

If disposing of this product, please recycle the paper.

As always, this book is dedicated to my loving wife. Her support and encouragement keep me moving forward and ensure that I get to the end of every project.

Preface

Cherish your doubts, for doubt is the handmaiden of truth. Doubt is the key to the door of knowledge; it is the servant of discovery. — Robert T. Weston.

I'm a vaccine advocate; I'll make no pretense about that. My background is in microbiology, specifically virology, and I've spent over 50 years working with and studying viruses. I am not a vaccine developer as my own work focused on understanding the biochemistry and molecular biology of certain viruses. However, creating vaccines against viruses is a major emphasis in the field of virology, and I've followed these efforts closely throughout my career. I've known several vaccine researchers well and met many others at conferences and scientific meetings. I've talked with them, heard their presentations, and read their published papers. I understand the science, and I've seen both the good and the bad in vaccines. There are clearly risks with every vaccine, but those risks are quite small and the benefits to the individual and to society from large-scale vaccination programs have been huge. Reducing viral infections through vaccination and preventing all the associated illnesses and deaths has been a triumph that science and medicine accomplished over the last 100 years. Millions and millions of lives have been saved, many of them children who otherwise would have died needlessly.

The average American receives between 50 and 100 vaccine doses in their lifetime, but few people know much about the nature of the vaccines they receive, the science behind them, or the history leading to their creation. I conceived this book many years ago as a tool to help educate the public; however, other obligations and projects always kept it as an idea rather than a reality. Contributing to the delay was the major inertia that must be overcome before finally embarking on a task this big and setting down the first words on a page. For years, whenever I thought about starting Conquest, the research and the writing seemed overwhelming on top of my already busy daily schedule. It was only after completing my more general book on viruses (*Viruses: Intimate Invaders*; Springer, 2022) that I felt emboldened to tackle this theme of viral vaccines. Not only did the successful completion of one book enhance my confidence about actually converting my idea into a finished product, but during the time writing *Intimate Invaders*, the world changed dramatically with the arrival of SARS-CoV-2 and the COVID-19 pandemic. That pandemic brought worldwide pain, misery, and death with the only real hope of controlling the virus being the successful development of vaccines. In a remarkable scientific tour de force a novel mRNA-based technology brought incredibly successful vaccines to the public in less than a year after the SARS-CoV-2 infections exploded around the world.

As the massive vaccine rollouts began in early 2021, I was initially overjoyed by what I thought was a pandemic-ending achievement. I believed that the public would vigorously embrace the vaccines and that many countries would achieve vaccine-induced herd immunity long before the end of 2021. However, as the year progressed my initial enthusiasm about quickly returning to normalcy dissipated and was replaced by incredulous disbelief and dismay over the widespread vaccine hesitancy. As an educator, I've studied vaccines and taught about them for decades to graduate students and medical students, and I know the value, safety, and efficacy of vaccines. Not only have vaccines saved countless millions of people from disease and death, they have even allowed us to eliminate the horrible smallpox virus from our planet so that no future human will ever experience that dreaded disease. Given the long and triumphant history of vaccines, I never would have anticipated so much resistance to a life-saving and pandemic-ending treatment. I was certainly aware that there were adamant anti-vaxxers, but I always thought they were the fringe few percent of the population, not the huge numbers that refused or delayed their COVID vaccination. While I understand and respect caution with any new medical advance, given the dire and dangerous nature of the pandemic, a small potential vaccine risk was vastly outweighed by the very real health risks of SARS-CoV-2 infection. Additionally, once the mRNA vaccine distribution began and millions were vaccinated worldwide it quickly became clear that adverse effects were extremely rare. Given the efficacy and safety data, I expected much of the initial hesitancy to vanish yet it did not. It puzzled me that many people were so afraid of a relatively simple vaccine whose mode of action was straightforward and well-understood while they seemingly weren't afraid of a complex, poorly understood, and decidedly lethal new virus. Many reasons for the remaining resistance to COVID-19 vaccination have been presented, from the political to the religious, but I suspect that a general lack of knowledge and understanding of vaccines underscores much of the concern that individuals have. Although not my initial intention, I now hope that this book can provide helpful information about vaccines that may alleviate unfounded concerns and replace dangerous misinformation with truthful evidence of the power and benefits of vaccines. By presenting the history of vaccine development and the sagas of the scientific pioneers who created these wonderful anti-viral prophylactics, I hope to imbue in the reader greater appreciation and respect for these incredibly potent additions to our modern medical armamentarium. I know that there will always be vaccine resisters, but maybe this book will help persuade some reluctant individuals that vaccines are just another effective medical treatment and are just as safe as any of the myriad drugs that people take for other diseases.

Bryan, TX, USA   Van G. Wilson

Competing Interests The author has no competing interests to declare that are relevant to the content of this manuscript.

Contents

1	**Vaccines in the Twenty-First Century**	1
	Let Us Begin	1
	The Nature of Viruses	2
	Our Immune Defenses	6
	A Vaccine Primer	10
	References	15
2	**Smallpox and Immunity: Success Precedes Science**	17
	The Speckled Monster—A Callous Killer	17
	Elusive Origins	20
	Variolation	24
	Cowpox and Vaccination	27
	Vaccinia Emerges	33
	The Death of Smallpox	35
	References	36
3	**Rabies: From Attenuation to Inactivation**	39
	The Microbial Paradigm of Disease	39
	The Birth of Vaccines	44
	The Mad Dog Disease	47
	The Rage Virus and Its Neural Death March	49
	Origin of the Rabies Vaccine	53
	Post-Pasteur—Rabies Vaccine Evolution	56
	References	59

Contents

4 Yellow Fever: America Goes to War — 63
- The Scourge of the Americas — 63
- A Disease That Flies — 67
- And a Vector Shall Carry It — 69
- The Agent and the Vaccine — 75
- The Virus and the Future — 80
- References — 82

5 Influenza: An Elusive and Evasive Foe — 87
- An Ill Wind — 87
- Influenza Hunters — 91
- Stumbling Towards A Vaccine — 96
- Drifting and Shifting—The Vaccine-Confounding Biology of Influenza Viruses — 100
- Eighty Years of Vaccine Evolution — 107
- A Final Note — 112
- References — 113

6 Poliovirus: An Insidious Plague — 117
- The Rise of Poliomyelitis — 117
- False Starts and False Hopes — 123
- The Path Forward At Last — 129
- Picornaviruses Revealed — 138
- The Salk Vaccine—Immediate Success and Unforeseen Problems — 141
- The Sabin Vaccine—Greater Success but a New Problem — 144
- The Poliovirus Endgame in the Post-Vaccine World — 148
- References — 150

7 Measles/Mumps/Rubella (MMR): The Childhood Trifecta — 155
- A Fabricated Controversy — 156
- Measles, The "First Disease" — 159
- Enders' Game — 160
- Handoff to Hilleman for a Touchdown — 163
- The Measles Endgame — 165
- A Measles Puzzle — 167
- Lumps and Mumps — 168
- Out of the Mouths of Babes — 169
- Close But Not Quite for Mumps — 173

	Rubella, The German Measles	175
	An Unsuspected Danger	176
	A Fateful Epidemic	178
	A Culture of Excellence	179
	References	183
8	**Hepatitis B Virus: Blood, Sex, and Drugs**	187
	A Persistent Enemy	188
	A Tale of Two Viruses	190
	A Mystery Down Under	194
	Chronic Carriers, Cancer, and a Vexing Causation	199
	A New Vaccine Paradigm	203
	Beyond Recombivax HB	207
	References	210
9	**Hepatitis A Virus: Feces, Food, and Fomites**	215
	The Other Hepatitis Virus	216
	Yellow Berets to the Rescue	217
	Polio's Cousin	222
	The Path to the HAV Vaccine	224
	HAV Vaccines in the Twenty-First Century	230
	References	232
10	**Herpes Varicella Zoster: The Other Pox (Chickenpox)**	237
	Chickenpox, Creeping Among Us	238
	The Duality of Herpes Viruses	240
	A Father's Love	246
	To Vaccinate or Not to Vaccinate	250
	Saint Anthony's Fire (Zoster)	256
	References	259
11	**Rotavirus: The Democratic Virus**	265
	A Wheel of Misfortune	266
	A Virus Runs Through Them	269
	RotaShield: Baby Steps and a Stumble	273
	Back into the Fray	279
	Pitfalls, Questions, and Opportunities—The Push for the Future	282
	References	285

Contents

12	**Human Papillomaviruses: An Ancient Enemy**	293
	Our Silent Passengers	294
	A Bumpy Past	298
	Sex, Cancer, and HPVs	301
	Did Our Cousins Give Us an STD?	307
	VLPs to the Rescue	309
	Ten Years On—The HPV Vaccine Success Story	315
	References	316
13	**SARS-Coronavirus-2: The Unexpected Plague**	323
	A New Threat	324
	A Spiking Disaster	326
	From Colds to Killers	329
	The Ascendency of RNA	332
	The mRNA Vaccine Revolution	337
	But Are They Safe?	340
	References	343
14	**Respiratory Syncytial Virus: A Shape-Shifting Adversary**	351
	The Silent Pandemic	351
	A Tragic Start	355
	About F'ing Time	358
	The Homestretch	363
	References	366
15	**The Fight Continues: Virus Without Vaccines**	373
	Three Centuries of Progress	374
	The Next Challenge	377
	The Other Herpesviruses	378
	The Other Hepatitis Virus	383
	A Persistent STD—Human Immunodeficiency Virus (HIV)	384
	The Tropical Diseases	386
	References	388
Epilogue		393
Index		397

1

Vaccines in the Twenty-First Century

Keywords Acquired immunity • Attenuated vaccine • Inactivate vaccine • Innate immunity • Subunit vaccine • Virion

Abbreviations

HPV	Human papillomavirus
MHC	Major histocompatibility complex
mRNA	Messenger Ribonucleic Acid
SARS-CoV-2	Severe acute respiratory syndrome-coronavirus-2
T_C	Cytotoxic T cells
T_H	T helper cells
T_{regs}	Regulator T cells
WHO	World Health Organization

Let Us Begin

This is a tale of pathogenic viruses and the men and women who tried to fight them through the creation of vaccines. The story of vaccines is complicated as it requires an understanding of both viral biology and human immunology, and that knowledge only slowly developed in the twentieth century. Science tries to be rational, but it isn't always linear. Sometimes critical advances stem from seemingly random or chance occurrences rather than thoughtful and cogent planning. A successful scientific story often includes a chaotic mixture of perceptive intelligence, perseverance, unrelenting effort, and occasionally good luck. Imagine the global body of scientific knowledge as an enormous and growing virtual book that contains the sum of all we've learned and discovered. Many disparate types of research, often from different countries and continents, will weave and intertwine as the tale unfolds. Multiple investigators pursue their individual plots without knowing how their efforts will contribute to the overall tome. Some scientists will write whole chapters in the book of knowledge whereas others may contribute only a few lines or words. And as in any book, some characters are heroes, and some are villains. Yet from these combined efforts ultimately emerges clarity, understanding, and a story that never before existed in the minds of men. Virologists, immunologists, molecular biologists, physicians, clinical trial volunteers, and many others all made important contributions to this tale, and we will explore vaccination from its crude beginnings to the sophisticated molecular vaccines that fought the COVID-19 epidemic. The chronicle of vaccines isn't finished yet, but we know the ending we seek, the conquest of all harmful human viruses through vaccination.

Sometimes when recounting a complicated story to a willing listener, it's better to begin with a brief overview of the ending before attempting to present the complex peregrinations of the tale; it can be easy to get lost in the details if you don't know where you are headed. Likewise, when explaining science, it's often best to start with existing knowledge before trying to describe the hazy beginnings of a field and the tangled paths that led to our current state of understanding. In the twenty-first century, we have a large and varied armamentarium of vaccine types, extensive knowledge of viruses, and a sophisticated understanding of our immune system, most of which accrued through major scientific advances over the last 150 years. However, before sharing the stories, the lives, and the discoveries that led to remarkable advances, brief introductions to viruses, the immune system, and vaccines are necessary. Hopefully, these primers will provide foundational support to help orient the reader and enhance appreciation of the remarkable backstories that led to our current world where many once-common viral scourges have been controlled or even eliminated.

The Nature of Viruses

For the uninitiated, a brief overview of the virus biology is critical to appreciate the accounts of vaccine development in subsequent chapters. Most people know that viruses are very small and that they cause certain diseases, but beyond those two concepts, knowledge of the true nature and properties of these amazing entities is likely limited. Some readers may believe that viruses are generally like bacteria or even our own cells, just smaller versions, and nothing could be further from the truth. Viruses are unique and distinct agents of disease that are nothing like cellular organisms. Indeed, some would argue that viruses are not even alive since they lack most of the physical and biochemical features of living organisms, and instead are simply self-replicating molecular machines. To fully grasp the issues and challenges of viral vaccine development requires an understanding of these intrinsic differences between viruses and all other cellular life forms.

Cells, from bacteria to humans, are complex structures that grow and divide to produce new cells. All cells possess a membrane structure that separates the external environment from the interior of the cell (the cytoplasm). Within every cell is the DNA genome of the organism that contains the genetic blueprint for all the cell's functions, including reproductive ability. Contained in the DNA sequence are the genes that provide the instructions to produce either proteins or molecules called RNA. Proteins are the workhorse of the cell and have many diverse functions. Some proteins form the structures of the cells, some are enzymes that catalyze biochemical reactions, some regulate the activity of other proteins or DNA, and some transmit signals within and between cells. Collectively these hundreds to thousands of proteins enable the cell to metabolize nutrients to produce energy and raw materials, use the raw materials to synthesize new components so that the cell can grow larger, and ultimately help the cell to reproduce by dividing into two daughter cells (a process called binary fission). RNA is related to DNA and both molecules are a type of compound called a nucleic acid. Like proteins, the RNAs encoded by the DNA are also quite diverse in function, but we'll only focus now on one type, messenger RNA (mRNA). mRNA is the intermediate between DNA and protein. When a gene for a protein is activated, the DNA sequence is copied into a mRNA molecule by a process called transcription. This mRNA is then transported to cellular structures called ribosomes where the RNA sequence is read and translated into the amino acid sequence of the protein. Thus, all living cells contain both types of nucleic acid, DNA and RNA, as each molecule provides essential functions to the cell. To summarize, cells are dynamic, membrane-bound entities that use DNA as their genetic material and can synthesize both RNA and protein. Cells perform metabolism to survive and/or grow larger, and with some limited exceptions, eventually divide to reproduce themselves.

In contrast to cells, viruses are inert particles. The simplest viruses are merely pieces of nucleic acid surrounded by a protective protein sphere known as a capsid, and we call an individual virus particle a virion. Like cells, some viruses do use DNA as their genetic material, but many viruses uniquely use RNA for their genomes. Consequently, viruses typically are categorized as either a DNA virus or an RNA virus depending on which type of nucleic acid is present in the virion. Because both DNA and RNA are relatively fragile molecules that can be easily damaged, capsids have evolved to provide a barrier that shields the nucleic acid from environmental assaults. Additionally, the capsid functions to help virions attach to and penetrate the cells that they infect. Larger and more complex virions can have hundreds of proteins associated with and within the capsid and may even have a membrane layer surrounding the capsid. Still, both the simple and the complex virions are merely passive entities that exhibit no activity when outside their target cell. Virions lack ribosomes so they cannot produce new proteins and they have no metabolic systems so they cannot generate energy or produce the raw materials for the synthesis of new virions. Without these abilities, virions in the environment do not reproduce and merely exist while they slowly degrade until they lose their infectious capacity.

Given all the things they lack, how do viruses reproduce? The key element is that viruses are obligate intracellular parasites and must be inside a susceptible cell to replicate themselves. While the details vary greatly from virus to virus, all viruses undergo a life cycle that includes seven steps: attachment, penetration, uncoating, gene expression, genome replication, packaging, and release. Attachment is the initial step where the virion forms a highly specific bond with an appropriate target cell. For the bond to occur, the cell must have a receptor on its surface that interacts strongly with a protein on the virion surface, referred to as the viral attachment protein. Cells lacking the receptor cannot be infected by that particular type of virus. If the cell has the appropriate receptor, then the viral attachment protein-cell receptor complex can form. This specific binding of the virion to the cell then triggers the subsequent penetration step. During penetration, the virion is internalized through the cell membrane and is deposited in the cytoplasm of the cell. The entry process culminates with the uncoating step which is the full or partial dissolution of the virion structure to release the virion nucleic acid. This uncoating is essential to release the nucleic acid from its sequestration within the virion where it is inaccessible to the cellular environment. Once released, the viral nucleic acid can now access the cellular biochemicals that it needs to produce viral mRNAs (the gene expression step). These viral mRNAs are translated into proteins via the cell's ribosomes. Some of the viral proteins are enzymes and regulatory factors that are needed for duplicating the viral genome. Once these critical viral proteins accumulate, they will use the incoming viral nucleic acid as a template to mass-produce identical new copies of the genome (the genome replication step). Other viral proteins being synthesized are classified as structural proteins and will assemble to form the new virions. Each newly formed virion will be packaged with a viral genome and these progeny virions accumulate within the cell, often reaching hundreds to thousands of virions per cell. Eventually, the cell typically lyses to release the progeny virions to go out and infect surrounding cells or be shed by the infected person and transmitted to other individuals.

As described above, the viral reproductive process is nothing like cellular reproduction. Living cells metabolize nutrients and use the energy and raw materials generated to both grow the cell larger and larger and to make a second copy of its genome. Once the cell is roughly doubled in size, the cell will divide into two daughter cells that each receive one genome copy. Unlike cells, individual virions do not enlarge in size and do not divide. Instead, viral reproduction is more akin to producing automobiles. Once it infects a cell, the virus takes over the cell and converts it into a virion manufacturing plant. The virus uses the cell's machinery to produce large quantities of its proteins and nucleic acid and then pieces them together like parts on an assembly line. The cell will continue to produce the viral components and construct new virions until it dies. In this way, a single virion entering a cell can produce thousands of offspring just like an automotive plant can produce thousands of cars.

The viral life cycle is intimately connected to the two major mechanisms of viral disease: cytopathogenesis and immunopathogenesis. Cytopathogenesis is the cellular injury caused directly by viral infection while immunopathogenesis refers to damage caused by our immune system. As the virus usurps the cellular functions during infection and redirects the cell's efforts to produce new virions, the cell becomes unhealthy and less able to perform its normal physiological function. As the number of infected cells grows during an illness, more and more cells are impaired, and eventually, organ function is impacted. As the cells die to release new virions this organ dysfunction is amplified as tens of thousands of cells succumb. For example, viral infection in the gastrointestinal tract can damage and kill cells lining the intestines that are critical for nutrient and water absorption, leading to pain and diarrhea. Similar viral-induced damage in other organ systems causes typical disease symptoms such as nasal congestion with respiratory viruses and skin lesions with herpes simplex virus. This cellular damage and destruction caused by viral reproduction constitutes viral cytopathogenesis and is a major contributor to the disease presentation of each virus. Coupled with this cytopathogenesis, many disease symptoms associated with viral infections stem from immunopathogenesis where our immune system attacks the infected cells.

Virus-infected cells display altered patterns that are recognized as foreign by our immune system leading to an aggressive defense. Part of the defense involves the production and release of soluble factors, for example, interferon which has strong antiviral activity. Unfortunately, interferon also causes side effects such as fevers and body aches that become part of the symptoms we feel during viral infection. Additionally, there are immune cells that can recognize and destroy viral-infected cells. This helps to eliminate these virion factories and is an important part of the recovery process. Unfortunately, this immune-mediated cell destruction also contributes to the overall loss of cells in an organ system. So whether cells die due to the virus directly or to our immune response, the net effect is organ damage that persists and will cause symptoms until the virus is controlled and new cells are regenerated.

This brief overview of virus structure, life cycle, and disease process is only intended to provide a modest foundation for understanding subsequent chapters. Since the year 1901, well over 200 human viruses have been identified and we continue to find an average of two new ones per year (Woolhouse et al. 2008). Every different virus has its own unique and complex biology that is a field of study unto itself. As we explore the development of viral vaccines, some of the complexities of individual viruses will be addressed, but much is beyond the scope of this book. For the interested reader, a more detailed presentation of general virology can be found in my previous book, *Viruses: Intimate Invaders* (Wilson 2022).

Our Immune Defenses

The underlying principle for all viral vaccinations is to use a harmless form of the pathogen to elicit an immune response that protects against the virulent form. This principle was first demonstrated in the late 1700s when deliberate infection with cowpox was first used to protect against deadly smallpox (Chap. 2). Without any comprehension of the immune system or how immunity worked, the concept that infection and recovery could provide future resistance to the same or related pathogens was already circulating in both the general public and the medical community. Early vaccine pioneers eagerly applied this concept, but their trial-and-error attempts were mostly stymied by their lack of knowledge about immunity and viruses. It wasn't until the 1900s when tools to study viruses were developed and our understanding of immunity exploded that there was real progress in viral vaccine development. We now know that our immune systems are amazingly complex and dynamic with the ability to protect us against everything from viruses to cancer. As we continue to unravel the secrets of how immunity works, the opportunities to apply this knowledge to new, safer, and more effective vaccines are enormous (Chap. 15).

The overall immune system is comprised of three levels: intrinsic, innate, and adaptive (also known as acquired) (Nicholson 2016). The intrinsic and innate levels provide important early defenses against various pathogens, including viruses. However, for vaccination, it is adaptive immunity that is the primary target (Fig. 1.1). The human adaptive immune response consists of two branches, the humoral and the cellular, and ideally, a vaccine would activate both branches for the maximum protective effect. For our purposes, discussion of the humoral immune system can be restricted to antibodies, the product of circulating white blood cells known as B lymphocytes. Antibodies are proteins, and each B cell makes a single unique type of antibody. Collectively, there are millions of B cells in our bodies, so our repertoire of different antibodies is huge. On each antibody is a region (the Fab domain) that can bind a foreign substance known as an antigen. Antigens can be different types of molecules, but one common type of antigen is proteins such as those found on the surface of viruses. This antibody-antigen interaction is highly precise so each antibody generally will recognize only one antigen (although sometimes an antibody will also recognize a closely related molecule in a process known as cross-reaction). This exquisite specificity between

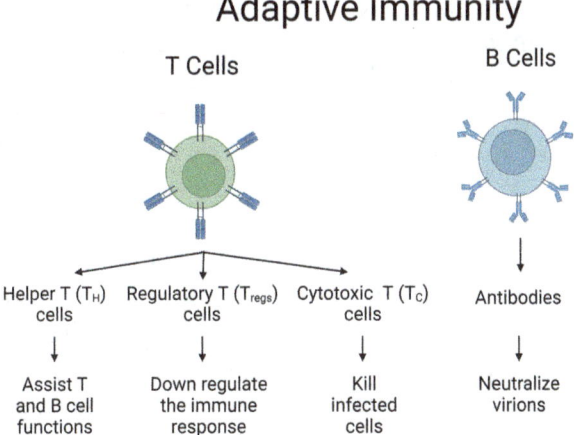

Fig. 1.1 Adaptive Immunity. Shown are diagrams of T and B cells illustrating their surface receptor proteins. Three subclasses of T cells are depicted, helper T cells (T_H), regulatory T cells (T_{regs}), and cytotoxic T cells (T_c), and their functions are listed for each cell type. Together these T cells comprise our cellular immunity. For B cells, their product is antibodies, one function of which is to neutralize virions. Antibodies constitute our humoral immunity

antibody-antigen binding makes antibodies like guided missiles that will seek out and bind only to their exact target.

Circulating B cells act as a surveillance system to detect foreign antigens. These so-called resting B cells have their cell surface studded with copies of their antibody. If these antibodies embedded on the cell surface encounter and bind their specific antigen, this activates that resting B cell to become a plasma cell. A plasma cell replicates to produce many identical copies of itself, and all of these copies now secrete the same antibody out into the bloodstream. These antibodies will spread throughout the body and will bind their specific antigen wherever they find it. Depending on the type of antigen, the formation of the antibody-antigen complexes can render the antigen inactive via several mechanisms. For illustrative purposes let's imagine that the triggering antigens are viral surface proteins. When an individual is infected with a virus, the virus replicates and virions begin to increase in number within the infected person. Resting B cells in circulation will happen upon the virions and any B cell that expresses an antibody capable of recognizing the virion surface proteins will bind the virion and become activated. The resultant plasma cells will secrete copious quantities of their antibodies, and these released antibodies will bind the virions, covering the entire virion surface with the antibody molecules. Remember that virions need a specific viral attachment protein on their surface to interact with a cell receptor as the first step in the viral life cycle. Antibodies that bind to the attachment protein will physically obstruct interaction with the receptor, much like wrapping a key in tape would prevent it from entering its keyhole. With the ability to bind their receptor blocked, these antibody-covered virions are no longer infectious and have been neutralized; those antibodies that prevent virion attachment are called neutralizing antibodies.

Neutralization is a critical mechanism for recovery from viral infection as it stops the spread of virions within the infected host and limits the cellular destruction that infection would cause. However, activation of B cells and the production of enough antibodies to neutralize all the accumulated virions is a somewhat slow process that takes 5–10 days to develop the first time that we encounter a virus. Consequently, on a first encounter with a new virus we often get sick for several days before the immune system ramps up enough to eliminate the virus and allow us to recover. As the infection wanes the plasma cells die off and the immune system returns to its resting state. Fortunately, a first exposure not only produces the transient plasma cells but also memory B cells. Memory B cells are long-lived and react to their antigen much more quickly than do naïve resting B cells. Upon a second or any subsequent encounter with their antigen (the virion in our scenario), these memory B cells immediately replicate and release antibodies in a matter of hours not days. This rapid production of neutralizing antibodies is usually sufficient to bind all the invading virions and prevent disease. All viral vaccines attempt to induce memory B cells that produce neutralizing antibodies against their target virus. This allows the vaccinated person to respond quickly and effectively should they ever encounter the virus, thereby protecting them from severe illness.

The second critical arm of immunity is cellular immunity. A primary mediator of cellular immunity is another type of white blood cell, the T lymphocytes. There are several classes of T cells, with the major three types being helper T cells (T_H), regulator T cells (T_{regs}), and cytotoxic T (T_C) cells. As their name implies, helper T cells assist other T cells, and even B cells, in responding to foreign substances and pathogens. The helper cells function both through direct cell-to-cell interaction with their target cells as well as by secreting soluble effector molecules. These effector molecules can disseminate widely and trigger responses in other cells that possess appropriate receptors for the effectors. The roles and contributions of helper T cells are critical for the coordination and regulation of the overall immune response to perceived threats, and without them, the entire immune system would go awry. Similarly, the T_{regs} also help regulate the immune response with an important role in shutting off the immune response once a threat has been eliminated. Since the immune reaction to pathogen-infected cells can also cause collateral damage to normal cells, deactivating the response and returning the immune system to the standby state is an essential function provided by the T_{regs}.

The third major type of T cell, the cytotoxic T cells, are the professional assassins of the cellular immune system. Like B cells, Tc cells use receptors to recognize foreign antigens, but in this case, they respond to antigens present on the surface of other cells. As viral proteins are synthesized within the cytoplasm, some protein molecules will be degraded into small pieces called peptides. These peptides are captured by the cell and transported to the cell surface in conjunction with an assemblage of cellular proteins known as the major histocompatibility complex (MHC). These viral peptides presented on the cell surface are now visible to the immune system and can be interrogated by circulating Tc cells. If a Tc cell has a receptor that can bind the viral peptide that binding event triggers the Tc cell to become a killer. While attached to the virus-infected cell, the activated killer Tc cell secretes several toxic proteins that attack and destroy the bound cell. Eradicating the infected cell may seem harsh, but by doing so that cell is stopped from producing new virions. The virus was going to kill that cell anyway and speeding up the process to prevent the synthesis and release of new virions that could infect hundreds of other cells is a worthy sacrifice. Collectively, the Tc cell population will kill off all the infected cells in the body. Removing these virion "factories" is an essential step toward eliminating the viral infection. Additionally, there are also memory T cells that will remember the viral antigens and respond very quickly to any future infections.

Together, the B and T cells provide a two-pronged attack that neutralizes free virions (via B cells producing antibodies) and destroys the virion-producing cells (via Tc cells producing toxins). Natural infection generates memory B and T cells that allow these responses to occur much more rapidly during subsequent exposures to the pathogen. Generally, any repeated infection will be stopped before the disease develops, and the person will usually not even know that they were exposed. The goal of a vaccine is to mimic the natural infection and elicit the appropriate memory B and T cells, thus rendering the recipient immune to the target pathogen. Ideally, this is accomplished without causing any disease. The difficult trick is finding the right material, the immunogen, that elicits robust and persistent protective immunity. Much of the early vaccine development utilized a slow trial-and-error approach that could take years to identify or create a suitable immunogen. In contrast, modern molecular science is now capable of more rational vaccine design as exemplified by the SARS-CoV-2 vaccine developed in under a year. In between those two extremes is a rich history of vaccine science that will be presented in the remaining chapters.

A Vaccine Primer

Most readers of this book have been vaccinated at some point in their lives and understand the basic purpose of vaccines if not their details. Vaccines are immunogens derived from microorganisms and include either the entire microorganism in a nonpathogenic form or just some critical portion of the microbe. These immunogens are introduced into our bodies to elicit an immune response that will protect us from serious diseases if we ever encounter the actual pathogen. Infants and children receive numerous vaccines and adults continue to receive various vaccines throughout their lives as booster shots or as new vaccines become available. Because of these remarkable treatments, many once prevalent viral diseases are now so rare that many people, including most young physicians, have never seen a case. Through vaccinations, smallpox was completely eliminated from the world and polio cases haven't occurred in developed nations in decades. Similarly, the childhood triumvirate of measles, mumps, and rubella (German measles) has been reduced from a yearly nationwide threat to rare and limited outbreaks. Collectively, vaccines are estimated by the World Health Organization (WHO) to save 2–5 million lives globally each year and to prevent many millions more cases of illness with attendant suffering and economic loss. No other single medical advance has had such a dramatic impact on the health of the world's population.

While all vaccines have the intent of producing an immune response that protects us from disease, they vary widely in their properties and immunogens. At the simplest level of classification, all vaccines can be categorized as either "live" or "inactivated" (Fig. 1.2) (Vetter et al. 2018). Live viral vaccines consist of the whole virus in an infectious form that can replicate and reproduce itself both in cultured cells as well as in the vaccine recipient. Although it may seem counterintuitive to protect someone by giving them a live virus, the key principle for these vaccines is that the virus must have reduced disease-causing ability while still retaining the capacity to elicit a protective immune response. The first vaccine against smallpox used exactly this strategy. In this case, the virus used for vaccination was a relative of smallpox called cowpox. Many viruses have adapted to a single primary host (plant or animal) and they

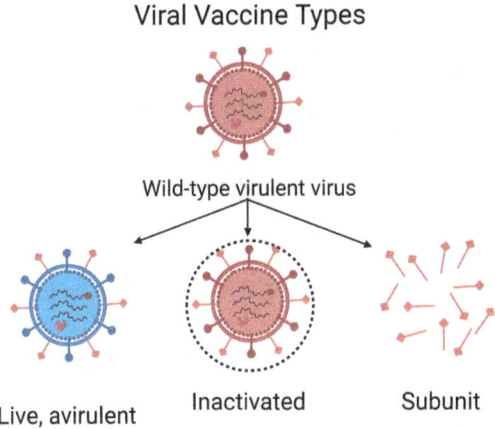

Fig. 1.2 Viral Vaccine Types. Depicted is a diagram of a wild-type, virulent virus and the three types of vaccines against that virus: live avirulent, inactivated, and subunit. Live, avirulent vaccines (also called attenuated vaccines) are viruses that have mutated to a form that has low virulence. Inactivated vaccines are made by chemically treating the wild-type virus so that it is no longer infectious ("killed"). For both the avirulent and the inactivated vaccines the inoculated viral particles still evoke an immunity that will protect against the wild-type virus. Subunit vaccines consist of only a portion of the wild-type virus, typically a viral surface protein. Inoculation with this protein (or a messenger RNA that encodes the protein) induces antibodies against the protein that will neutralize the wild-type virus and render it non-infectious

may have a restricted ability to infect other species. Cowpox is no exception, and being a disease of cattle, the virus is inherently poor at infecting humans. This makes cowpox naturally attenuated for humans and only able to cause a very mild disease with skin lesions at the site of infection. Fortunately, even this trivial cowpox infection is sufficient to generate a strong immune response that is cross-reactive for smallpox and provides excellent protection against the ravages of the virulent virus. We now know that simply injecting a small amount of cowpox (or vaccinia—see Chap. 2) into the skin of the recipient caused a small, localized infection that stimulated both the B and T cell branches of the immune system, resulting in a robust protective immunity against smallpox.

Although the use of a non-human virus as the vaccine for smallpox was highly effective, this has not generally been a viable strategy for live vaccine development. In most cases, there is not an animal virus relative to a human virus that can work analogously to the cowpox-smallpox situation. Instead, most other live vaccines are based on converting a virulent human virus into an attenuated form that can be used as the immunogen. Typically, this is done by serially passaging the virus through some type of non-human host such as mice, chicken eggs, or cells in culture. After the first round of infection, any newly made virions are harvested and used to infect a second round of the same type of host. The process is continued for multiple rounds, and with luck, the virus eventually becomes fully adapted to the new host, loses its virulence for humans, and becomes suitable for use as a human vaccine. As with cowpox, since the attenuated virus can still replicate weakly in humans it will generally evoke both a B and T cell response, providing a powerful and long-lasting immunity.

Because the serial passage approach is usually successful at attenuation, requires little or no knowledge about the molecular properties of a virus, and generally produces a very effective vaccine, making live attenuated viruses was the earliest technology adopted and still accounts for many viral vaccines. Yet these vaccines are not without concerns. Since they do replicate to at least a limited extent in the vaccine recipients, they can cause some mild disease symptoms. While this minor vaccine-induced illness is not generally an issue for healthy people, it can progress to severe disease in individuals who have an underlying immunodeficiency. Additionally, because the replicating vaccine virus can enter the bloodstream there is a theoretical concern that the vaccine virus could cross the placenta in a pregnant woman and cause fetal harm. Consequently, live virus vaccines cannot be given to pregnant women or immunocompromised people.

A potentially more problematic issue with live virus vaccines is the reversion to virulence. During the serial passage attenuation process, human viruses adapt to non-human hosts through replication errors in the viral genome. Every time a virus replicates its DNA or RNA there can be random copying mistakes which are mutations. The rate at which these replication mutations occur varies widely from virus to virus but is generally much higher for RNA viruses than for DNA viruses. If these random genetic changes enhance the reproduction of the virus in the passaging host then they will be selected, and after multiple rounds of passage, the virus becomes adapted to the new host and hopefully attenuated for humans. Unfortunately, the mutations that make a human virus attenuated for human infection are not permanent. After vaccination with a live, attenuated virus there will still be limited viral replication in the vaccine recipient. During this replication period, there can be additional mutations that may revert the virus to the human virulent form. In practice, this has not proven to be a significant concern for live vaccines except for the oral polio vaccine (Chap. 6). This virus has a reversion rate high enough that it does occur in the rare vaccine recipient leading to severe disease. Due to this problem, the live polio vaccine has been discontinued in many countries, including the United States which stopped using this vaccine in the year 2000.

To avoid the issues with live attenuated vaccines, inactivated or "killed" vaccines have been developed for some viruses. As the terminology implies, such vaccines contain noninfectious virions created by treating the viruses with inactivating chemicals (for example, formaldehyde or β-propiolactone). Because the virions are no longer infectious after the chemical treatment, when injected into the recipient there is no viral replication, and without replication, there is no reversion issue. Furthermore, without viral reproduction no viral disease is possible. Consequently, there are not even mild disease symptoms in healthy recipients and these vaccines are safe for pregnant women and the immunocompromised. Importantly though, the treated virions must retain their shape, structures, and antigenicity if they are to be suitable as vaccines. When injected, the body recognizes the inactive virions as foreign antigens and produces antibodies against the virion surface molecules. Among the multitude of antibodies produced will be ones with virus-neutralizing capability. It is the neutralizing antibodies (or more specifically the memory B cells that can produce the neutralizing antibodies) that defend against subsequent virus infection. There may also be some T cell immunity invoked, although this is typically less robust for inactivated vaccines, and it is usually the B cell response that provides the bulk of the protective immunity.

Inactivated vaccines are safe because the lack of viral reproduction means that there is no chance of the vaccine causing the viral disease in the recipient. However, this also means that there is no amplification of virions and their surface antigens. As a virus reproduces and spreads during an actual infection, the large quantity of viral antigens that accumulate ensures that a strong immune response is evoked. In contrast, the only viral antigens the recipient is exposed to with an inactivated vaccine are the small quantity in the delivered dose. Because of this limited amount of antigen, inactivated vaccines can require between 2 and 4 doses to generate sufficient immunity. Furthermore, the immunity induced by inactivated vaccines is more likely to wane over time compared to live vaccines and subsequent booster shots may be required years after the original vaccination to maintain adequate protection. While these issues are not serious impediments to the use of inactivated vaccines, these inherent constraints do provide ample incentive for the scientific community to develop more potent inactivated vaccines with enhanced antigenicity and increased duration of protection.

A final caveat with inactivated viruses is the complexity of the antigens present in the vaccine. The whole virion can be a very intricate molecular structure with numerous proteins and other biomolecules decorating the surface of the virion. Each of these surface molecules is a foreign antigen that can elicit antibodies, however, many of the antibodies generated are not capable of neutralizing the virions so are not protective. The B cell population simply responds to any foreign antigen without knowing which resultant antibodies will be effective at controlling the invading virus. Thus, the process is inherently a shotgun approach and a significant portion of the immune response to an inactivated whole virus is wasted effort. Ideally, rather than using the whole virion in the vaccine, it would be better to use only the viral surface molecule(s) that is(are) the target for neutralizing antibodies. Such vaccines that use only a part of the virion are called subunit vaccines. The first viral subunit vaccine was the vaccine against hepatitis B (Chap. 8). This vaccine consists of a hepatitis B virus surface protein called the S antigen. The cloned gene for the protein is expressed in yeast where the S protein can be harvested and purified for the vaccine. Injecting this single protein elicits antibodies that effectively bind to virion S proteins and neutralize the virions. Since the vaccine consists of only this purified viral protein it cannot cause hepatitis B and is safe for both pregnant women and immunocompromised individuals. Other currently available viral subunit vaccines include the human papillomavirus (HPV—Chap. 12) vaccine and the SARS-CoV-2 messenger RNA (mRNA) vaccines (Chap. 13).

This brief overview was intended only to introduce the general concepts of live, inactivated, and subunit vaccine types. Each virus is unique and presents its own challenges for vaccine development. In some cases, the nature of the virus and the disease it causes may provide clues as to which vaccine type would be most feasible or effective. For other viruses, the choice of which type of vaccine to pursue is not often clear, and multiple types and approaches may be tried before discovering the most effective vaccine strategy. Some issues can only be resolved through systematic testing and tweaking of vaccine candidates, a process that can take years to accomplish. Our cadre of modern vaccines is a remarkable testament to the skill and dedication of many researchers who strove to fight disease and rid the world of dangerous viruses. As we explore the history and development of viral vaccines in the subsequent chapters, you will learn the intricacies of individual vaccines, confront their diverse issues, and hopefully develop a greater appreciation for these lifesaving and world-changing medical advances.

References

Nicholson LB (2016) The immune system. Essays Biochem 60(3):275–301. https://doi.org/10.1042/EBC20160017

Vetter V, Denizer G, Friedland LR, Krishnan J, Shapiro M (2018) Understanding modern-day vaccines: what you need to know. Ann Med 50(2):110–120. https://doi.org/10.1080/07853890.2017.1407035

Wilson VG (2022) Viruses: intimate invaders. Springer. https://doi.org/10.1007/978-3-030-85487-4

Woolhouse ME, Howey R, Gaunt E, Reilly L, Chase-Topping M, Savill N (2008) Temporal trends in the discovery of human viruses. Proc Biol Sci 275(1647):2111–2115. https://doi.org/10.1098/rspb.2008.0294

2

Smallpox and Immunity: Success Precedes Science

Keywords Cowpox • Edward Jenner • Vaccination • Vaccinia • Variola virus • Variolation

Abbreviations

BCE	Before common era
CE	Common era
CMLV	Camelpox virus
CPXV	Cowpox virus
TATV	Taterapox virus
VACV	Vaccinia virus
VARV	Variola virus
WHO	World Health Organization

The Speckled Monster—A Callous Killer

We live in a viral world. We are continually exposed to viruses in our environment, and some viruses are even part of our own genomes. Fortunately, most viruses are innocuous although some can sicken us, some can cripple or scar us, and some can kill us with frightening destruction of our bodies. Among the many types of viruses that infect humans, their physical and biological properties vary greatly. Yet they all have one thing in common—once they infect us they are difficult to treat. In many cases, the only option is to provide supportive or palliative care and hope that our immune system defeats the viral invader before we are ravaged or killed by the infection. Unlike the widely available antibiotics for bacterial infections, until recent decades there were no effective drugs to treat any viral infections. Much progress has been made in antiviral drug development in recent years, yet we still lack antivirals against many of our most important viral pathogens. Our only defense against many viruses is to prevent infections from happening through prophylactic vaccination. Consequently, much of twentieth-century virological research was devoted to developing vaccines against dangerous endemic viruses that routinely swept through the global population each year. These efforts were an unqualified success and many once common viral diseases virtually disappeared from any nation capable of widely vaccinating its populace. Vaccines are widely acknowledged as one of the most successful medical advances in the twentieth century, preventing hundreds of millions of deaths worldwide.

One theme that will become obvious is that the temporal order of vaccine development paralleled the seriousness and/or frequency of the viral disease. Initial vaccine efforts focused on the most severe or impactful diseases while vaccines for many milder viral infections weren't tackled until the greater threats were vanquished. By the criteria of severity and impact, smallpox was a prime target for any kind of treatment or prophylactic. Most estimates suggest that smallpox alone killed hundreds of millions of people throughout its history, with possibly as many as 500 million in the twentieth century alone (Koplow 2003). Even the survivors were not untouched as smallpox often left behind severe scarring from the pox lesions and blindness from lesions in the eyes. Smallpox's enormous cost in human lives and its negative influence on societies across every continent from Asia to South America are almost incalculable. From the death of at least 10 Eurasian sovereigns (Habsburg Emperor Joseph I, Queen Mary II of England, Czar Peter II of Russia, King Louis XV of France, King Luis I of Spain, William of Orange, Queen Eleonora of Sweden, Princess Louise Hippolyte of Monaco, Emperors Higashiyama and Go-Kōmyō of Japan, and both the Shunzi and Tongzhi Emperors of China) to the decimation of the native populations in both North and South America, smallpox took a horrendous toll everywhere it appeared. As penned by Thomas Macaulay in the 1800s, "*the smallpox was always present, filling the churchyards with corpses, tormenting with constant fears all whom it had not yet stricken, leaving on those whose lives it spared the hideous traces of its power, turning the babe into a changeling at which the mother shuddered, and making the eyes and cheeks of the betrothed maiden objects of horror to the lover*" (Macaulay 1881). Given its many centuries of inflicting death and illness, it's not surprising that the smallpox virus attracted intense scrutiny and became the first virus with a successful vaccine.

2 Smallpox and Immunity: Success Precedes Science

The smallpox virus is formally known as the variola virus (VARV). VARV belongs to the Poxviridae family, a group whose members have large, double-stranded DNA genomes that encode dozens of viral proteins. Initiation of new infections is usually via inhalation of VARV particles released by a cough or sneeze from an infected person (Chertow and Kindrachuk 2020). Alternatively, the virus is also present in bodily fluids, including the characteristic pox lesions on the skin, and exposure to these fluids can also lead to infection. When introduced into the nose or mouth, the virus infects mucosal cells, spreads to the nearby lymph nodes, and eventually into the bloodstream. Viruses in the bloodstream (an event called a viremia) are disseminated throughout the body where these microscopic invaders establish infections in the major internal organs and the skin. This process of replication at the site of initial infection, followed by dissemination and subsequent systemic replication, is relatively slow so that symptoms such as fever don't begin until 1–2 weeks after exposure to the virus. Typically 3–4 days after the fever starts the highly characteristic rash appears, starting as raised bumps (called papules) that evolve to pustulant, virus-filled blisters over 5–7 days. These pervasive lesions can be so extensive that they can cover the entire body in painful, suppurating sores. In survivors, these lesions eventually heal over but often leave horrid permanent, pitted scars, hence the moniker for this disease as the speckled monster. Elsewhere in the patient, virus-induced destruction in the respiratory system triggers coughs and sneezes, both of which produce and disperse small, aerosolized droplets that carry virus particles to the next victim. Along with lung infection, viral replication in various organs causes widespread systemic damage that is deadly in a high percentage of cases (Martin 2002). In naïve populations that have never before encountered VARV, the fatality rate approached 80–90%, a level of mortality almost unimaginable in modern times. Even in populations accustomed to the disease, typically 25–40% of the patients succumbed to the infection, 5–9% of survivors developed ocular complications that often led to blindness, and many more were permanently scarred (Semba 2003). The precise mechanisms by which VARV elicited its killing effects remain uncertain as the disease was eliminated over four decades ago, long before many modern molecular analysis techniques were available. Nonetheless, historical records leave no doubt about the ferocious morbidity and mortality associated with VARV infections as illustrated in this passage from the seventeenth century (Anselment 1989):

> Uppon the 29th of September … began my daughter Katte with a violent and extreme pain in the backe and head, with such scrikes and torments that shee was deprived of reason, wanting sleepe, nor could she eate anything. For three

> daies she contineuded, to my great affliction, not knoweing what this distemper would be. At last the smale pox appeared, breaking out abundantly all over; ... She was all over her face in one scurfe, they running into each other. Her extreamity beeing soe great, crieing night and day, that I was forced to be removed, though very weake, into the scarlett chamber, for want of rest.

Given the endemic nature of smallpox and its high level of morbidity and mortality, the entire world was desperate for any treatment or preventative that could ameliorate its devastating effects on communities and whole countries. This desperation fueled many attempts to understand the disease and find some measure of protection. While the concept of a virus and an understanding of the nature of these entities was still far in the future, this didn't stop inquiring minds from observing the disease and trying to learn its secrets. Over the centuries, many cultures reached the same important conclusion that ultimately formed the basis for all vaccination: surviving smallpox made the recovered patient resistant to any future return of the disease. Without knowing how the disease was caused or how it was spread, it still became clear that recovery provided some type of lasting protection. Somehow the body reacted against the disease agent, learned how to fight it off, and remembered that lesson to prevent any future illness if reexposed to the pathogen. We now refer to this process as immunity, an incredibly complex and powerful defense system that constantly surveils our bodies to identify and attack foreign substances, including viruses. This concept of induced immunity stemming from the smallpox work, and its broad application to all infectious diseases, is a legacy of the smallpox vaccine perhaps even more important than fighting the disease itself. From these early and unsophisticated observations arose the fields of immunology and vaccinology that now keep many of our most fearsome viruses well controlled. The creation of vaccines over the last 150 years has reduced viral disease incidence to only a fraction of what it was in previous centuries.

Elusive Origins

So where did this vicious pathogen come from and why did it become a prevalent disease that terrified humans for thousands of years? Poxviruses are widespread in nature with 83 known species that infect different vertebrates and invertebrates (e.g. insects, arachnids, and crustaceans). This broad distribution of poxviruses among diverse species indicates an ancient origin for this virus family that preceded the evolutionary split between vertebrates and invertebrates which occurred over 500 million years ago. Yet despite the historically distant origin of the poxvirus family, the VARV appears to be a much more recent entry into human populations, occurring thousands rather than millions of years ago. VARV is a specific human pathogen that is highly adapted to our species and does not naturally infect any other species. However, there is no definitive answer about where and when the VARV arose as reported estimates of its entry into the human population vary widely. To determine viral origins, paleovirologists use multiple approaches, including historical records of disease outbreaks, attempts to isolate viral material from ancient corpses, and comparing genomic DNA sequences from different human and animal poxviruses to create evolutionary (phylogenetic) trees. Sadly, the historical writings about potential VARV cases are difficult to validate and little physical evidence remains of ancient human infections. Written descriptions of outbreaks such as the Plague of Athens (430 BCE) and the Antonine Plague (170 CE) have been postulated as examples of smallpox, as have historical writing from 1100 century BCE China. However, none of these reports is definitive as each outbreak could have a non-VARV cause (Babkin and Babkina 2015). The first reliable written evidence of historical smallpox cases comes from more recent Chinese and Indian texts created in the first few centuries of the Common Era (CE). For example, the Chinese physician Ko Hung wrote in 340 CE that *"some people have suffered from seasonal epidemic sores which attack the head, face and trunk. In a short time they spread all over the body. They look like red boils, all containing some white matter. The pustules arise all together, and later dry up about the same time. If the severe cases are not treated immediately many will die. Patients who recover are left with dark purplish scars the colour of which takes more than a year to fade"*. Such a detailed and accurate description confirms that these early physicians were well acquainted with the disease of smallpox, so we know VARV is at least roughly 2000 years old (Theves et al. 2016). Additional accepted accounts of smallpox began to appear in the early centuries of the Common Era, first in the Arabian Peninsula (Syria in 302 CE and Saudi Arabia in 570 CE). The sixth century CE saw widespread dissemination of the disease with reports coming from Asia (Japan), Africa (Egypt), and Europe (France and Italy). Ultimately, the Crusades (1100–1300 CE) fueled the pervasive introduction of smallpox to Europe and by the 1500s this disease was endemic throughout the European nations.

With only minimal written documentation, scientists have looked for more direct evidence of historical smallpox by examining preserved cadavers. Unfortunately, similar to the uncertainty in written records, attempts to isolate VARV from ancient human remains have not yielded definitive results.

Presumptive smallpox-like lesions are found on several Egyptian mummies, including Ramses V who died in 1157 BCE. However, efforts to visualize viral particles or isolate viral DNA from the Ramses mummy and a second Egyptian mummy from the same period were unsuccessful (McCollum et al. 2014). Our inability to detect molecular evidence of VARV from millennia ago prevents a direct dating of the earliest cases of smallpox.

Without unequivocal proof from either historical observations or human remains, genetic analysis is the major tool for dating the origin of VARV. Attempts to confirm the presence of VARV in more recent cadavers have been successful with partial or full viral sequences detected in several mummies from the 1600s or later (Ferrari et al. 2020). Having VARV DNA sequences from these earlier times reveals evolutionary changes that can help define the origin of this virus. DNA is composed of four nucleotides: A, C, G, and T. You can think of these nucleotides as the letters of the genetic alphabet. Combinations of these four nucleotides spell out the genes just as the 26 letters of the English alphabet spell out words. During the replication of poxvirus genomes (and the genomes of every other life form), synthesis mistakes (mutations) can occur. A replication error can cause a permanent change in the DNA sequence of the newly made genome, for example, putting an A in the position that should have a G nucleotide. Such replication errors are the major source of genetic variation in viruses.

Poxviruses have a low synthesis error rate, yet mutations do gradually accumulate in their genomes. By comparing genomes isolated from different time periods, scientists can estimate the mutation rate of poxvirus genomes, i.e. the length of time required to acquire one DNA sequence change. DNA sequences from human and animal poxviruses obtained from different global locations and periods of history are compared to determine the number of sequence differences between any two viruses. The greater the number of sequence differences the longer the time since the isolates diverged. Using this approach one can construct phylogenetic trees showing the degree of relatedness between different poxviruses (Fig. 2.1) and can estimate the timing of branch divergence. Importantly though, these approaches have limitations and some assumptions have to be made concerning both the genome changes and the timing when a particular poxvirus type reaches its geographic location. Because of these inherent uncertainties, the estimates for when VARV first entered human populations range from 68,000 years ago (Li et al. 2007) to the more generally accepted 3000–4000 years ago (Babkin and Babkina 2015; Hughes et al. 2010). It is this more recent emergence that is consistent with two other types of evidence. First, epidemiologists estimate that it requires a fairly large population base (>100,000 people) to maintain smallpox in an

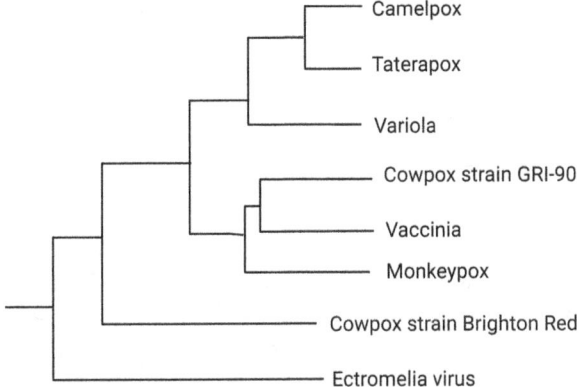

Fig. 2.1 Phylogenetic Tree for Poxviruses. A phylogenetic tree depicts the genetic relatedness between different organisms. Shown are eight different poxviruses, with Variola being the smallpox virus. The closer two organisms are branched, the more closely related they are. For example, Camelpox and Taterapox are closely related but are only very distantly related to the Ectromelia virus

endemic form. Large, dense populations only began to arise after the introduction of agriculture 7000–10,000 years ago. The lack of sufficient population densities before that time would likely have prevented VARV from being established and maintained in human civilizations. Second, the two closest relatives of VARV are the camelpox virus (CMLV) and the taterapox virus (TATV), a poxvirus specific for the naked sole gerbil (see Fig. 2.1). The close genetic relatedness of these three viruses implies that they all arose from a common ancestor at around the same time. It is presumed that an unknown animal poxvirus with a broad host range concurrently infected humans, gerbils, and camels and adapted to each new host to become the three distinct modern poxviruses. The naked sole gerbil is only found in Africa, suggesting that this continent is where the ancestor of human poxvirus existed. Humans have been in Africa for many millennia, but we know the concurrent viral jump into humans, gerbils, and camels couldn't have occurred until 3500–4500 years ago when camels were introduced to Africa. The period for the arrival of camels to the African continent is consistent with the phylogenetic estimate of VARV's emergence 3000–4000 years ago. Why an animal poxvirus endemic to Africa suddenly acquired three new hosts is unknown, but such host switching by viruses typically arises from disruption of the ecosystem of the host animal. Anything that affects the original host animal's habitat can result in an infected animal now intermingling with new host species, providing the virus with an opportunity to jump into one or more new susceptible species. Somehow the incursion of humans and camels into the region of the naked sole gerbil disturbed the ecosystem and facilitated the passage of the endemic ancestral poxvirus into all three species. While the exact timing and cause will likely never be confirmed, we know that the outcome of this ancient event was for an animal poxvirus to change hosts and become a deadly human pathogen.

Variolation

The Anglo-Saxon word "smallpox" arose in the sixteenth century in Britain, however, this disease was well-known in Middle and Far Eastern countries for many centuries before it appeared in Europe. Some writings even suggest that smallpox was present in China more than 3000 years ago where it was called tai-tou. Given its presumed presence in ancient Asia, it is not surprising that the earliest attempts to avert this disease, known as variolation, have Chinese or Indian origins (Boylston 2012). A key precept behind the various preventative techniques was that smallpox survivors never incurred a second bout of the disease. This observation was likely made independently in many diverse cultures and became folkloric wisdom. However, given that disease in ancient times was often regarded as caused by evil spirits or angry gods, how and why this protective effect worked must have been quite puzzling. Still, even without any mechanistic explanation for this phenomenon, intrepid individuals across the centuries came to a similar conclusion—if one could induce a mild case of smallpox it would protect the patient from suffering severe disease during any future smallpox outbreak. Without realizing it, these early medical practitioners identified and relied on the concept of adaptive immunity. Survivors of smallpox, whether it was acquired naturally through community infection or via some prophylactic intervention, developed immunological memory against VARV. This immunologic recall is mediated by two types of long-lived immune cells, memory B and T cells. Once formed, these cells remain in circulation for decades and quickly respond to the virus if it is ever encountered again, thus preventing a recurrence of the disease. Although this biological phenomenon of adaptive immunity wouldn't be understood until the twentieth century, the observation that recovery from a disease could render the individual resistant to any further instances of that disease underpinned the development of all vaccines.

The oldest purported approach for smallpox prevention comes from China at around 1000 CE. This initial method, known as variolation, involved isolating and drying scabs obtained from people with mild cases of smallpox. The dried material was ground into a powder and blown into the nostrils of children. Since the VARV virus particles are unusually stable and resistant to drying, these powders would have contained viable viruses that initiated an infection in the recipient. This ancient tradition was not well documented until the mid-1500s, so the efficacy and extent to which this procedure was used are not known. From the later written accounts, many recipients developed full-blown smallpox and died, a not unexpected result considering that live virus was being introduced into its natural portal of entry, the nose. Nonetheless, those who survived were indeed protected from future VARV infection.

From the 1500s on there is more documentation of variolation from other nations though one thing is clear: the Chinese practice of nasal variolation (also known as inoculation or insufflation) did not spread widely. Failure of other cultures to adopt the Chinese variolation practice was probably due to the high incidence of systemic smallpox and high mortality associated with this procedure. Instead, an intradermal variolation method became common throughout Asia, the Middle East, and eventually Europe. The application of the intradermal method varied somewhat in different locales but remained fairly consistent in principle. Typically, pustulant material from a smallpox patient was collected, and then a needle or other sharp object was dipped into the pus and used to stab the smallpox material into the skin of the recipient. The number and location of the stabs depended on the local customs, but all variations provided fairly consistent smallpox protection. Introducing the virus through an unnatural route, the skin, tended to produce a more localized infection with only a few pox lesions at the sites of inoculation. These lesions were usually accompanied only by mild systemic effects (fever, muscle aches, malaise) rather than generalized smallpox that could result from the inhalation method, thus making the intradermal method much safer than the nasal route. This greatly improved risk-to-benefit ratio is likely the reason that intradermal variolation became the more widely accepted smallpox prophylactic. Without realizing it, the purveyors of the skin inoculation method stumbled upon a key tenet of many future vaccines, a process called attenuation. The attenuation strategy tries to induce immunity to the target pathogen by exposing the recipient to a weakened or nonvirulent derivative of the pathogen. This attenuated version causes either no or very mild disease yet presents to the immune system just like the virulent parent organism. Consequently, the attenuated pathogen invokes the immune system to produce a lasting immune response that protects against the virulent virus. Although variolation didn't use an attenuated version of VARV, introducing the virulent virus through the skin rather than the respiratory system resulted in an attenuated infection that still generated protective immunity. No one knows what inspired some bold soul to conceive this idea of skin inoculation, but it clearly protected those who survived the procedure. Centuries later, Louis Pasteur and others would seize upon the attenuation concept to begin the modern vaccine era (Chap. 3).

An interesting note about the intradermal method is that its origin is uncertain. It is not known if there was a single place where this form of variolation arose which then spread to other regions or if this approach was independently developed by multiple cultures. Numerous historical letters and writings indicate that by the 1600s intradermal variolation was practiced in Africa, Turkey, India, parts of the Arab world, and even in Wales (Boylston 2012). Some historians argue that the similarity in the procedure and terminologies used by practitioners of variolation in these various countries favor a common origin for the procedure which then spread via trade and commerce. However, even if its origin remains obscure, this wide geographic distribution suggests that variolation was perceived as a valuable tool to reduce smallpox infections and prevent the horrid scaring and tragic deaths so common with this disease. Regardless of culture or country, people everywhere were anxious to protect their children and themselves from this ravaging disease. With no effective treatments for smallpox cases, the concept of a preventative measure that warded off the disease must have seemed miraculous to those whose minds weren't closed against new ideas.

While we are unlikely to ever know the first inventor or inventors of intradermal variolation, the spread of this procedure into eighteenth-century Europe is more clearly delineated in the historical records. Lady Mary Montagu receives and deserves much credit for introducing variolation to England although she was not the first to present this method to the British (Grundy 2000; Dinc and Ulman 2007). The first report of variolation in England was published in 1714 in Philosophical Transactions, the journal of the Royal Society. Submitted by Emmanouel Timonius, the report details his use of variolation in Constantinople (now Istanbul) during a smallpox outbreak. Two years later, Philosophical Transactions received and published another account of variolation in Turkey provided by the Greek physician Jacobus Pylarinos. Neither report appears to have had an impact on the British medical community. Instead, acceptance of variolation and implementation of the procedure can be traced to the persistence of Lady Montagu. As a young adult, Lady Montagu suffered smallpox-induced facial scaring and lost her beloved brother to the disease, so she was very familiar with and very fearful of this contagion. Consequently, when she later lived in Constantinople with her husband, the British Ambassador to Turkey, she was very receptive to the protective power of variolation. Ultimately, in 1718 she had her 5-year-old son Edward variolated by the Embassy physician, Charles Maitland.

Three years after the Montagus returned to England, London was in the throes of the great smallpox epidemic of 1721. Now fearful for her 4-year-old daughter Mary, Lady Montagu again sought out Maitland to perform the variolation. However, sentiment in the medical community was against this procedure and proof of safety was requested before proceeding with Lady Montagu's daughter. Six convicts and eleven orphans were variolated as a safety trial (a practice certainly not allowed under modern informed consent policies). With no adverse effects observed in these seventeen test subjects, Maitland was allowed to perform the variolation on Mary. Young Mary had an uneventful variolation and never contracted smallpox. Lady Montagu's fearless example in variolating her children inspired other members of the upper class and even the royal family to undergo variolation. She also continued to advocate for variolation in personal encounters and letters for years thereafter. Her efforts helped the procedure spread slowly in England and then into Europe and the New World colonies, although it was mostly adopted by the upper classes and the free thinkers who had positive views of scientific advances. Variolation of the masses never became a widely accepted practice due to a combination of issues, including religious objections, cost and availability, rejection by some of the established medical community, and a degree of danger as some recipients did develop fatal smallpox from the procedure. Nonetheless, the idea that it was possible to prevent a serious disease by intentionally inducing a less severe form of the disease was an important intellectual advance that set the stage for the introduction of vaccination. Without Lady Montagu championing variolation and spreading this novel preemptive protection concept, a procedure that would become known as immunization, eighteenth-century society might have been much less receptive to the subsequent work of Edward Jenner and his contemporaries.

Cowpox and Vaccination

After variolation, the next advance toward modern vaccines was the use of another form of attenuated infection. Variolation used low doses of virulent VARV delivered by an unnatural route (intradermal rather than respiratory) to create a mild yet protective infection. However, since live VARV was still involved in this procedure some people did develop full-blown smallpox from variolation, and there was a death rate of 1–2% among the treated individuals. While much safer than contracting VARV naturally, variolation was not an ideal prophylactic, and a much less dangerous method was sorely needed before there would be widespread acceptance of a preventative treatment for smallpox. Fortuitously, a benign and effective alternative was soon discovered in the inaccurately named cowpox virus (CPXV). Historically, cowpox was a mild skin disease sporadic in cattle throughout Eurasia, hence the name. However, modern genetic analysis indicates that rodents are the likely authentic host species for this virus with bovine infection being an incidental cross-species transmission (Carroll et al. 2011). The disease in milking cows was very innocuous and manifested as a few pustulant lesions on the teats. This location made the virus easily transferred to the hands of milkers (usually women by social convention). Infected humans generally had localized skin lesions on the hands and only mild if any systemic symptoms. Some of the first recorded reports to suggest a connection between cowpox and smallpox were letters to the Medical Society of London in the mid-1700s. Two physicians (Drs. Fewster and Rolph) independently noted that persons who previously had cowpox did not respond to variation and failed to develop either pox at the site of injection or the typical systemic symptoms such as fever (Barquet and Domingo 1997). Fewster wrote up his observations in a paper entitled "Cow Pox and its Ability to Prevent Smallpox" that was submitted to the Medical Society of London which declined to publish it. Similar though less well-documented reports of the negative effects of cowpox infection on variolation attempts were noted by several other individuals both in England and Europe. These observations that puzzled the physicians of that era are now easily explained as immunologic cross-reaction. CPXV and VARV are related viruses whose surface proteins share similar sequences and structures. Because of these similarities, infection with CPXV elicits B and T cells that will also recognize and attack VARV. Thus, CPXV acts like an attenuated form of VARV in that infection with CPXV usually causes only mild disease yet provides complete protection against the more virulent VARV. Because of this cross-protection, variolation produced no noticeable effects in cowpox-positive individuals as the inoculated VARV was immunologically neutralized before it could establish a productive infection in the skin.

In addition to the early noted observations about cowpox and variolation, by the latter half of the 1700s, there was widespread folklore across England and Europe suggesting that milkmaids who contracted cowpox were resistant to smallpox. A corollary to this folk wisdom was the observation that milkmaids generally had nice complexions because they never suffered from the scarring ravages of a smallpox infection. This belief that cowpox infection could protect from smallpox apparently inspired some adventuresome individuals to conduct their own primitive scientific experiments. Towards the end of the 1700s, there are a few poorly detailed accounts from different countries of individuals trying to inoculate themselves or family members with pustulant material from cowpox lesions in hopes of preventing smallpox. Among these attempts, perhaps the best-documented case is that of the English cattle farmer, Benjamin Jesty (Behbehani 1988). Jesty was known to have had cowpox as a child and likely was well aware of its supposed protective benefit against smallpox. This belief was further supported by the observation that two of his servant girls with prior cowpox infections remained smallpox-free even after multiple exposures to VARV. During the smallpox outbreak of 1774, Jesty used material directly from a cow's cowpox lesion to inoculate his wife and two young sons (ages 2 and 4) by scratching their arms with a needle covered with the pox material. While none of them contracted smallpox from this inoculation, Mrs. Jesty developed a serious infection at her scratch site (likely bacterial) and had to be treated for this side effect. Both the local physician and the community at large were outraged by this familial experiment, and a chastised Jesty refrained from any additional exploration of cowpox as a preventative for smallpox. However, 15 years later Jesty had both of his sons variolated, and consistent with early reports, neither son responded, suggesting that they were both fully protected from VARV infection.

Similar to the Jesty story, using cowpox as a smallpox prophylactic was also attempted by a Dutch tutor, Peter Plett, working in what is now Germany (Barquet and Domingo 1997). He too used material taken from a cow's lesion and used it to inoculate three children, two of his employer's daughters, and one other local child. As with the Jesty effort, one of the children developed severe inflammation at the inoculation site and Plett never attempted cowpox inoculation again. There is no evidence that any of the three children were subsequently variolated, so the direct efficacy of the cowpox exposure was not tested. However, 3 years after their inoculation a severe smallpox outbreak swept through their region, and all three cowpox-treated children survived while many others perished.

Despite all the above evidence that there was widespread folklore about cowpox infection giving protection from smallpox, Edward Jenner is often credited as the inventor of vaccination. Jenner was an English physician and naturalist who had wide-ranging scientific interests from botany to human anatomy (Baxby 1981). During his training, Jenner apprenticed under John Hunter who is acknowledged as one of the leading physician-scientists of his era. Under Hunter's tutelage, Jenner learned the scientific method of conducting experiments to answer questions rather than just proposing ideas without concrete proof. He also learned to make and record careful observations on the things that he studied so that there would be a permanent record. Jenner made good use of these skills and by the age of 39 was made a Fellow of the Royal Society in reward for his astute scrutiny of Cuculidae birds published as the "Natural History of the Cuckoo".

As diverse as his interests were, Jenner was always intrigued by smallpox and the possible protective effects of cowpox infection. As a child, he nearly died from his own variolation with smallpox which may have stimulated his interest in finding safer solutions to prevent smallpox infections. Like others from his era, Jenner certainly knew of the folklore surrounding milkmaids, cowpox, and smallpox resistance. Like prior physicians, Jenner noted during his medical practice that patients with a history of cowpox infection never responded to variolation, supporting cowpox's protective ability. Although not unique observations, Jenner's insight and contribution were to turn folklore into fact by conducting direct, if primitive, clinical trials while carefully documenting his experiments. Starting in 1789, Jenner began to test the effects of cowpox on a series of patients, many of them children, a horrific and unethical approach by modern standards. His general technique was to use a needle, knife, or another sharp object that was dipped in pustulant material from a cowpox lesion. This tainted instrument was then used to make scratches or small cuts, typically on the arm, that would introduce the cowpox material into the skin of the recipient. In essence, the process was simply a modified variolation procedure using cowpox rather than smallpox as the inoculum.

The first test subject was his 10-month-old child, also named Edward (Behbehani 1988). Edward's nurse developed a mild pox illness, presumably cowpox, and Jenner used fluid from her pustules to inoculate young Edward and two young servant women from a neighboring family. After an incubation period, Jenner variolated both his son and his son's nurse and found no reaction to the variolation procedure, consistent with the inoculation providing resistance to smallpox. This "challenge" test was an important aspect of Jenner's studies and is an approach still used today in some vaccine testing. Rather than waiting to see if the trial subjects developed smallpox from community exposure, the challenge provided rapid results. Not only did this greatly speed up the evaluation process, but it was also a much more definitive test. Failure to develop smallpox during a community outbreak could be due to the protective effect of cowpox or could just be due to a lack of exposure, so negative results were at best anecdotal. Using the variolation challenge added an immediate and unambiguous assessment of the experiment which considerably enhanced Jenner's credibility in documenting the protective efficacy of cowpox.

Jenner's next attempt wasn't until 1796 when he inoculated an eight-year-old boy named James Phipps. In this case, the inoculation material came from a pox lesion on the hand of a local milkmaid named Sarah Nelmes. Again, after an incubation period of several weeks, the young boy was variolated with no response. Jenner's experiments with cowpox inoculation, along with his previous observations on smallpox resistance in patients who had naturally acquired cowpox, were submitted to the Royal Society in 1797. However, the medical community was still resistant to the idea of cowpox inoculation and refused to publish the study, citing insufficient data. Interestingly, two different copies of his original manuscript still exist, one with the Wellcome Institute for the History of Medicine and one with the Royal College of Surgeons of England.

Rejection of research studies by journals and scientific societies is commonplace and is intended to ensure that only the highest quality and most sound manuscripts are published. While rejection is painful, it is usually accompanied by critical commentary that indicates the shortcomings of the work and areas that need to be improved before the work is acceptable. Not deterred by the verdict of the Royal Society, Jenner was determined to conduct further experimentation but was transiently stymied by the lack of cowpox cases in his community. It wasn't until early 1798 that cowpox was again detected on cows at a local farm. Over the next weeks, using lesion material either directly from infected animals or from human cowpox cases, Jenner inoculated several individuals, mostly children. One significant concept that he tested in these studies was the feasibility of multiple person-to-person transmission of the cowpox protective effect. Waiting for natural cowpox to appear in local animals was a haphazard and inefficient way to garner the material for his inoculations into people. Jenner and others tried methods to preserve cowpox lesion material as dried samples, but the effectiveness of these preparations was short-lived and highly variable. However, Jenner already knew that using lesion samples from a human directly infected by an animal provided adequate protection for the recipient, so could human-to-human transmission be extended and still provide suitable material for protection? In his new studies, Jenner passed the cowpox lesions from person to person and then tested the final recipients by variolation. The protective effect was still observed, confirming that there was no loss of activity during human passage. This was an important observation since it reduced the need for animals and provided a method to transport the inoculum material over time and distance. A remarkable implementation of this approach was used by King Charles IV of Spain to send cowpox to the Spanish holdings in the New World in 1803. In exchange for education and support, twenty-two orphans were enlisted to form a living chain of cowpox infection (Franco-Paredes et al. 2005). At the start of the voyage from Spain, several of the orphans were variolated with cowpox. As they developed lesions, their pustulant material was used to inoculate another set of orphans, and the process was repeated until the ship reached its final destination, thus bringing the life-saving material to a new world of people.

Having thoroughly convinced himself of the safety and efficacy of his cowpox variolation technique, Jenner designated his method variolae vaccinae, vaccinae being derived from the vacca, the Latin word for cow. It would be nearly a century later that Louis Pasteur converted Jenner's vaccinae to vaccine and used it to mean any material used to immunize against disease. Jenner self-published his results in a 1798 monograph entitled "*An Inquiry into the Causes and Effects of the Variolae Vaccinae, a Disease Discovered in Some of the Western Counties of England, Particularly Gloucestershire, and Known by the Name of the Cow Pox*". As he stated in a later publication, "*what renders the Cow Pox virus so extremely singular, is that the person who has been thus affected is forever after secure from the infection of the Small Pox*". Perhaps as important as the strong protective effect, Jenner's studies also demonstrated that cowpox variolation was generally safe, making it a superior substitute for smallpox variolation. The detailed and well-documented reports that Jenner produced quickly attracted both interest and support from influential persons such as well-known London physicians, Drs. William Woodville and George Pearson, who both promoted his findings. Although cowpox vaccination initially had many strident opponents, Jenner's ardent proselytizing coupled with the backing of numerous converts from traditional smallpox variolation quickly established his method as the preferred approach for smallpox prevention. Within less than a decade after his 1798 publication, the variolae vaccinae procedure was widespread in both Europe and the Americas. While Jenner did not originate the idea of using cowpox as a surrogate for smallpox, his systematic approach and enthusiastic endorsement of this alternative strategy helped turn folklore into medical practice (Jenner and Miller 1983). These accomplishments propelled him into a favored status in the history of vaccination and explain why he is often credited as the inventor of this therapeutic advance. As Francis Darwin (Charles Darwin's third son) noted: "But in science the credit goes to the man who convinces the world, not to the man to whom the idea first occurs". Regrettably, Jenner's seminal work on smallpox prevention would not be replicated for any other infectious illness until several generations later. In his era, disease itself was still a very mysterious phenomenon as was the body's response to disease. The scientific and medical communities of his time were ignorant about how bacteria and viruses caused sickness and about the role of human immunity in fighting microorganisms. This lack of knowledge precluded the easy adaptation of his vaccination concept to other prevalent ailments. Additionally, no other human viral disease had an obvious animal counterpart like cowpox that could be used as an attenuated vaccine component. This combination of issues kept vaccination unique to smallpox for decades until the brilliant and egotistical Louis Pasteur's triumph over rabies.

Vaccinia Emerges

As Jenner's cowpox vaccination protocol gained acceptance and dispersed to countries around the globe, the need for cowpox inoculum grew. Jenner provided many samples to colleagues who could then initiate their own supplies through person-to-person transmission and harvesting of pox material. In other instances, physicians or laypersons simply used pox material from their local cattle or sometimes horses. For most of the nineteenth century, no oversight, regulation, or record-keeping tracked the usage, quality, or nature of the material being used for vaccination against smallpox. During this period, vaccination remained a cottage industry handled at the local level of individual towns and cities. There were efforts in some European countries to encourage universal smallpox vaccination, although most wide-scale inoculation efforts were only in response to local outbreaks of the disease. For example, after the smallpox outbreak of 1837–1840, the British passed the Vaccination Act of 1840. This regulation outlawed variolation with smallpox and mandated Jenner's cowpox variolation free of charge to infants. As England and other countries began to make efforts to inoculate large numbers of people, a major handicap for vaccination programs was the inconsistent source of cowpox material. Spontaneous outbreaks of cowpox in cattle or other farm animals were sporadic and unable to provide the volume of material required for numerous vaccinations. Ultimately, a procedure developed in Italy in the early 1800s became the dominant method for vaccine production. This technique involved making multiple, long scratches in the flanks of calves and then seeding these wounds with cowpox material. The extensive rows of pox lesions that arose provided significant amounts of pustulant material for vaccine doses. However, while providing an ample source of vaccine material, this method didn't eliminate the uncleanness problems inherent in Jenner's vaccination protocol. As you might imagine, harvesting pus from a sore on a cow's skin did not provide a pure sample of the cowpox virus. Along with cowpox, this material was frequently contaminated with other viruses and bacteria that caused illness and death in some of the human recipients. Definitely not the ideal vaccine, but in the absence of an alternative this approach persisted into the twentieth century.

From the latter half of the nineteenth century into the early decades of the twentieth century, attempts were made to enhance the quality and consistency of the smallpox vaccine. Freeze-dried preparations were developed that greatly increased the stability, storage life, and ease of transportation of the vaccine, allowing vaccine distribution into many parts of the world that were previously excluded. In many countries, the production of the smallpox vaccine became more centralized as manufacturing capabilities for the vaccine were developed. Different countries and regions utilized independent versions of the CPXV for their vaccine manufacturing purposes, and all the vaccine strains became known collectively as vaccinia virus (VACV). Unfortunately, the provenances and propagation histories of these vaccinia "strains" are poorly documented as strains were often shared, mixed, and produced with little or no record keeping. Ultimately, in 1939 it was shown by immunological analysis that two major VACV strains in use for vaccination were not even CPXV (Downie 1939). Every type of analysis since that time has confirmed that VACV and CPXV are related but distinct viruses. Even different vaccinia strains that were widely used in twentieth-century vaccination programs turn out to be substantially different at the genome level. Modern DNA sequence evaluation of VACV genomes reveals that vaccinia strains have a complex and uncertain lineage (Qin et al. 2015). Interestingly, current VACV strains have closer ties to horsepox than cowpox, a fact consistent with the historical records that Jenner and others used horsepox as well as cowpox as a vaccine source. Another important conclusion from sequencing data is that VACV is not a natural virus that ever existed as an infection of some animal species. Instead, VACV appears to be an artificial hybrid virus resulting from a mixing of various progenitor virus types (animal poxviruses and human VARV). When various individuals, organizations, and countries prepared and shared smallpox vaccines there was ample opportunity for different types of poxviruses to be introduced into the samples used to inoculate people or production animals. During mixed viral infections, the viral DNA genomes can exchange pieces (called recombination) and undergo genome rearrangements such as deletions and duplications of some DNA regions. Without any understanding of these processes or any methods for detecting these changes, the resulting vaccine viruses accumulated significant changes over time. It is safe to conclude that modern VACV is a man-made virus resulting from this haphazard mixing and propagating of different poxviruses that occurred over a hundred-plus years of smallpox vaccine preparation. Yet even with this dismayingly vague and convoluted history, VACV proved to be an incredibly effective vaccine agent that facilitated the complete eradication of VARV as a human pathogen.

The Death of Smallpox

Several detailed accounts have been written about the twentieth-century campaign to eliminate smallpox (Tucker 2001; Henderson 2009; Behbehani 1988), so the reader can refer to those works for more extensive coverage of the demise of smallpox. For our purposes, the endgame began roughly 150 years after Jenner revolutionized smallpox protection with cowpox vaccination. It was at the Third World Health Assembly of the World Health Organization (WHO) in 1950 that Dr. Fred Soper proposed that smallpox could be eliminated in the Americas through an aggressive mass vaccination program. This strategy was dependent upon several factors unique to smallpox. First, VARV existed naturally only in humans. With no animal reservoirs for this virus, once all human cases were absent from an area then smallpox was gone unless reintroduced by an incoming infected person. Second, smallpox was a dramatic and acute disease with no asymptomatic cases. Therefore, it was easy to recognize infected individuals and the disease didn't spread silently like many other viral infections such as hepatitis C (Chap. 15) or SARS-CoV-2 (Chap. 13). This obvious disease presentation allowed infected persons to be identified rapidly and then quarantined to prevent spread while their contacts and fellow community members were quickly vaccinated. Lastly, there was the highly effective VACV vaccine, still produced in cows, that could be stored and transported as a dried powder to be rehydrated for administration at the final destination (Belongia and Naleway 2003). With a convenient and stable vaccine formulation, healthcare workers could carry the vaccine for days to reach remote areas that lacked refrigeration. This combination of factors proved phenomenally successful and by the late 1950s smallpox cases no longer occurred in North America or Europe, although there were still outbreaks in South America.

The successful elimination of smallpox from several nations on two continents led the WHO to propose an even more ambitious plan to completely eradicate smallpox from the world. In 1959, the WHO launched its first global attack on smallpox, but progress was slow due to a lack of sufficient funding, vaccine shortages, and only half-hearted commitment from some countries. It wasn't until the mid-1960s that the WHO made a renewed commitment to exterminating smallpox. That commitment, along with significant donations of vaccine doses and funding from the United States and the

U.S.S.R., was the turning point. This influx of resources, along with a new global dedication to the collective goal, finally produced the expected results. Using vaccination, smallpox was abolished in South America by 1971, followed by Asia in 1975, and Africa in 1977. After waiting several years to see if any smallpox cases would appear anywhere in the world, in 1980 the WHO declared victory over smallpox, and a centuries-old scourge was finally abolished. This is the first and still only human virus eliminated from the world, although poliovirus may soon be next (Chap. 6). It is fitting that the first viral vaccine, developed long before any scientific understanding of viruses or immunity, resulted in the first elimination of a human disease. The work of Jenner and many others established a new paradigm for preventing infections and started humanity on the pathway of controlling pathogens rather than being at the mercy of the microbial world.

References

Anselment RA (1989) Smallpox in seventeenth-century English literature: reality and the metamorphosis of wit. Med Hist 33(1):72–95. https://doi.org/10.1017/s0025727300048912

Babkin IV, Babkina IN (2015) The origin of the variola virus. Viruses 7(3):1100–1112. https://doi.org/10.3390/v7031100

Barquet N, Domingo P (1997) Smallpox: the triumph over the most terrible of the ministers of death. Ann Intern Med 127(8 Pt 1):635–642. https://doi.org/10.7326/0003-4819-127-8_part_1-199710150-00010

Baxby D (1981) Jenner's smallpox vaccine: the riddle of vaccinia virus and its origin. Heinemann Educational Books, London

Behbehani AM (1988) The smallpox story: in words and pictures. University of Kansas Medical Center, Kansas City

Belongia EA, Naleway AL (2003) Smallpox vaccine: the good, the bad, and the ugly. Clin Med Res 1(2):87–92. https://doi.org/10.3121/cmr.1.2.87

Boylston A (2012) The origins of inoculation. J R Soc Med 105(7):309–313. https://doi.org/10.1258/jrsm.2012.12k044

Carroll DS, Emerson GL, Li Y, Sammons S, Olson V, Frace M, Nakazawa Y, Czerny CP, Tryland M, Kolodziejek J, Nowotny N, Olsen-Rasmussen M, Khristova M, Govil D, Karem K, Damon IK, Meyer H (2011) Chasing Jenner's vaccine: revisiting cowpox virus classification. PLoS One 6(8):e23086. https://doi.org/10.1371/journal.pone.0023086

Chertow DS, Kindrachuk J (2020) Influenza, measles, SARS, MERS, and smallpox. In: Highly infectious diseases in critical care. Springer, pp 69–96. https://doi.org/10.1007/978-3-030-33803-9_5

Dinc G, Ulman YI (2007) The introduction of variolation 'A La Turca' to the West by Lady Mary Montagu and Turkey's contribution to this. Vaccine 25(21):4261–4265. https://doi.org/10.1016/j.vaccine.2007.02.076

Downie AW (1939) The immunological relationship of the virus of spontaneous cowpox to vaccinia virus. Br J Exp Pathol 20(2):158–176

Ferrari G, Neukamm J, Baalsrud HT, Breidenstein AM, Ravinet M, Phillips C, Ruhli F, Bouwman A, Schuenemann VJ (2020) Variola virus genome sequenced from an eighteenth-century museum specimen supports the recent origin of smallpox. Philos Trans R Soc Lond Ser B Biol Sci 375(1812):20190572. https://doi.org/10.1098/rstb.2019.0572

Franco-Paredes C, Lammoglia L, Santos-Preciado JI (2005) The Spanish royal philanthropic expedition to bring smallpox vaccination to the New World and Asia in the 19th century. Clin Infect Dis 41(9):1285–1289. https://doi.org/10.1086/496930

Grundy I (2000) Montagu's variolation. Endeavour 24(1):4–7. https://doi.org/10.1016/s0160-9327(99)01244-2

Henderson DA (2009) Smallpox: the death of a disease: the inside story of eradicating a worldwide killer. Prometheus Books, Amherst

Hughes AL, Irausquin S, Friedman R (2010) The evolutionary biology of poxviruses. Infect Genet Evol 10(1):50–59. https://doi.org/10.1016/j.meegid.2009.10.001

Jenner E, Miller G (1983) Letters of Edward Jenner, and other documents concerning the early history of vaccination. The Henry E Sigerist supplements to the Bulletin of the history of medicine, vol new ser , no 8. Johns Hopkins University Press, Baltimore

Koplow DA (2003) Smallpox: the fight to eradicate a global scourge. University of California Press, Berkeley

Li Y, Carroll DS, Gardner SN, Walsh MC, Vitalis EA, Damon IK (2007) On the origin of smallpox: correlating variola phylogenics with historical smallpox records. Proc Natl Acad Sci USA 104(40):15787–15792. https://doi.org/10.1073/pnas.0609268104

Macaulay TBM (1881) The history of England from the accession of James the Second. Seaside library, vol 48, no 976. G. Munro, New York

Martin DB (2002) The cause of death in smallpox: an examination of the pathology record. Mil Med 167(7):546–551

McCollum AM, Li Y, Wilkins K, Karem KL, Davidson WB, Paddock CD, Reynolds MG, Damon IK (2014) Poxvirus viability and signatures in historical relics. Emerg Infect Dis 20(2):177–184. https://doi.org/10.3201/eid2002/131098

Qin L, Favis N, Famulski J, Evans DH (2015) Evolution of and evolutionary relationships between extant vaccinia virus strains. J Virol 89(3):1809–1824. https://doi.org/10.1128/JVI.02797-14

Semba RD (2003) The ocular complications of smallpox and smallpox immunization. Arch Ophthalmol 121(5):715–719. https://doi.org/10.1001/archopht.121.5.715

Theves C, Crubezy E, Biagini P (2016) History of smallpox and its spread in human populations. Microbiol Spectr 4(4). https://doi.org/10.1128/microbiolspec.PoH-0004-2014

Tucker JB (2001) Scourge: the once and future threat of smallpox. Atlantic Monthly Press, New York

3

Rabies: From Attenuation to Inactivation

Keywords Anthrax • Cell culture • Germ theory • Robert Koch • Louis Pasteur • Zoonotic disease

Abbreviations

CNS	Central nervous system
HDCV	Human diploid cell vaccines
PCECV	Purified chick embryo cell vaccine
PDEV	Purified duck embryo vaccine
PEP	Post-exposure prophylaxis
PHKCV	Primary hamster kidney cell vaccine
PVRV	Purified Vero cell rabies vaccine
RABV	Rabies virus

The Microbial Paradigm of Disease

Decades passed following Jenner's promotion of cowpox vaccination as an effective preventative for smallpox, and science slowly began to recognize and appreciate the connection between microorganisms and disease. The concept that bacteria, viruses, and other microorganisms can infect humans and produce specific diseases is so entrenched in our modern lives that this concept almost seems intuitive. In contrast, at the start of the 1800s, the concept that microbes could cause diseases was largely unknown and foreign to the prevailing beliefs. Ancient societies mainly attributed disease to supernatural causes, viewing illness as a punishment or tribulation imposed by gods and spirits. As civilizations advanced and became more sophisticated, emerging physicians and philosophers argued and postulated about the nature and causes of illness. While diseases were still mostly attributed to mystical entities, early Greek writing proposed that illnesses arose from an imbalance or corruption in the four humors: blood (sanguine), black bile (melancholic), yellow bile (choleric), and phlegm (phlegmatic) (Lagay 2002); these four humors also corresponded to the natural elements of air (blood), earth (black bile), fire (yellow bile), and water (phlegm). This idea that disease stemmed from real-world factors rather than supernatural causes, was an important conceptual advance even if it would take centuries before there was an actual scientific understanding of diseases. Ascribing disease to causes that were at least somewhat under human control meant that tangible therapies could be developed as an alternative to simply praying to some deity. Thus, adherents of the humoral theory could try to cure illnesses by re-establishing the correct equilibrium among these elements. For example, fevers ("hot blood") were often treated by reducing "excess" blood using leeches or by making cuts in the patient. While mostly ineffective, and some perhaps even harmful, the humoral therapies at least encouraged medical practitioners to try something to help patients and provide relief for various ailments. This spirit of active intervention rather than passive acceptance of disease underlies all the progress that eventually led to modern medicine.

Although the humoral model persisted well into the 1800s, it competed with the miasma theory as an explanation for how disease occurred and spread (Karamanou et al. 2012; Kannadan 2018). Also originating in ancient Greece, the miasma theory attributed illnesses to "bad" air such as the emanations from decaying corpses and plants, and even from the exhalations of diseased people. These invisible, noxious vapors, especially ones with putrefying odors, were believed to invade the body and cause various illnesses. Gaining popularity from the Middle Ages on, the miasma theory permeated throughout Western medicine until the mid-1800s and even influenced the name of a well-known disease, malaria. The word malaria derives from the Italian *mala aria* and literally means bad air, reinforcing the errant belief in the eighteenth and early nineteenth centuries that this illness was acquired via inhalation of inert toxins. We now know that malaria is spread by infected mosquitoes, but it's understandable how a disease transmitted by these small, ubiquitous, flying insects could easily seem like it was simply floating in the air.

At the beginning of the 1800s, healthcare practitioners and the general public largely believed that miasmas were responsible for most common maladies. Swamps, marshes, and stagnant waters, along with offal, were often indicted as sources of disease outbreaks. While civic attempts to clean up these presumed culprits may have generally improved sanitation and indirectly helped prevent some diseases, such as water-borne or mosquito-transmitted infections, these improvements had little impact on most common ailments of that period. Additionally, this erroneous focus on miasmas was so ingrained that many in the medical community were hostile and resistant to the growing evidence in the second half of the nineteenth century that invisible microorganisms were the real etiological agents of infectious diseases. As with any widely held and popular belief, challenges to the miasma belief were met with skepticism and a refusal to consider alternative explanations. Nonetheless, the so-called "germ theory" slowly gained converts and was eventually irrefutably proven by scientific experimentalists.

Jenner's success with smallpox prevention proved to the world that it was possible to thwart a disease even if the etiology was not understood. Yet the smallpox story remained an isolated case that was not generalizable to other common ailments. Without a cowpox equivalent to use as an inoculum for other human diseases, there was no obvious source of vaccine material. And just as importantly, as long as illnesses were ascribed to intangible and immaterial miasmas or humors, there was no intellectual or practical approach for exploring disease causes and transmission mechanisms. Citizens of that era thought that diseases just happened, like inclement weather, and trying to prevent illness seemed no more feasible than trying to stop or control storms. Overcoming this conceptual barrier and providing a scientific framework to study all infectious diseases is the great contribution of nineteenth-century scientists. Prominent among these pioneers were two individuals that we recognize as giants in microbiology, Robert Koch (1843–1910) and Louis Pasteur (1822–1895). Pasteur and Koch were contemporaries, rivals, and major intellectual forces whose work firmly established the microbial basis for infectious diseases. The detailed backstory of each scientist can be found in Paul De Kruif's classic 1926 book, *Microbe Hunters* (De Kruif 1926). For our purposes, it will suffice to focus on a few of their most significant achievements relevant to the development of germ theory and vaccines.

Louis Pasteur trained as a chemist and made his first important scientific contributions concerning the structure of tartaric acid crystals, a substance found as a byproduct of wine fermentation (Berche 2012). These early studies embued him with a life-long interest in fermentation that led him into the microbial world. While working on production problems in the French beer industry he detected tiny structures in the fermenting fluid that could only be seen by microscopy. Months of careful, even obsessive, observation eventually convinced him that these structures were actually living organisms (yeast) whose presence was essential for the fermentation process. Years of additional painstaking work on fermentation in milk, wine, and vinegar confirmed his impression that all fermentation required either yeast or bacteria to proceed, but led to a new question—where did these minuscule life forms come from? The prevailing theory for the origin of new life was spontaneous generation, the ability of living organisms to arise from inanimate material. People believed that new life would arise de novo from putrefying organic material such as dead plants, animals, or other decaying refuse. This belief seemed obvious since such dead material always began to rot and become infested with tiny insects without any apparent source of these small creatures. Pasteur designed elegant experiments to test this hypothesis and showed that organic material that was sterilized and then protected from the air never rotted or became infested. His series of studies finally changed public opinion and convincingly disproved this long-held certainty in spontaneous generation. Instead, Pasteur proved that new microorganisms only arose from pre-existing microorganisms and that the pre-existing microbes were widely prevalent in air, soil, and water where they could easily contaminate decaying substances. These findings radically changed the perception of the microbial world and created the framework for a new science of microbiology. Armed with these new concepts, Pasteur was next called to investigate diseases in silkworms which were an important industry in France. Several years of dedicated inquiry resulted in his demonstration that two prevalent silkworm diseases were caused by two different bacteria, one of the first direct links between bacterial infections and specific diseases. From his decades of study on bacteria and yeast, Pasteur became convinced that these microscopic creatures were more generally the cause of various illnesses. Pasteur's so-called germ theory, posited first in an early form in an 1861 paper, would have a profound effect on Robert Koch who became familiar with it early in his medical training.

Unlike Pasteur who performed most of his early studies on fermentation and agricultural problems, Koch was trained as a physician and focused exclusively on human diseases (Blevins and Bronze 2010). Koch was very receptive to the germ theory and ultimately spent his life researching bacteria. He is credited with many innovative and significant advances that helped establish bacteriology as a precise and reproducible discipline. Among his most notable achievements are the development of culture media for growing bacteria (including the invention of agar plates where visible bacterial colonies could form), the invention of staining methods to visualize and distinguish bacteria, and the refinement of many microscopic techniques for viewing individual bacteria. Using his considerable skills and expertise, Koch attacked the anthrax problem in Germany where hundreds of people and thousands of cattle died each year. Following Pasteur's lead, he rapidly established through careful experimentation that this was a transmissible disease caused by a rod-shaped, spore-forming bacterium (subsequently named *Bacillus anthracis*). Koch's anthrax finding was a turning point in medical history as it was the first study to compellingly link a human disease with a specific bacterial organism. Importantly, just as Pasteur's germ theory had influenced Koch, Koch's publication of his anthrax results in 1876 would profoundly impact Pasteur's later research efforts.

The combined work of Pasteur and Koch, along with contributions from many of their contemporaries, finally convinced the world that microorganisms not only existed but were responsible for many diseases. Although Pasteur and Koch together only identified the specific microbes associated with a small number of illnesses, their studies suggested that this microbial paradigm for disease would be broadly applicable to other human maladies. Their concept that each infectious disease resulted from a single, unique microbial agent stimulated a frantic quest to isolate and identify the causative agents for every human illness, a search that continues to this day.

On a side note, as often happens in science, certain ideas that turn out to be correct surface before their time and fail to gain traction in the world. This can be due to the lack of sufficient evidence or just the unwillingness of the public to embrace a new concept. One such example belongs to the sixteenth-century Italian polymath, Hieronymi (aka Girolamo) Fracastoro (Pesapane et al. 2015). He made contributions to several fields from literature to medicine, including coining the word syphilis for the rampant disease that we now know is caused by a sexually transmitted bacterium called *Treponema pallidum*. In 1546, three centuries before Koch and Pasteur, Fracastoro published a work called "De Contagione". In this book, he added substance to the miasma theory by proposing the existence of invisible particles that spread disease. He called these particles "spores" or "seeds" and postulated that they floated through the air to infect people. He also conjectured that these particles could contaminate surfaces where they could be picked up on hands and transferred to people unwittingly. These nearly 500-year-old speculations are amazingly prescient about how many viruses spread via respiratory droplets that are inhaled by nearby people or are picked up on hands after the droplets settle on surrounding objects. Fracastoro even appears to be the first to use the term "fomite" to describe surfaces and objects contaminated with infectious material, a term still used today in science and medicine. Yet while fundamentally correct, Fracastoro's theory remained largely ignored by medical practitioners until Koch and Pasteur finally exposed the error of the prevalent dogmas. With their irrefutable experimental evidence, they quashed existing beliefs and decisively established that microorganisms, not imbalanced humors or some vague "bad air", were the cause of infectious diseases. Koch is also remembered for his simple, yet elegant, recipe describing how to establish the connection between a particular organism and a specific disease. The eponymous Koch's postulates state that for proof of causality the following conditions must be met (Grimes 2006):

- The same organism must be present in every case of the disease.
- The organism must be isolated from the diseased host and grown in pure culture.
- The pure isolate must cause the disease when inoculated into a healthy,
- susceptible animal.
- The organism must be reisolated from the inoculated, diseased animal.

These rules have guided investigators ever since Koch proposed them, although we'll see that fulfilling these four steps is not always easy or feasible for some viruses.

The Birth of Vaccines

Koch's anthrax studies provided solid confirmation of Pasteur's long-held belief that germs caused human illness, not just agricultural problems. Such identification of human pathogens was critically important, but Pasteur immediately realized that the real prize was using knowledge about microbes to save lives. Pasteur was always about practical science (what today we would call applied science) that used research to solve real-world problems, and now there were opportunities to apply his skills to farm animals and humans, not just plants and insects. Consequently, Pasteur altered his research interests to focus on human disease-causing bacteria with an intent to find new ways of treating or preventing these infections. Taking up this new challenge, Pasteur's attention turned first to the anthrax problem in France as he began to study diseased cattle with the same zest and insight that he brought to his former studies.

However, Pasteur was not a physician and lacked both medical knowledge and the technical skills for working with animals and humans, leading him to recruit young physicians as his assistants. Among these assistants was Pierre Paul Émile Roux who joined the lab in 1878 and played a prominent role in Pasteur's later rabies vaccine work.

One of Pasteur's early experiments with anthrax involved inoculating four cows with live anthrax and testing a purported cure touted by a veterinarian named Louvrier (De Kruif 1926). Two cows were given the cure treatment and two were left untreated. In both groups, one cow died and one cow survived. By the standards of modern science, this was not a large enough sample size to have statistical significance, and today these results would be considered worthless. Nonetheless, Pasteur claimed that the cure was ineffective since the test group fared no better than the control group, a conclusion that happened to be true even if his experiment did not rigorously prove this assertion. More importantly, Pasteur used the surviving cows in another experiment where they were injected with a more potent batch of anthrax to see what would happen. Neither cow developed any sign of anthrax, not even lesions at the site of injection, a response equivalent to the cowpox-treated people not responding to smallpox variolation. As with the cowpox work from the previous century, Pasteur intuited that one infection with anthrax imbued surviving animals with immunity to subsequent infection with this organism. Now the solution became apparent, he just needed to find some way to produce a mild, survivable anthrax infection and the recipient would be safe from serious disease if exposed in the future. Without yet naming it or understanding how and why it worked, Pasteur was starting to develop the general principle of attenuation, a principle that would form the basis for many future vaccines. The fundamental theme that would emerge from his work was simple: the administration of a weakened form of a microbe (the attenuated form) would generate protective immunity against the virulent form of that same microbe. As with the cowpox/smallpox situation, we now understand why attenuated microbes work as vaccines. Each type of microorganism has a specific size and shape, and its surface is covered with a unique variety of biomolecules such as proteins and carbohydrates. When a pathogenic organism enters our body, these surface molecules (collectively called antigens) are detected as foreign by our immune system and we initiate an immune response that targets these surface molecules and helps control or destroy the pathogen. These same surface antigens are also present on the attenuated microorganisms, so exposure to the attenuated strain evokes the same array of immune responses as the virulent strain yet without producing serious disease. The resultant immune memory responses allow for a rapid, effective, and highly protective immune attack on that pathogen should it ever be encountered again. Still, although attenuation is simple in concept, the difficulty for Pasteur and future vaccine developers was finding a process for generating safe, reproducible, attenuated organisms that retained their surface antigens in a native form.

While struggling with anthrax, Pasteur's research group was also working on a different disease problem, chicken cholera, caused by a recently discovered tiny bacterium (subsequently named *Pasteurella multocida*). In an 1854 talk delivered to the University of Lille, Pasteur stated that "in the fields of observation chance favors only the prepared mind", and his mind was certainly prepared for the fortuitous accident that was about to occur with the cholera research. Cultures of the cholera bacteria were routinely grown in Pasteur's lab and tested for virulence by injecting them into chickens; such infected birds typically died within 24–48 hours. On one occasion a weeks-old culture was used for the inoculation rather than a fresh culture and while the infected animals all got sick they eventually recovered and survived. Similar to what he observed with anthrax, when these recovered birds were subsequently given a lethal dose of fresh cholera culture they did not become ill. These observations established that aging the cholera cultures produced an attenuated material that could be injected into chickens to provide immunity. Without any knowledge of bacterial genetics, Pasteur inadvertently discovered the malleable nature of bacteria. Rather than having permanently fixed properties, bacteria can change their characteristics through mutation and other genetic processes. Because bacteria reproduce in just minutes, growing cultures for days represents hundreds of generations, plenty of time for genetic changes to arise and become established. We now know that one such common change in pathogenic bacteria is the loss of virulence when they are grown for prolonged periods on artificial media rather than reproducing in susceptible animals. Pasteur could not have conceived of nor understood the mechanisms at play, but he was quick to appreciate that changing bacterial growth conditions was a practical solution for creating attenuated organisms. In honor of Jenner's work with variolae vaccinae, Pasteur called his attenuated cholera material a vaccine, and the word "vaccine" has become a general term for any substance used to stimulate protective immunity against a disease. Eventually, this strategy of finding growth and propagation conditions that created attenuated strains would also be applied to viruses as they are just as genetically protean as bacteria.

As active, productive, and triumphant as Pasteur's lab was, they were not the only ones working on vaccine development. A countryman of Pasteur, Jean Joseph Henri Toussaint, is now credited with the first successful anthrax vaccine development. Rather than finding appropriate growth conditions to attenuate the organism, Toussaint explored physical and chemical treatments to reduce virulence. Ultimately he demonstrated that heating the culture to 55 °C in the presence of a small amount of phenol produced a safe and effective vaccine material. This approach is a splendid example of the second major category of vaccines, the inactivated or killed organism vaccines. For this type of vaccine, the challenge is to define conditions that kill the organism without disrupting its antigenic presentation. If the killed bacteria retain their critical surface antigens intact then injecting these dead microbes will still elicit protective antibodies that can prevent future infections. As we will see in subsequent chapters, many viral vaccines in use today are either attenuated or inactivated type vaccines that were created using the concepts established by early investigators such as Toussaint, Pasteur, and their contemporaries.

The Mad Dog Disease

While Pasteur's group was grappling with chicken cholera, they were also beginning studies with rabid dogs and would soon make history again with an anti-rabies vaccine. Rabies was endemic in Europe during Pasteur's lifetime, and even today, few infections elicit more terror and horror than this dreadful disease (Wasik and Murphy 2012). The ghastly and unique constellation of symptoms associated with rabies plus the inevitably fatal outcome once symptoms appear, make this infection one of our most feared diseases. Fortunately, rabies is extremely rare in developed nations and few of us will ever personally encounter a victim. Nonetheless, fictional accounts (movies, television, books) are widespread and have acquainted most of us with the basic features of this illness. From Old Yeller's tragic death in the eponymous 1956 novel (Gipson 1956) to the slavering Saint Bernard in Stephen King's Cujo (King 1981), rabid dogs haunt popular culture. Even when these fictional presentations are a bit hyperbolic, they often capture many of the truly devastating consequences of a rabies infection. The symptoms can vary but generally present as either an encephalitic form (furious or classical) or as paralytic (dumb) rabies (Scott and Nel 2021). Victims with rabies initially exhibit nondescript symptoms such as fever, chills, malaise, vomiting, and headache, and they may show signs of intolerance to light and noise. Soon after these early symptoms, behavioral changes may develop and present as depression, insomnia, or irrational anger. About 80% of infected people will progress to the encephalitic form where they display more pronounced neurological dysfunctions that can include hallucinations, confusion, agitation (thrashing and biting), hyperactivity, seizures, and hydrophobia (fear of water)—all the signs of madness that are classically associated with rabies. Along with these outward manifestations of the infection, serious alterations are occurring in organ functions, particularly the heart, as well as progressive muscle paralysis. In the other 20% of patients, delirium, hydrophobia, and other neurological symptoms are absent and the patients just progress slowly to the paralytic stage. For both the encephalitic and paralytic forms, death results from muscle paralysis that causes heart failure and/or suffocation as the respiratory muscles fail, a slow and unpleasant way to die. Even with all the weapons of modern medicine we still lack an effective therapy for this disease once symptoms have begun, and only a few rare individuals have ever survived rabies.

Unlike smallpox whose historical legacy is vague until the early centuries CE, rabies is prominently featured in writing as far back as Sumerian texts from roughly 2000 BCE (Yuhong 2001). Ancient civilizations around the world, including Greece, Egypt, Persia, and China all left written descriptions of a malady that must be rabies based on the symptoms unlike those of any other disease (Tarantola 2017; Dalfardi et al. 2014). For example, madness (bizarre behavioral changes) and hydrophobia were widely recognized as hallmarks of this illness, as was its fatal and incurable outcome. As stated in the ancient Vedic text Sushruta Samhita: "If the patient becomes exceedingly frightened at the sight or mention of the very name of water, he should be understood to have been afflicted with Jala-trsisa (hydrophobia) and be deemed to have been doomed" (Suraweera et al. 2012). Hydrophobia was also noted by the Roman philosopher, Pliny, who stated that "rabies … is dangerous to human beings … as it causes fatal hydrophobia" (Pliny et al. 1961). Interestingly, many of these texts mention dogs and dog bites as the source of human cases, so the canine connection was apparent long ago even if the nature of the contagion wouldn't be established until the modern era. For example, the Greek physician Philumenos noted in the second century CE that the rabid dog "does not eat or drink, has saliva flowing from its mouth, and appears irritable and aggressive" (Theodorides 1984). There is even a threat from Marduk, the patron god of Babylon, that invoked dogs and rabies as punishment (Yuhong 2001): "Dogs will become rabid and bite people. All the persons whom they shall bite will not survive but will die." Whether ascribing the illness to a deity's punishment, an imbalance of humors, or some other mysterious source, the written depictions of rabies from widely distributed cultures indicate that the disease was well established in dogs across Europe, Asia, and the Middle East thousands of years ago.

The presence of rabies in dogs is important because dogs are such an integral part of society, from pampered domestic pets to feral creatures that roam cities and the countryside (Wang 2019). Their frequent proximity to humans and human habitats makes them a prime vector for rabies transmission into people. Because of our familiarity with and acceptance of dogs, we generally think of them as friendly neighbors rather than potential disease carriers. This perspective tempers our natural wariness of animals and can prevent us from recognizing aberrant dog behavior until too late to avoid an attack. However, as important as dogs are for human cases of rabies, dogs are not alone as rabies carriers (Gilbert 2018). Rabies is an endemic zoonotic disease that circulates in wild animal populations. Certain mammals, particularly carnivores, are susceptible to this disease. Foxes, raccoons, skunks, bats, and monkeys remain prominent sources of rabies in different regions of the modern world. Unfortunately, many other wild animals, as well as domestic animals can also harbor rabies. Cats, cows, horses, sheep, goats, and pigs can all be infected with rabies and pose a danger for transmission to humans. Infected animals, wild or domestic, shed the rabies virus in their saliva, making bites the primary method of transmission. Self-grooming activities by an infected animal may deposit virus-infested saliva on the paws and claws, so scratches can also transmit the disease. Even licks from infected animals can transmit the rabies agent if the saliva contacts mucosal surfaces (eyes, inner nose, and mouth) or open wounds in the skin. Since infected animals (and humans) produce viruses in their saliva in advance of the obvious symptoms, any contact with a wild animal that is a known carrier species or bites/scratches from an unvaccinated susceptible domestic animal must be treated as a potential threat for rabies infection. On a positive note, rodents (rats, mice, squirrels, hamsters, gerbils, etc.) and lagomorphs (rabbits and hares) rarely if ever carry rabies. Human rabies cases from these animals have never been documented so there is no need to undergo rabies treatment for contact with or even bites from these animals (although other diseases can result from their bites).

The Rage Virus and Its Neural Death March

While Pasteur and his contemporaries could only speculate about the nature of the rabies agent, research from decades later established that the causative agent is an enveloped RNA virus that in scientific nomenclature is designated *Rabies lyssavirus* (species name) in the genus *Lyssavirus* of the family Rhabdoviridae. The Latin word "rabies" originated from Sanskrit where "rabbahs" meant "to do violence" (Fu 1997), an apt descriptor for animals or humans afflicted with this terrible illness. Similarly, Lyssa in Greek mythology was a spirit that embued rage, frenzy, and madness in animals, and the *lyssavirus* name acknowledges both the zoonotic hosts of this virus and the effects of the virus on the host's behavior. As RNA viruses go, the rabies virus has a fairly average genome size of approximately 11,000 nucleotides that only encode five proteins (G, L, M, N, and P). Five proteins may seem a paltry number for a virus that produces such catastrophic outcomes in its victims, yet these five enable the virus to reach the central nervous system (CNS) and reproduce effectively where the pathogenic damage occurs. Although neither the genome size nor the protein capacity is unusual for an RNA virus, the virus does have one obvious and remarkable feature, its shape (Fig. 3.1). The rabies viral particle resembles a bullet, being a rod-shaped cylinder that is flat on one end and rounded on the other. This shape is unique to the rabies virus and its relatives in the *Rhabdoviridae* family (Rhabdo coming from "rod" in Greek) as most other viruses generally have spherical shapes.

In addition to the rabies virus (RABV), the genus *Lyssavirus* contains numerous members, many of which are highly related to RABV. Modern lyssaviruses arose tens of thousands of years ago, most likely in the Palearctic region (Eurasia together with North Africa and the temperate part of the Arabian peninsula) with subsequent spread throughout the world (Hayman et al. 2016; Fisher et al. 2018). Bats may be the original host species for lyssaviruses as bats throughout the world are the primary carriers of these viruses. When and how RABV evolved and entered other animals remains unclear. The historical record of disease description remains our best evidence that rabies and therefore some type of RABV has been plaguing humans for at least 4000 years.

While the origin of RABV remains uncertain, its effects on humans are well characterized. Once the rabies virus is introduced into a victim via a bite, scratch, or lick, the virus infects muscle cells at the site of entry and slowly begins to replicate (Begeman et al. 2018; Scott and Nel 2021). Viruses can persist in this location for days to months until some viral particles (virions) eventually enter the peripheral nerves, often through neuromuscular junctions called motor endplates. After entering the peripheral nerves, the virus

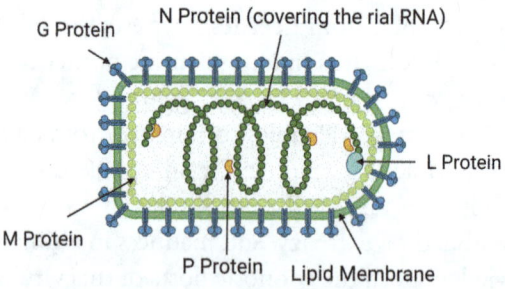

Fig. 3.1 The Rabies Virion. Shown is a single rabies virus particle (a virion) with its distinctive bullet shape. The location of the five viral proteins is indicated. G is the glycoprotein that is embedded in the lipid membrane coat, M is the matrix protein organized inside the lipid membrane, L is the viral polymerase that functions to synthesize new RNA copies, N is the nucleocapsid protein that coats the viral genomic RNA, and P is a phosphoprotein that acts with the L protein to facilitate replication. The coiled structure is the viral RNA genome entirely covered by bound N proteins

continues to replicate as it slowly ascends the nerve fibers (a process called retrograde axonal transport) on its inexorable journey to the CNS where the critical pathology will begin. Along the way, the virus must evade immune detection and destruction during its prolonged travels, and multiple viral proteins are critical for shielding the virus from immune attack (Davis et al. 2015). Innate immunity, particularly the interferon response, is an early defense system against viral infections that is critical for controlling viral spread before adaptive immunity (B and T cells) develops. Several rabies viral proteins can inhibit or disrupt aspects of the innate immune response resulting in a reduced ability of this system to fight the viral infection. The combined efforts of these proteins in disabling early immune functions are likely important in allowing the virus to establish itself and persist in muscle and nerve tissue for long periods.

After successfully thwarting the immune response and traversing along a nerve fiber, the virus finally arrives at its target, the central nervous system. Once the virus enters the CNS there are additional less well-understood immune avoidance mechanisms at play (Scott and Nel 2016), but an important element is simply the location. The CNS is considered an immunologically privileged site where immune responses are more tightly controlled. By infecting and exerting its major pathogenic effects in the CNS, the rabies virus is targeting an organ less able to defend itself than other sites in the body. Upon reaching this final destination, the rabies virus spreads and proliferates within the brain. Given the dramatic and horrid consequences on neurological function, it is somewhat surprising that the virus does not cause massive destruction of brain cells (Scott and Nel 2021). Instead, there is mild inflammation and only modest evidence of visible damage to nerve cells. Because the brain cells are not physically destroyed and remain mostly normal in appearance this suggests that the virus must be exerting its effects by disrupting cellular functions rather than directly killing cells. A variety of molecular processes that may contribute to the viral effects are being investigated and should eventually reveal more about pathological mechanisms in the future (Scott and Nel 2016). As the viral infection of the brain expands, the neurological dysfunction intensifies over 1–2 weeks until the patient finally succumbs. During this period of brain infection, newly made viruses begin to move back down the nerve fibers to various organs where localized viral replication and concomitant cellular dysfunction may also contribute to some of the clinical symptoms. Significantly, this dissemination also carries viruses into the salivary glands where they populate the saliva for transmission to the next victim. This cycle of infection, transmission, and death occurs relentlessly in animal populations across the globe to maintain rabies endemically as a pervasive zoonotic disease. Humans are not part of the endemic cycle but remain at risk due to incidental contact with infected animals. The World Health Organization estimates that there are approximately 60,000 human deaths from rabies annually year after year. Even if the true molecular nature of the viral pathology remains elusive, it is certainly easy to understand the clinical and social impact that this frightful virus continues to exert.

One final important feature of rabies infection is the long incubation time before disease symptoms appear, a reflection of the slow process of virus movement from the body surface to the CNS. The time between initial infection and clinical symptoms averages 2–3 months but has a wide range from a few weeks to even years in some cases (Boland et al. 2014). These extreme differences seen in individual incubation periods may stem from three factors. First, there is likely variability in the time of viral entry to the peripheral nerves. In some individuals, the virus may linger for prolonged times in the muscles where the virus was initially seeded before the peripheral nerves ever become infected. Second, the rate at which virions progress up the nerves and into the brain may not be equivalent in every person, and differing movement rates would impact time to disease. However, we do know that distance also matters during the incubation period. Infections that occur on the head and neck generally proceed to disease faster than those on the extremities such as hands and feet. This appears to be a simple function of the distance that the viruses have to travel through nerve fibers to reach the brain; shorter distances to travel yield shorter incubation times. Lastly, strain-specific differences in the virus likely contribute to both the pace of infection and the type of presentation, encephalitic or paralytic. Fortunately, all of these factors prevent the virus from reaching the brain quickly, allowing successful medical intervention after infection. As first demonstrated by Pasteur, because of the long incubation period for this disease it is possible to vaccinate after exposure to the virus. Between the initial infection and entry into the brain, there is sufficient time for the vaccine to train the adaptive immune system and produce a protective immunity that prevents a fatal neurological attack by the virus. This so-called post-exposure prophylaxis (PEP) is effective for rabies and has become the standard of care. However, PEP is a rarity because most other common viruses have much shorter incubation periods and their vaccines must be administered before infection to be protective.

Origin of the Rabies Vaccine

People in nineteenth-century Europe were well acquainted with rabies and its transmission by rabid dogs, and they were just as familiar with the lack of any effective treatment for the poor souls attacked by rabid animals. What remained a puzzling mystery was the causative agent of the disease and how to combat it. Nonetheless, even without the ability to visualize, isolate, or grow the rabies agent, some progress was being made toward understanding this disease. In 1804, Georg Zinke took saliva from a rabid dog and used a small brush to spread the saliva into wounds he made on a Dachshund's leg. The Dachshund subsequently developed rabies confirming that the causative agent was present in the saliva. Using the same approach, Zinke also induced rabies in cats, rabbits, and chickens, thus establishing the broad host range of this agent (Wilkinson 1988). Within a few years, Zinke's experiment would be duplicated by Francois Magendie using saliva from a human rabies victim. Magendie, a prominent French physiologist, showed that the saliva from this patient would produce rabies when inoculated into a healthy dog. He didn't publish his results until many years later in 1821, but this work was additional evidence that rabies was a saliva-borne agent that could be transmitted from an infected animal or human. In addition to saliva, work by others was noting the connection between rabies and the brain, even going so far as to show that pieces of the sciatic nerve from a rabid cat, when placed in a fresh wound, caused rabies in a dog (Tarantola 2017). As important as these early observations were, the next several decades saw little progress in defining the rabies agent as rabies remained refractory to methods that identified bacteria and linked them to specific illnesses. Unlike other diseases whose causes were being exposed by intrepid researchers, the rabies agent remained invisible and resistant to propagation in culture media. It would take the careful studies of the insightful Pierre Victor Galtier to finally establish workable conditions for handling rabies in the laboratory.

Galtier, a veterinarian and professor at the National Veterinary School of Lyons in France, made two important observations about rabies that would influence Pasteur's seminal studies on this disease. His first advance was to build on Zinke's work with rabbits. Galtier showed that not only could rabies be transmitted from dogs to rabbits, but also from rabbit to rabbit using the cutaneous injection of saliva, observations that he published in 1879. The ability to passage the disease sequentially in rabbits was advantageous because rabbits were smaller and easier to maintain in the lab than dogs and the disease also had a shorter incubation period in rabbits of 2–3 weeks. Both these factors allowed for more convenient and rapid experimentation, as well as providing a steady source of infectious material for the studies, and Pasteur would soon incorporate rabbits into his rabies work. Going further, Galtier tried and failed to induce rabies in sheep by intravenous injection of saliva, giving a tantalizing hint that rabies did not reach the brain via the bloodstream and must be reaching there by some other route. Importantly, when he tried to induce rabies in these treated sheep with standard inoculations they were now resistant to the disease. Galtier perceptively realized that he had inadvertently discovered a method of vaccination, and this method was even used for a time to protect sheep. Similar to smallpox variolation into the skin, introducing live RABV into the bloodstream was an unnatural route that induced immunity while not allowing the virus to reach the CNS and cause disease. Still, the concept of injecting virulent rabies into the bloodstream was likely not an attractive idea for protecting people or even animals, and Galtier's work never progressed to a generally acceptable vaccine.

By the time Galtier published his sheep study in 1881, Pasteur and his collaborators had already begun their initial studies with rabies in dogs, and they quickly adopted rabbits as an additional model for the disease. As with cholera and anthrax, Pasteur's intent was not just to understand the basic features of rabies but always included a single-minded focus on finding treatments or preventative measures. Drawing from his bacterial studies, attenuation seemed to be the key that might tame rabies, whatever the agent might be. Jenner was lucky that naturally occurring cowpox acted like an attenuated version of smallpox while Pasteur had no such mild version of rabies to utilize in his experiments. Instead, he needed to invent a method for reliably attenuating the actual RABV, and it was his success in this process that has history acclaiming him as the inventor of the rabies vaccine. One of the first innovations developed by Pasteur and his collaborators was to use brain material rather than saliva as the inoculum. As you can easily imagine, collecting saliva from a live rabid dog or other animal is fraught with danger for the collector. Instead, an euthanized animal's brain stem could be dissected, minced, and used to infect another animal with less risk to the investigator. They also found that rabies could be induced by the direct introduction of infected brain material into the brain of the recipient animal. This intracranial inoculation procedure was much more reliable and produced disease in virtually all the test animals with a shorter incubation period than inoculation into the skin or muscle. Using the rabbit model, consecutive passage through 21 rabbits yielded an adapted rabies strain that consistently produced disease in 8 days, what Pasteur referred to as a "fixed" strain. Now, even without being able to grow the rabies agent in a culture like a bacterium, this was a method to maintain the agent in the lab in a constant form that would make future experimentation less variable.

With the tools developed for careful and trustworthy studies, the next step was finding an attenuation strategy (Wasik and Murphy 2012). In a long series of trials, Pasteur's assistants, particularly Emile Roux, explored host adaptation as a means for attenuation. Passage of the rabbit "fixed" rabies strain through different animals did alter its virulence in some cases but failed to produce a fully attenuated strain suitable for a vaccination trial. Whether or not Pasteur's rabbit strain was less virulent in humans is unknown as direct testing on humans was impossibly dangerous and could not be done. Seeking an alternative method, Roux tested desiccation as a means to inactivate the virus. Spinal cord material from rabid rabbits was dried under sterile air for different lengths of time then emulsified in sterile liquid and injected into the brains of dogs. Over the 5 years from 1880 to 1885, Pasteur's group tested and optimized dried spinal cord material as a source of a rabies vaccine. Their studies established that a drying time of 14 days yielded a material that did not cause rabies and protected the dogs from a subsequent challenge with fresh rabies virus. Importantly, this vaccine worked even if dogs were first inoculated with virulent rabies and then given the spinal cord material, establishing the post-exposure protection concept. While Pasteur would not know the basis for this inactivation, his result is not surprising to modern virologists. The outer layer of RABV is a membrane envelope, a lipid structure that you can think of as a soap bubble, that contains the surface antigens. Many viruses have membrane envelopes and these envelopes are very sensitive to drying. Drying typically destroys the membrane which eliminates virus infectivity while leaving the surface antigens intact. It likely took Pasteur's RABV 14 days to inactivate because it was protected within the moist spinal cord material, and it took longer for that material to completely dry. The virions within the dried material, while no longer infectious, would have retained their antigenicity and been capable of eliciting protective antibodies against the virus.

Pasteur's success in protecting dogs from rabies attracted widespread attention that led to a momentous human trial of his vaccine. In 1885, 9-year-old Joseph Meister from a small town in northeastern France was mauled by a rabid dog. Faced with his certain death, Joseph's family approached Pasteur and implored him to use his rabies vaccine. The historical records indicate that Pasteur had his trepidations about using the vaccine on a human (Wasik and Murphy 2012): "I have not dared to attempt anything on men, in spite of my own confidence ….I fear too much that a failure might compromise the future." Yet despite the dangers and his personal concerns, the absence of any other treatment overcame his hesitation. On July 6, 1885, approximately two and half days after the dog attack, Joseph began a series of inoculations with the dried rabbit spinal cord material. One milliliter of emulsified material was injected under the skin of his abdomen, an unpleasant delivery protocol that would persist until the 1980s and which made many people fearful of this vaccine. On the first day, he received material dried for 14 days, and over the next 10 days, he was given 12 more injections of material dried for successively less time. These later injections had increasing amounts of virulent RABV, but whether or not this increasing exposure to live virus was necessary to provide protection is unclear. What is clear is that Joseph survived and lived a long, normal life until his death during World War II. Preventing rabies in this child was a landmark achievement that solidified Pasteur's stature in history and finally gave hope to all the victims of rabid animal attacks. Through meticulous study and careful experimentation, Pasteur's group did what others could not and crafted an effective methodology for preventing disease in people exposed to rabies. Like Jenner, without any real concept of viruses or how they differed from bacteria, Pasteur's insight and determination yielded a practical solution to a societal problem. Within a few years, Pasteur had successfully treated hundreds of patients, his technique was rapidly adopted in countries around the world, and the age of manmade vaccines had begun.

Post-Pasteur—Rabies Vaccine Evolution

As exciting and revolutionary as Pasteur's vaccine was, it was not without problems. First and foremost was the consistency of inactivation. Drying biological material, such as the rabbit spinal cord, was a lengthy and hard-to-standardize procedure with the technology of that era. Inadequate preparation could not only fail to protect the bitten patient but could even introduce virulent viruses. Since not all dog bite victims were necessarily exposed to RABV, some unfortunates were likely given rabies by the vaccine rather than their animal wounds. To eliminate the residual live virus, several investigators in different countries tried variations of the Pasteur protocol while others turned to chemical inactivation. One of the first reported chemical attempts was in 1908 by Claudio Fermi working in Italy. Using aqueous suspensions of RABV-infected brains from sheep and goats, he showed that room-temperature treatment with phenol (also called carbolic acid) created a material suitable for vaccine use even though some live viruses remained in his preparation. Building upon Fermi's work, 3 years later David Semple at the Central Research Institute in Kasauli, India incubated his rabbit brain extracts in phenol at 30 °C for 2–3 days and found that the virus was completely inactivated while still retaining its vaccine potency (Chakrabarti 2010). Extensive utilization of his formulation in India showed great success in preventing rabies and convinced other nations to adopt his protocol. Semple's version, along with derivatives (for example switching to sheep brains rather than rabbit brains), remained the most common rabies vaccine worldwide until the end of the twentieth century. Importantly, Semple's research marked an abrupt shift from attenuated vaccines to a fully inactivated type of vaccine. Demonstrating that a live virus was not necessary to invoke a protective immune response was a critical advance, and researchers for the next 100 years would use both the attenuation and inactivation strategies to develop new viral vaccines.

Removing the virulent virus from rabies vaccines improved but did not completely render these vaccines safe. Being made from either brain or spinal cord material meant that these preparations were rich in myelin, a protein prevalent in these tissues. Injection of the rabies vaccines introduced significant amounts of myelin which caused adverse reactions and neurological complications in some patients, including allergic encephalomyelitis and even demyelination in the CNS. Over the next five decades, various strategies were developed to reduce or eliminate myelin contamination (Wu et al. 2011). Embryonic and newborn animals were found to have less myelin in their nerve tissue and some vaccine producers switched to these systems to produce their vaccine stocks. Alternatively, RABV strains were adapted to grow in chicken and duck embryos so that the virus could be passaged in eggs for vaccine production. For example, a duck embryo-derived rabies vaccine was used in the United States from the 1950s to the 1980s. The vaccine consisted of a suspension of RABV-infected embryos inactivated with beta-propiolactone (a chemical that modifies viral nucleic acids and proteins and renders them nonfunctional). While a great improvement over the nerve tissue-derived vaccines, the egg-derived vaccines suffered from inconsistent potency and were not ideal for industrial-scale production.

The next phase in rabies vaccine development was the introduction of cell-culture-adapted strains of RABV. Cell culture is the technique for growing isolated animal cells in artificial media, either in suspension solutions or attached to surfaces such as plates or flasks. Although similar in concept to growing bacteria in broth or on agar plates, animal cells proved more fastidious and the development of conditions for their growth in culture was a long, slow process. And unlike bacteria which grow and divide indefinitely, normal animal cells will typically only divide a limited number of times before they senesce and die (an observation called Hayflick's lament after the researcher who first described this issue). Consequently, it took decades of work before human cells were established in culture, and these early cell lines were all derived from tumor tissue and were not suitable for vaccine production. It wasn't until 1961 that Leonard Hayflick created the first "normal" human cell culture (a so-called diploid cell) from fetal lung tissue that he named WI-38 (Hayflick and Moorhead 1961). RABV was quickly adapted for growth in WI-38 cells in hopes of producing the virus in quantities sufficient for vaccine production and without any myelin contamination. However, getting regulatory approval in the United States for the use of a new cell line to produce vaccines was a slow process. While this work was progressing, France approved a rabies vaccine in 1974 that was produced in a different fetal lung cell line called MRC-5. The vaccine consisted of RABV grown in and purified from MRC-5 cells then inactivated with beta-propiolactone, and it was finally approved for use in the United States in 1980. Rabies vaccines produced in WI-38, MRC-5, or similar human cells were generically called human diploid cell vaccines (HDCVs).

As important as the HDCVs are for their safety and consistency, they are not the only type of rabies vaccine currently available. Alternative vaccines available in different countries include the purified chick embryo cell vaccine (PCECV), the purified Vero cell (Vero cells come from African green monkeys) rabies vaccine (PVRV), the primary hamster kidney cell vaccine (PHKCV), or the purified duck embryo vaccine (PDEV). Each of these vaccines has its strengths and weaknesses, but all contain inactivated viruses and lack myelin, making them vastly safer than the original Pasteur vaccine. Additionally, these improved vaccines no longer need to be administered into the abdomen or need as many doses. Instead, the recipient typically is given only 4–5 injections over 2 weeks, and the shots are into the arm muscle as for other routine vaccinations. While the modern vaccines are not as terrifying or dangerous as the ancestral versions, data about how long the protective effects persist in humans is limited and the rabies exposure risk is small in many countries. Consequently, rabies vaccines remain primarily as post-exposure treatments for human victims of rabies exposure rather than for use in the mass vaccinations that are common for other viral diseases. Instead of widespread human vaccination, the United States and much of the world focus on vaccinating animals, primarily cats and dogs, to reduce rabies transmission to people. Some countries have even inoculated wild animal populations using food spiked with an oral rabies vaccine. The treated food is spread in the targeted animal's habitat for consumption and vaccination. These strategies, made possible through Pasteur's pioneering efforts, have eliminated rabies in some countries and reduced its incidence to only a few cases per year in many regions. With its huge zoonotic reservoir in bats and numerous other animal species, it may never be possible to eliminate the monstrous RABV from the world, but vaccines have turned rabies exposure from a nightmarish and fatal illness into a completely survivable event. The pioneering work with rabies introduced the concepts of both attenuation and inactivation as feasible strategies for vaccine creation, and the twentieth century would see these approaches implemented to control many human viruses.

References

Begeman L, GeurtsvanKessel C, Finke S, Freuling CM, Koopmans M, Muller T, Ruigrok TJH, Kuiken T (2018) Comparative pathogenesis of rabies in bats and carnivores, and implications for spillover to humans. Lancet Infect Dis 18(4):e147–e159. https://doi.org/10.1016/S1473-3099(17)30574-1

Berche P (2012) Louis Pasteur, from crystals of life to vaccination. Clin Microbiol Infect 18(Suppl 5):1–6. https://doi.org/10.1111/j.1469-0691.2012.03945.x

Blevins SM, Bronze MS (2010) Robert Koch and the 'golden age' of bacteriology. Int J Infect Dis 14(9):e744–e751. https://doi.org/10.1016/j.ijid.2009.12.003

Boland TA, McGuone D, Jindal J, Rocha M, Cumming M, Rupprecht CE, Barbosa TF, de Novaes OR, Chu CJ, Cole AJ, Kotait I, Kuzmina NA, Yager PA, Kuzmin IV, Hedley-Whyte ET, Brown CM, Rosenthal ES (2014) Phylogenetic and epidemiologic evidence of multiyear incubation in human rabies. Ann Neurol 75(1):155–160. https://doi.org/10.1002/ana.24016

Chakrabarti P (2010) "Living versus dead": the Pasteurian paradigm and imperial vaccine research. Bull Hist Med 84(3):387–423. https://doi.org/10.1353/bhm.2010.0002

Dalfardi B, Esnaashary MH, Yarmohammadi H (2014) Rabies in medieval Persian literature - the Canon of Avicenna (980-1037 AD). Infect Dis Poverty 3(1):7. https://doi.org/10.1186/2049-9957-3-7

Davis BM, Rall GF, Schnell MJ (2015) Everything you always wanted to know about rabies virus (but were afraid to ask). Annu Rev Virol 2(1):451–471. https://doi.org/10.1146/annurev-virology-100114-055157

De Kruif P (1926) Microbe hunters. Harcourt, New York

Fisher CR, Streicker DG, Schnell MJ (2018) The spread and evolution of rabies virus: conquering new frontiers. Nat Rev Microbiol 16(4):241–255. https://doi.org/10.1038/nrmicro.2018.11

Fu ZF (1997) Rabies and rabies research: past, present and future. Vaccine 15(Suppl):S20–S24. https://doi.org/10.1016/s0264-410x(96)00312-x

Gilbert AT (2018) Rabies virus vectors and reservoir species. Rev Sci Tech Oie 37(2):371–384. https://doi.org/10.20506/rst.37.2.2808

Gipson F (1956) Old yeller. Harper, New York

Grimes DJ (2006) Koch's postulates—then and now. Microbe Magazine 1(5):223–228

Hayflick L, Moorhead PS (1961) The serial cultivation of human diploid cell strains. Exp Cell Res 25:585–621. https://doi.org/10.1016/0014-4827(61)90192-6

Hayman DT, Fooks AR, Marston DA, Garcia RJ (2016) The global phylogeography of Lyssaviruses - challenging the 'Out of Africa' hypothesis. PLoS Negl Trop Dis 10(12):e0005266. https://doi.org/10.1371/journal.pntd.0005266

Kannadan A (2018) History of the miasma theory of disease. ESSAI 16(1)

Karamanou M, Panayiotakopoulos G, Tsoucalas G, Kousoulis AA, Androutsos G (2012) From miasmas to germs: a historical approach to theories of infectious disease transmission. Infez Med 20(1):58–62

King S (1981) Cujo. Viking Press, New York

Lagay F (2002) The legacy of humoral medicine. Virtual Mentor 4(7). https://doi.org/10.1001/virtualmentor.2002.4.7.mhst1-0207

Pesapane F, Marcelli S, Nazzaro G (2015) Hieronymi Fracastorii: the Italian scientist who described the "French disease". An Bras Dermatol 90(5):684–686. https://doi.org/10.1590/abd1806-4841.20154262

Pliny RH, Jones WHS (1961) Natural history. The Loeb classical library. Harvard University Press, Cambridge

Scott TP, Nel LH (2016) Subversion of the immune response by rabies virus. Viruses 8(8). https://doi.org/10.3390/v8080231

Scott TP, Nel LH (2021) Lyssaviruses and the fatal encephalitic disease rabies. Front Immunol 12:786953. https://doi.org/10.3389/fimmu.2021.786953

Suraweera W, Morris SK, Kumar R, Warrell DA, Warrell MJ, Jha P, Million Death Study C (2012) Deaths from symptomatically identifiable furious rabies in India: a nationally representative mortality survey. PLoS Negl Trop Dis 6(10):e1847. https://doi.org/10.1371/journal.pntd.0001847

Tarantola A (2017) Four thousand years of concepts relating to rabies in animals and humans, its prevention and its cure. Trop Med Infect Dis 2(2). https://doi.org/10.3390/tropicalmed2020005

Theodorides J (1984) Rabies in Byzantine medicine. In: Symposium on Byzantine Medicine. Dumbarton Oaks, pp 149–158

Wang J (2019) Mad dogs and other New Yorkers: rabies, medicine, and society in an American metropolis, 1840–1920. Animals, history, and culture. Johns Hopkins University Press, Baltimore

Wasik B, Murphy M (2012) Rabid: a cultural history of the world's most diabolical virus. Viking, New York

Wilkinson L (1988) Understanding the nature of Rabies: an historical perspective. In: Campbell JB, Charlton KM (eds) Rabies. Springer US, Boston, pp 1–23. https://doi.org/10.1007/978-1-4613-1755-5_1

Wu X, Smith TG, Rupprecht CE (2011) From brain passage to cell adaptation: the road of human rabies vaccine development. Expert Rev Vaccines 10(11):1597–1608. https://doi.org/10.1586/erv.11.140

Yuhong W (2001) Rabies and rabid dogs in Sumerian and Akkadian Literature. J Am Orient Soc 121(1):32–43

4

Yellow Fever: America Goes to War

Keywords Filterable agent • Mosquito vector • Walter Reed • Max Theiler • Viremia • 17D strain

Abbreviations

mRNA   Messenger ribonucleic acid
YEL-AND Yellow fever-associated neurotropic disease
YEL-AVD Yellow fever-associated viscerotropic disease

The Scourge of the Americas

It will likely come as a surprise that the third viral vaccine created was against yellow fever, a disease that is virtually unknown to the American public in the twenty-first century. While yellow fever remains endemic in parts of South America (particularly Brazil) and Africa, North America hasn't seen a case in over 100 years. Given its long absence, it's hard for contemporary society to imagine that yellow fever was once the most feared illness in the Americas, throwing entire cities into a panic when cases materialized. Unlike rabies that came from animal bites, yellow fever was mysterious as it was never certain when or where it would appear. In the United States, it was known as a disease that typically began in the spring and early summer and persisted until the first frost, but its occurrence was haphazard. Cities might have a terrible outbreak one summer and then not see cases again for many years. Theories abounded about where it came from and how it spread, most of them completely wrong, and it wouldn't be until the early 1900s that this disease was finally understood. The inability to fathom the source of this disease, predict when and where it might strike, or take effective preventative measures contributed to the fearfulness toward yellow fever. It was as if this illness was a random curse that descended without warning upon ill-fated communities. If cases did begin to appear, many city residents would immediately flee to escape the disease and avoid being trapped in areas that were often quarantined to prevent further spread. Given that there were no effective treatments and the fatality rate could be as high as 50% (Patterson 1992), fleeing seems like a very prudent response.

Yellow fever was originally an African illness where the disease has been known for at least seven centuries (Chippaux and Chippaux 2018), and the viral agent is likely much older than that (Bryant et al. 2007). Yellow fever in the Western hemisphere appeared roughly 400 years ago, coincident with the burgeoning African slave trade in the Caribbean (Pierce and Writer 2005; Crosby 2006). By the mid-1600s there were reports of disease outbreaks resembling yellow fever on several Caribbean islands including Barbados, St. Kitts, and Martinique, as well as the Yucatan peninsula (Patterson 1992). The timing of these historical observations is consistent with the amount of genetic divergence between African and New World strains of yellow fever (Bryant et al. 2007). However, it is not certain that all these early reports were truly yellow fever as other endemic diseases share some of the same symptoms. Perhaps the first well-documented yellow fever cases were in Cuba from around 1648 to 1655. Surprisingly, after that outbreak, the disease disappeared from the island and wasn't seen again for 100 years at which point it became endemic to Cuba. Nonetheless, over the next century as European nations extensively occupied other Caribbean islands and expanded the slave trade the disease slowly spread with sporadic but deadly outbreaks on many islands.

As the disease was introduced to new territories, it was originally known by its Spanish name, el vomito negro (black vomit). This name arose because victims, particularly those likely to succumb to the illness, bled into their stomachs and regurgitated dark, coagulated blood with the consistency and color of coffee grounds. In contrast, in English-speaking colonies, the term yellow fever was adopted to emphasize the yellowish skin produced by jaundice occurring in many patients. Eventually, this English descriptor became the commonly accepted name for the sickness although other designations abounded. For example, yellow fever was commonly referred to as "yellow jack" which was a reference to the yellow quarantine flag flown on ships and towns with yellow fever outbreaks. Other slang terms included yellowjacket, bronze john, the saffron scourge, yellow plague, and dock fever, the last referring to the prevalence of the disease in ports and port cities. Regardless of the name, by the late 1600s, yellow fever was well known throughout the Caribbean and was making its first appearance in America as the slave trade spread to the North American continent. There would even be outbreaks in European ports including Wales, France, and Spain as trade ships carried the disease back and forth across the Atlantic Ocean.

Yellow fever is a puzzling disease as many people do not experience symptoms and many others only present with mild symptoms that resolve in 3–4 days (Monath and Barrett 2003). The initial incubation period is 3–6 days at which point many patients experience chills, fever, and headaches. Some individuals resolve these symptoms and recover quickly while about 15% of patients progress to a severe acute illness characterized by a constellation of ailments including myalgia, vomiting, photophobia, intense back pain, and neurological features such as restlessness, irritability, and dizziness. During this acute stage, the patients typically exhibit a high fever (102–103 °F) which can soar to 105 °F, with the extreme temperatures associated with poor survival. This acute phase is followed by a period of remission lasting 1–2 days where the symptoms dwindle and the patient is improving. Most infected individuals continue to recover and return to full health, but unfortunately, 25% of the patients at this stage relapse and enter the most dangerous intoxication phase where half will die within 10–14 days (Roukens and Visser 2008). This final stage is the so-called hemorrhagic fever phase where high fever returns along with internal bleeding throughout the body as well as from the eyes, nose, and mouth. Bleeding causes damage and dysfunction in the liver, kidneys, heart, and digestive system, and this combined internal damage is the cause of death. In addition, the liver injury resulted in jaundice with the distinctive yellowing of the victim's skin while bleeding into the digestive tract produced the telltale black vomit that gave this disease its original name, both signs of a dangerously ill patient. As one eighteenth-century observer wrote "the black vomiting, which was generally a mortal symptom, and universal yellowness,… were constant symptoms, and coma came on very speedily" (Carey 1794). Even today effective therapeutics are lacking at the late stages of the disease and only supportive and palliative care is available.

Given its origin in tropical Africa and its subsequent Western hemisphere localization in the tropical Caribbean, yellow fever's initial North American appearance in Boston was unanticipated (Pierce and Writer 2005). Nonetheless, Boston's outbreak would foreshadow the American experience for the ensuing two hundred years where yellow fever was a threat to port cities from New Hampshire to Texas. In 1693, a British fleet headed by Admiral

Joseph Wheeler arrived in Boston from the West Indies triggering the first confirmed yellow fever epidemic in North America (Patterson 1992). That same year, cases were reported in the port cities of Philadelphia and Charleston, and the outbreaks in these three initial cities initiated the start of a two-hundred-year medical problem. Throughout the next century, yellow fever would repeatedly strike up and down the Atlantic coast, carried in on trading ships from Africa and the Caribbean. No major port city was immune to suffering and death from yellow fever, and Boston, Philadelphia, and Charleston would be repeatedly hit by outbreaks, along with other prominent port cities such as New York, Baltimore, Norfolk, Portsmouth, and Providence. Some years saw no disease, but others brought high mortality with thousands of deaths. The year 1793 was particularly significant historically as a massive yellow fever epidemic occurred in Philadelphia killing roughly 5000 people which was nearly 10% of the city's population. With no real idea of how the disease spread, residents either fled or tried to avoid contact with other people. One resident wrote that "the old custom of shaking hands fell into such general disuse, that many were affronted at even the offer of a hand" (Pierce and Writer 2005), a response similar to what occurred 200 years later with SARS-CoV-2 (Chap. 13). Importantly, at that time Philadelphia was the capital of the United States and the seat of our government. However, the devastation from yellow fever and the threat of future outbreaks forced the relocation of the capital inland to what is now Washington, D.C., just one example of how this disease helped shape American history.

By the early 1800s, the slave trade was diminishing in the northern portion of the United States and concomitantly the location of yellow fever outbreaks shifted. The last reported yellow fever cases in northern cities were in Baltimore and New York in 1822. After that date, yellow fever became a Southern problem with all the outbreaks confined to ports in slave trading states. Galveston (Texas), New Orleans (Louisiana), Mobile (Alabama), Savannah (Georgia), and Charleston (South Carolina) became the major epicenters of yellow fever in the 1800s although even smaller ports in the Gulf region and Florida saw periodic yellow fever attacks. As in the previous century, outbreaks were unpredictable in timing, location, and severity, contributing to the disconcerting nature of this disease. Some ports saw yellow fever return regularly each year for many consecutive years while other ports only had scattered and random outbreaks. When cases emerged it was never known if only a few citizens would be stricken and die or if the disease would rampage through the community killing hundreds to thousands. For example, the epidemic of 1878 started quietly in New Orleans but silently spread north via the Mississippi River trade causing outbreaks in river towns all the way to

Memphis, Tennessee. Yellow fever roared into Memphis that summer ravaging the city and killing 10% of the population (Crosby 2006). More than half the residents fled as the disease inexorably spread throughout the city, leaving much of the community eerily deserted and quiet with the bodies of dead families lying untended in their homes. Many of the city's doctors struggled heroically to treat the sick, but their remedies based on superstitions, anecdotes, and misconceptions did nothing to stem the illness and often were more harmful than helpful. As one local physician stated in a letter to his wife, "I am good for nothing, death coming in upon the sick in spite of all that I can do" (Crosby 2006). By the time the cool weather of fall finally ended the infections, the 1878 outbreak killed nearly 20,000 people from New Orleans to Memphis and would be recorded as the worst American yellow fever epidemic of the nineteenth century. Fortunately, while scattered American outbreaks would continue for another 25 years, all were modest compared to the 1878 scourge. Eventually, the work of Walter Reed and his collaborators in the late 1800s established preventative measures that halted disease spread and the last yellow fever cases on American soil occurred in New Orleans in 1905, bringing to an end 200 years of yellow fever misery.

A Disease That Flies

In 1878 while yellow fever was devastating New Orleans, Memphis, and other Mississippi river cities, Pasteur, Koch, and other Europeans were pushing science forward by developing the germ theory. These innovative researchers were visualizing bacteria by microscopy, learning to grow these microbes, and proving that different bacteria were the causative agents for specific diseases. In contrast, American medicine languished and adhered to old ideas about noxious gasses and mysterious effluvia as the cause of infectious diseases. Some citizens even continued to ascribe disease to divine retribution for various sins. For yellow fever, there were opposing schools of thought about what caused it and how it spread. Many believed that it was a toxic agent spontaneously generated by the filthy conditions of cities where garbage and sewage were strewn haphazardly in residential areas. One nineteenth-century physician even specifically blamed yellow fever on fumes from putrefying piles of vegetables and coffee that were commonly found on the docks of port cities (Pierce and Writer 2005). Others were convinced that natural effluvia from wetlands and swamps were the causative factor. Regardless of whether the source was natural or manmade, inhaling poisonous vapors from sewage and decaying matter was widely postulated as initiating the yellow fever symptoms. This model fitted well with the observed rise of yellow fever in the summer when the odor of decay and sewage peaked and with the disease's disappearance in colder weather when decay subsided. Based on this thinking, many cities opted to improve public sanitation through garbage collection and improved sewage systems. While these laudable public works contributed to overall better health and well-being for the citizenry, they had no appreciable effect on controlling yellow fever.

In contrast to the noxious gases theory, the other major school of thought was that yellow fever spread on inanimate objects (fomites) such as clothing, bedding, or anything else handled by diseased individuals. Fomite supporters believe that coming in contact with contaminated materials would spread the disease to susceptible individuals, although what the actual agent was and how it entered the body were less clear. Proponents of this theory argued that the disease was carried into port cities by contaminated cargo and infected crew or passengers on ships arriving from the Caribbean where the disease was prevalent. Their solution was to enforce quarantines on arriving vessels to ensure that all disease dissipated before any cargo or people were allowed onshore. Although fomites would prove to be entirely irrelevant to yellow fever transmission, this approach was moderately successful as the quarantine period allowed the disease to run its course through the crew and passengers. Without actively infected individuals coming ashore there was no risk to the local population. Unfortunately, quarantines were poorly tolerated by both the shippers and the local merchants. The shippers hated the lost time and expenses incurred while sitting idly in the harbor and the merchants disliked waiting for badly needed supplies and materials. Consequently, quarantines were often poorly enforced or done away with completely if yellow fever was absent for a few years. This inconsistent approach may have prevented some disease spread but never eliminated the constant threat that was lurking in every incoming vessel.

A major flaw with both the noxious gas and fomite theories was that neither accurately reflected the pattern of disease transmission in community epidemics. Once outbreaks began the occurrence of cases was maddingly unpredictable. Had fomites been involved, then those closest to the patients, such as family members and those providing care and treatment, would have been the most likely next victims. Yet in practice cases didn't necessarily cluster in families and many doctors and clergy administered to hundreds of patients without ever becoming ill. Similarly, if inhalation of some agent from decaying matter or sewage was involved, then areas of the city closest to these sources should have had the highest caseload. Instead, cases often hopped around the city with no discernible pattern. It was as if the disease could fly, skipping randomly over homes and businesses to strike unsuspecting victims no matter how hard they tried to avoid the yellow plague. When yellow fever struck a city, no one was safe, and nothing provided any effective protection from the disease.

And a Vector Shall Carry It

Mystifying and terrifying, the spread of yellow fever defied simple explanations, and the solution to this puzzle needed careful, rational, and scientific analysis, not baseless conjecture without supporting data. While various physicians pursued their solitary studies on yellow fever and some local communities implemented measures that they hoped would reduce disease transmission, a coordinated investigative strategy was lacking. It took 200 years of casualties from yellow fever before the first real national attempt to understand and combat this illness was implemented. In response to the 1878 epidemic along the Mississippi Valley, the United States Congress assembled a Yellow Fever Commission and sent them to Cuba, a known endemic area for yellow fever. Although this initial investigative group provided little new insight into yellow fever, one member of the team, Dr. Carlos Finlay, became convinced that mosquitoes were the transmitting vector for this disease. He noted that mosquito geographic and temperature ranges correlated well with the distribution of yellow fever cases (Clements and Harbach 2017), and while he could not prove this connection he steadfastly maintained his belief that mosquitos carried yellow fever even as the medical community and the general public mockingly dubbed him the "Mosquito Man". Undissuaded by the disdain his theory generated, Finlay would become an unquestioned mosquito expert whose talents ultimately helped solve the mechanism of yellow fever transmission.

In the decade after Finlay's original mosquito work, Cuba again became the focal point for yellow fever, this time for the United States Army. The 1800s had numerous examples of yellow fever severely disrupting European armies in the Caribbean, starting with Napoleon's forces. In 1801 Napoleon sent a massive army to Haiti to suppress a rebellion, but yellow fever decimated his troops, killing over 80% of his men and causing him to abandon his interest in the Americas (Marr and Cathey 2013). Parenthetically, the dismal outcome of the Haiti campaign also induced him to sell his Louisiana Territory holdings to the United States, thus ensuring the continued westward expansion of the new nation. Similarly, British and Spanish troops stationed in their Caribbean possessions routinely suffered greatly from yellow fever, making the disease a much greater risk to troops than actual combat. The history of the yellow fever problem was not lost on the United States military which in the 1890s was monitoring the ongoing conflict between Cuba and Spain. Cuba declared independence from Spain in 1894, launching a years-long war for control of the island. America was a watchful observer of this war as the U.S. had extensive business investments on the island and the U.S. government was concerned about protecting those interests. However, the U.S. was well aware that thousands of Spanish troops were dying of yellow fever each year and that many more were being incapacitated, greatly crippling the Spanish war efforts. By one estimate, nearly 4000 Spanish soldiers were dying annually in Cuba of yellow fever, and by 1898 over 75% of the troops were incapable of fighting due to illness (Pierce and Writer 2005). When the U.S. was drawn into what became the Spanish-American War and eventually took control of Cuba in 1899, it became crucial to finally understand yellow fever and develop protocols to protect American troops and civilians.

The 1900s would ultimately become a century of fabulous scientific progress, including the birth of modern evidence-based medicine, but the year 1900 saw no abatement of yellow fever in Cuba despite American efforts to improve general sanitation in the cities. Seeing no progress against this illness, on May 23, 1900, the Army Surgeon General, George Sternberg, appointed a team to deploy to Cuba and address "the etiology and prevention of yellow fever" (Clements and Harbach 2017). This team would become the Yellow Fever Commission and be headed by Walter Reed, an Army surgeon with a long history of investigating infectious diseases. Other members were Dr. James Carrol who would oversee bacteriological work, Dr. Jesse Lazear who would perform microscopic pathology, and Dr. Aristides Agramonte who would conduct autopsies and gross pathology. While all the members of the commission were seasoned and expert physicians, the designation of Dr. Reed as the group leader was particularly fortuitous as he was a thorough and meticulous researcher. The 1800s saw many attempts to explain the cause of yellow fever, and although some proposals contained a modicum of truth, none could be confirmed experimentally and all remained suspect. Reed was determined to avoid these past failures that he attributed to poorly controlled, poorly designed, and haphazard experimental work.

One of the many confounding issues with yellow fever was the inability to visualize the pathogen in samples from infected individuals. The germ theory of Pasteur and Koch had become the rage in the closing decades of the 1800s with researchers around the world vigilantly searching for disease-causing bacteria that they could visualize in the microscope and grow in the laboratory. If you could isolate a bacterium from a patient with disease X, propagate that microbe in the lab, and then cause disease X by infecting animals or humans with the laboratory organisms (Koch's postulates), then the etiology of a disease could be established. By the time Reed and his team began their work in Cuba, not only had bacteria been shown to be responsible for a variety of common diseases, but parasites were also proven to cause diseases such as malaria, filariasis, and trypanosomiasis. However, viruses were still vague and poorly understood disease agents that could not be grown and propagated like bacteria. Additionally, being much smaller than bacteria and parasites, they were invisible to light microscopy so there was no means to visualize them in patients' samples. Consequently, the initial extensive search by the Yellow Fever Commission for some visible microbe present in yellow fever victims failed. The only consolation was that the Commission did eliminate some candidate microbes as the yellow fever agent. Their careful examination of blood and tissue from yellow fever patients confirmed that several proposed causative bacteria were irrelevant passengers and were not responsible for the disease. After exhaustive studies, they were forced to conclude that the causative agent could not be determined and must be highly elusive. Without an agent in hand and no further means to identify one, the mechanism of disease transmission became the focal question.

As noted in the previous section, the transmission of yellow fever was a perplexing puzzle that intrigued and frustrated physicians for centuries. Was it airborne, was it transmitted via food or water, could it be carried on fomites, or was it spread directly from person to person? All of these mechanisms had been proposed, but none were backed by any rigorous proof and none were entirely consistent with known patterns of outbreaks. What was needed were carefully controlled transmission experiments to investigate these possibilities, however, there were no animal models of yellow fever, and humans appeared to be the only species susceptible to this illness. With no animal models, human experiments would be critical for gathering conclusive data but these would be dangerous studies as yellow fever could potentially be fatal for test subjects. Reed understood the danger and was troubled by the risks yet was convinced "that only can experimentation on human beings serve to clear the field for effective future work" (Crosby 2006). Knowing that the consequences of their experiments could be dire, the Commission implemented certain conditions to ameliorate risk as best they could (Bean 1983; Cutter 2016). First, all subjects would be adult volunteers who would be paid a generous fee to participate and paid more if they became sick; there would be no using prisoners, children, or the mentally ill as was often done in prior eras. Second, only healthy young adults could participate as yellow fever was known to be more severe in older adults. And lastly, the Commission created an informed consent document to ensure that volunteers understood what they were undertaking and the risks involved. The concept of informed consent would become the bedrock of twentieth-century human research but was largely unheard of in Reed's time and showed his remarkable sensitivity towards the brave individuals that he desperately needed for his research efforts.

In addition to the standard routes of disease transmission to be investigated, the Yellow Fever Commission slowly began to consider a highly novel yet long-suggested method of transmission, the mosquito as a carrier of yellow fever. Up until the late 1800s, the idea that insects could be vectors for human diseases was rarely considered and mostly ridiculed. However, by 1900 there was convincing proof that mosquitoes and ticks did transmit diseases such as malaria and Texas fever (babesiosis, a cattle disease), respectively, so the possibility of an insect vector for yellow fever couldn't be dismissed without examination. Importantly, several previous investigators suggested mosquitos as the cause of yellow fever. As far back as 1807, John Crawford of Baltimore published an account, without any real evidence, claiming that mosquitos carried yellow fever and other diseases (Oldstone 1998). And of course, the unappreciated Carlos Finlay was a strong proponent of the mosquito-yellow fever connection based on his correlations of mosquito habitats with yellow fever cases (Clements and Harbach 2017). After unexpected success with some desultory mosquito experiments, Reed and his team decided that all the possibilities, including mosquitos, must be rigorously tested (Mehra 2009). Tragically, Jesse Lazear contracted yellow fever and died early in the Commission's studies, although whether his case was acquired by accidental exposure or by a self-inflicted infection with an infected mosquito remains uncertain.

The full details and drama of the Yellow Fever Commission's experiments are beyond the scope of this book but are wonderfully described in The American Plague (Crosby 2006) and Yellow Jack (Pierce and Writer 2005) and will only be covered briefly here. True to his scientific nature, Reed was a thorough researcher who planned and recorded his work meticulously. Where prior yellow fever investigators often relied on incidental observations to derive their flawed conclusions, Reed and his collaborators conducted active trials designed to eliminate confounding variables. Instead of poorly designed tests with serious flaws, the Commission set up carefully constructed experiments with rigorous controls to assess each type of transmission mechanism. All the subsequent work was conducted at a new and specially chosen site named Camp Lazear in honor of their fallen colleague. New facilities were erected that were planned for specific transmission studies. For example, to test transmission from contaminated objects the Commission used an isolation barrack known as the "fomite house" (Ploth 2019). The building was designed to be insect-free and was kept at 90 °F to simulate summer conditions. The test subjects spent up to 3 weeks isolated in this barrack sleeping on bedding and wearing clothing that were both intentionally contaminated with bodily fluids, vomit, and feces from active yellow fever cases. The bedding, clothing, and other contaminated objects were replaced daily to create a pungent, noxious environment that would mimic the "foul atmosphere" that some pundits maintained was the cause of yellow fever. The condition must truly have been horrific, but no cases of yellow fever ever developed in these heroic volunteers, finally ending the erroneous belief that either fomites or some toxic fumes were responsible for spreading yellow fever.

In a related experiment, a second barrack was constructed with a wire screen separating two living areas. As for the fomite house, building number 2 was insect-proof and was thoroughly disinfected before conducting the transmission test. On one side of the screen was an active yellow fever patient and on the other side healthy volunteers. Although they lived in close contact, ate the same food, and breathed the same air, again the volunteers remained free of yellow fever, supporting the conclusion that yellow fever did not spread freely through the air. Having ruled out fomites and airborne routes of transmission, the mosquito vector became increasingly attractive as the actual carrier for yellow fever. Again using volunteers, both soldiers and civilians, the Reed team put striped house mosquitos (*Aedes aegypti*) in glass tubes and placed these tubes over the arms of yellow fever patients to let the mosquitos feed. The tubes were then transferred to the arms of the health volunteers so that the mosquitos could bite these new victims and potentially transmit the disease. Care was taken to ensure that the recipients were isolated and unlikely to be exposed to yellow fever from any natural sources. Based on some previous suspicions that there might be a necessary incubation time in the insect, the interval between the mosquitos feeding on a yellow fever patient and feeding on the uninfected persons was varied. This proved extremely insightful as they discovered that at least a 12-day interval was necessary to produce disease in the recipient, likely a very mysterious phenomenon to Reed and his collaborators. We now know that this reflects the time required for the virus to replicate in the insect gut and then make its way to the salivary glands where it can be released when the mosquito bites a new victim. In addition to this long incubation time in the insect, they also discovered that only mosquitos feeding on patients in the first few days of disease symptoms were effective at transmitting the disease. This latter observation is due to the timing of viruses in the bloodstream (viremia) of the patient. After infection, the virus replicates rapidly and reaches high levels in the bloodstream as symptoms emerge in the initial acute phase. Because of the high level of viremia, a mosquito feeding during this time has a good chance of ingesting viruses and becoming a potential disease spreader. As the acute phase wanes, virus levels in the patient's blood diminish and remain low for the duration of the illness, making it less likely that a mosquito feeding on these later-stage patients would pick up and transmit viruses. Discovering these critical parameters was key to finally proving yellow fever's dependence on mosquitos. Prior investigators, including Carlos Finlay, failed in their yellow fever studies with mosquitos because they never established these necessary conditions for efficient transmission. Reed's group, using acute-stage patients and the proper incubation period for the mosquitos, could consistently produce yellow fever in their test subjects. The first mosquito-transmitted case they caused was in an enlisted volunteer, Private John Kissinger who described his illness as feeling "as if every bone in my body had been crushed … and as if my head was going to burst open" (Pierce and Writer 2005). As horrible as the results were in some recipients, at last, the mechanism of yellow fever's spread was clear: the disease really could fly through the air although it needed the mosquito as a helper.

Once Reed's group discovered that only early-stage patients were suitable for transmission studies they reasoned the disease agent, whatever it was, must be present in the patient's blood at this stage. To confirm this conclusion they did direct blood transfers from acute state patients into volunteers and were able to transmit yellow fever without the mosquito (Bean 1983; Reed 1902). Convinced at last that the infectious agent existed in the bloodstream and could be transferred by mosquitos (Reed et al. 1900; Reed and Carroll 1901), they were still unable to identify the culprit. No examination of blood or tissues ever revealed visible bacteria or parasites as the yellow fever microbe, suggesting that some much smaller, invisible organism must be present (Reed 1902). In the decade preceding Reed's studies, European science was defining just such an infectious agent. In 1892 the Dutch botanist, Martinus Willem Beijerinck, adopted the word "virus" to describe the agent of tobacco mosaic disease (Bos 1999). His working definition of the virus was that it was a filterable agent—something too small to be seen with light microscopy and able to pass through filters that retained all cells and bacteria. This working definition would remain the state of the art until the 1930s when the electron microscope was invented and virus particles could finally be visualized. Using Beijerinck's filtration approach, Loeffler and Frosch showed in 1898 that the agent of hoof-and-mouth disease in cattle must also be a virus (Brown 2003). Aware of these previous studies, Reed conducted similar experiments by passing yellow fever-infected blood through bacteriological filters. The resultant "sterile" filtrate still caused yellow fever in volunteers, confirming the existence of a minute microbe that we now acknowledge as the first human virus defined by this experimental approach (Reed 1902). While lacking the scientific tools to elucidate the agent any further, Reed correctly postulated that the yellow fever problem could be solved not by fighting the virus, but by fighting the mosquito. He carefully delineated strategies for mosquito reduction and eradication (Reed and Carroll 1901) that once applied led to a rapid and dramatic decline in yellow fever cases throughout the Western hemisphere. Within a few years, the dreaded disease disappeared from North America, thus ending the great American scourge that had lasted for over two centuries. Sadly, shortly after publishing his ground-breaking yellow fever research, Walter Reed died of appendicitis in 1902 and never lived to see the tremendous impact of his work.

The Agent and the Vaccine

Walter Reed's contributions to medicine were enormous as he solved the mystery of yellow fever's spread and delineated the public health measures that would reduce mosquito numbers and control the disease in urban areas. Still, the disease remained endemic in sub-Saharan Africa and tropical regions of South America with sporadic outbreaks in other areas. Eliminating mosquitos would never be possible and even reducing exposure to mosquitos was difficult in some parts of the world. What was sorely needed was a protective vaccine. Given that yellow fever patients who survived their illness appeared to have life-long immunity, a vaccine seemed feasible but that would take another 35 years to develop. One of the continuing impediments to yellow fever research in general and vaccine development specifically was the absence of an animal model for the disease. Susceptible animals were urgently needed for serial propagation of the suspected viral agent so that it could be maintained in the laboratory without the constant need for reacquisition of samples from infected humans, similar to what Pasteur did for rabies virus. Ideally, blood from an infected human would infect a subject animal that would become ill. Blood from the ill animal would be used to inoculate a second animal, and then the second animal's blood used to infect a third animal. This procedure could be repeated indefinitely from animal to animal to provide a constant stock of infected blood, thus eliminating human patients from the process. Animals were also needed for studies on the disease pathology as well as for assessing therapeutics and potential vaccine candidates. Without an appropriate test animal, humans remained the only vulnerable species, severely handicapping efforts to understand and defeat the virus.

In the early 1900s, the global challenge of yellow fever became a major focus for the eponymous Rockefeller Foundation which was founded in 1913 by the Rockefeller family. Headquartered in New York, the Foundation aspired to enhance science, public health, and medical education, both nationally and globally. The Foundation hired and supported several prominent researchers specifically to work on yellow fever and organized several international commissions to address yellow fever control around the world. Adrian Stokes and colleagues, investigators supported by the Foundation's 1925 West African Yellow Fever Commission, finally made the critical breakthrough in 1927. Local African primates were not vulnerable to yellow fever so they explored different species for yellow fever susceptibility. Fortuitously, they discovered that non-African monkeys, such as Asian rhesus macaques, became ill and often died from this agent (Bryan 1997; Adrian Stokes 1928). It wouldn't be elucidated until much later, but local primates were not actually resistant to yellow fever. Instead, yellow fever is endemic in many African and New World primates. What confounded early researchers was that these seemingly resistant primates actually had silent infections and just didn't become clinically ill so it wasn't apparent that they were being infected by the virus (Haddow 1967). In nature, the silently infected primates serve as the actual reservoir for yellow fever, and the mosquito spreads the virus from animal to animal just as it can spread the disease from human to human (Chippaux and Chippaux 2018). Humans aren't truly part of the natural life cycle of this virus, and we are merely incidental hosts when we happen to encounter an infected mosquito. Parenthetically, much later it was established that the yellow fever virus could also be transmitted directly from infected mosquitos to their offspring in the wild (called vertical transmission) (Fontenille et al. 1997). In addition to sequential human-to-human transmission, mosquito-to-mosquito transmission may have contributed to the ability of the virus to persist for months aboard the ships sailing from Africa to the New World.

Stokes' demonstration that rhesus monkeys were susceptible to yellow fever enabled him to establish the first permanent yellow fever isolate from a Ghanaian patient named Asibi. Asibi's blood was used to inoculate a rhesus monkey then that monkey's blood was used to inoculate the next animal and so on, creating a perpetual chain of transmission. The so-called Asibi strain of the virus was propagated for years and was critical for the eventual development of a successful yellow fever vaccine. Just as important as the establishment of a viral strain, having a susceptible animal model now allowed candidate vaccines to be tested for safety and efficacy in primates rather than human volunteers. Initial vaccine attempts focused on viral inactivation by chemical procedures similar to those used for rabies vaccines. First published by Edward Hindle, virus-infected liver or spleen extracts treated with formalin or glycerol-phenol produced a vaccine that protected rhesus monkeys from a yellow fever challenge (Hindle 1928). His procedure was adopted in several countries and put into human clinical trials with limited success. Lacking methods to ensure a constant amount of virus in the extracts or to assess the extent of the inactivation process, batch-to-batch potency and safety were too inconsistent for human use. As poor results accumulated, these early inactivated vaccines fell into disfavor and were never widely adopted for general public use.

Stokes' work with rhesus monkeys was seminal in jump-starting laboratory-based research on yellow fever and finally allowing the isolation of viral strains from patients. However, monkeys are expensive to obtain and maintain as well as difficult to handle and work with—it is no surprise that smaller animals such as rodents and rabbits eventually became the mainstay of animal research. By 1930, mice were already a commonly used laboratory species, and there were even published reports showing that mice could support some human viral infections. Max Theiler at Harvard University was working with a second strain of yellow fever known as the Dakar strain and was seeking a simple and convenient way to propagate and study the virus. Turning to mice he quickly determined that yellow fever was fatal when injected intracerebrally (Theiler 1930b). Under these conditions, the disease was limited to the brain and the animals didn't develop the typical liver infection common in humans. Nonetheless, even though mouse infection via this route did not produce a disease reflective of the human response, it created a powerful tool for diagnosing and assessing yellow fever. For example, blood from patients or monkeys suspected of having yellow fever could simply be inoculated into mouse brains and tested for lethality and characteristic pathology. Similarly, patients or test animals could be assessed for anti-yellow fever antibodies by mixing their serum with a known lethal dose of yellow fever virus. The mixture was injected into the mouse brain and if the mouse died the subject lacked protective antibodies. Conversely, if the mouse survived then the subject had antibodies that inactivated the virus and prevented the lethal infection, and such protective antibodies became known as neutralizing antibodies. This simple approach revolutionized the field as researchers could now rapidly test and quantify both the relative amount of virus and the immune response. Tests using serum antibodies to detect foreign antigens, primarily pathogens such as viruses and bacteria, are collectively known as serology, and this technique remains widely used in clinical medicine as a diagnostic tool.

Impressed with Theiler's breakthrough, the Rockefeller Foundation hired him away from Harvard and backed his further studies on yellow fever. As important as the mouse model was for diagnostics and research, the major public health need for a safe and effective vaccine remained. The disappointing results with the inactivated yellow fever vaccines encouraged researchers, including Theiler, to consider attenuation as an alternative strategy for vaccine development. Pasteur's experiments many years before had shown that passage of the canine rabies agent in alternative hosts could change its virulence properties, leading Theiler to explore the long-term passage of yellow fever in mice as a means of attenuation. These early attempts to alter viruses were based purely on empirical observations without any understanding of the underlying genetic mechanisms of change. Modern science appreciates that most viruses are highly adapted to their natural host species and may reproduce poorly in other species. Forcing a virus to propagate in a foreign host puts the virus under tremendous selection pressure to adapt to the new environment. Under this selection pressure, the many random genomic mutations that continuously arise in viral populations are fertile fodder for viral evolution. Progeny viruses bearing chance mutations that favor better replication and transmission in the new host will thrive and eventually dominate the viral population, leading to new strains with properties different from the original parental virus.

Max Theiler could not have known what was happening to the virus at the molecular level, but he must have been hopeful that a prolonged mouse passage might produce a strain of yellow fever suitable for vaccine usage (Monath 2010; Norrby 2007). After 29 passages of the Asibi isolate in mice, extracts made from infected mouse brains protected rhesus monkeys from a challenge with virulent yellow fever (Theiler 1930a), suggesting that this strategy was promising. However, the early passage strain was still highly variable in its effect on mice so more extensive passaging was performed. After 100 passages the mouse brain-adapted strain was extremely stable ("fixed" in Pasteur's terminology) and consistently caused lethality in mice 4 days after intracerebral injection. The original Asibi strain was viscerotropic (attacked the liver and other organs), and Theiler now had a neurotropic strain (brain-adapted) that had lost its viscerotropic virulence.

As Theiler pursued his studies, other groups conducted similar vaccine development leading to the first human trial. A Harvard-Institut Pasteur collaboration produced a parallel mouse brain-adapted (neurotropic) strain starting from a different human isolate of yellow fever (the Dakar isolate). This so-called "French vaccine" underwent initial human testing in 1932 and showed effective protection against yellow fever leading to its implementation as the first yellow fever vaccine. Throughout the rest of that decade, modified versions of this vaccine were widely used in French West Africa. As the French vaccine went into use, many yellow fever researchers had doubts about using a neurotropic virus as a vaccine. While the early small trials in monkeys and humans didn't reveal any serious issues with the French vaccine, there were still theoretical concerns that the mouse neurotropic virus might be able to attack human brains in some vaccine recipients, a concern that was finally demonstrated decades later. By the 1950s, there was mounting evidence that this neurotropic vaccine could cause nervous system complications, including death, especially in young children. Ultimately the use of this vaccine was restricted to persons over 15 years of age, and eventually, the French vaccine was discontinued and replaced by the vaccine developed by Theiler.

Struggling with the neurotropic concern, Theiler embraced the emerging field of tissue culture (Ambrose 2019). Beginning in the early 1900s, researchers developed techniques for maintaining animal tissues in nutrient media that allowed for examination of and experimentation with isolated cells and organs rather than whole animals. Returning to the original Asibi strain that was never passaged through mouse brains, Theiler showed that the yellow fever virus could be propagated in chicken and mouse embryo tissue cultures. Trying multiple passage strategies, he eventually developed a strain called 17D that lacked both viscerotropic and neurotropic properties. To produce 17D, the Asibi strain was passaged 18 times in whole mouse embryos followed by 58 passages in minced chicken embryos. However, the key to eliminating the neurovirulence was additional passages in minced chicken embryos whose nervous tissues had been removed. After 100 serial passages in these depleted chicken embryo cultures (176 total passages of the original Asibi isolate), 17D arose as a stable strain that was no longer virulent in mice and fully protected monkeys from the normally lethal Asibi strain. Field trials in Brazil in 1937 confirmed that 17D was safe and provided effective human protection against yellow fever. 17D was licensed for human use in 1938 and rapidly became a global vaccine as an alternative to the French vaccine. Between mosquito abatement and aggressive vaccination programs, yellow fever in endemic areas diminished from a massive killer to a well-controlled disease. In the early 1980s, the French vaccine was finally discontinued due to its neurological issues and 17D became the sole yellow fever vaccine. Remarkably, while there have been some modest modifications of the 17D vaccine over the years, mostly concerning manufacturing and storage formulations, 17D and its derivatives remain the only yellow fever vaccines available worldwide (Roukens and Visser 2008). In 1951, Max Theiler received the Nobel Prize in Physiology or Medicine for his development of this life-saving vaccine, and his recognition remained the only Nobel Prize given for a viral vaccine until the SARS-CoV-2 mRNA vaccine (Norrby 2007).

The Virus and the Future

Even as yellow fever cases dwindled around the world, there was still much science to be done on the yellow fever virus. Max Theiler introduced methods to propagate the agent and assess immune responses to the virus, but his advances didn't purify or characterize the virus itself. He showed that infected mouse brains or tissue culture cells could be used to make virus-containing extracts for research purposes, for further virus passage, or to produce a vaccine, but these preparations were far from pure virions. These virion extracts were cell-free solutions that also contained undefined cellular biochemicals and any other small cellular components that could pass through the bacteriologic filters. While the virions present in these extracts were functional and infectious, nothing was known about the size, shape, or molecular properties of the virions themselves. In the 1930s, most of the tools needed for viral characterization were lacking, but over the next 50 years, the biology and molecular features of the yellow fever virus would slowly emerge. It wasn't until 1953 that virions were first visualized by electron microscopy and shown to be spherical particles (Reagan and Brueckner 1953). Subsequent analysis revealed that within the viral particle, the viral genome consisted of a single strand of RNA, and this RNA genome of roughly 11,000 nucleotides was entirely sequenced in 1985 (Rice et al. 1985). This single-stranded RNA genome is designated a positive-sense genome meaning that the viral genome functions as a messenger RNA (mRNA). Because it biochemically resembles our own mRNAs, upon delivery of this viral RNA into human cells it is immediately recognizable by our ribosomes and is translated into 10 viral proteins. These viral proteins initiate viral genome replication and progeny virion production with the attendant pathogenic effects leading to disease and sometimes death (Chambers and Rice 1987).

As the biochemical and molecular features of the yellow fever virus were elucidated it became the prototype member of a new viral family called *Flaviviridae* from the Latin flavus meaning yellow. Currently, there are 4 genera and over 100 species within this family and they all share similar molecular features with the yellow fever virus (de Bernadi Schneider et al. 2020). These related viruses have a common virion size and appearance, have the same genome type (single-stranded, positive-sense RNA), and possess comparable genes organized in the same order along the genome (Fig. 4.1). Interestingly, among the closest relatives of yellow fever are several other human pathogens including dengue virus, West Nile virus, Zika virus, and St. Louis encephalitis virus. Like the yellow fever virus, each of these related

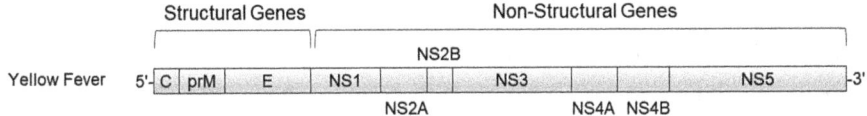

Fig. 4.1 Genome Organization of the Yellow Fever Virus. Shown are the 10 viral genes of the yellow fever virus oriented along the genome from the 5' end to the 3' end. The 3 structural genes encode proteins that make up the virion while the 7 non-structural genes encode proteins required for viral reproduction. The other flaviviruses such as dengue, Zika, and West Nile virus have equivalent genes organized in the same order along their genomes

viruses is also spread by mosquitos and doesn't generally transmit via fomites or the respiratory route. Although it is not a formal taxonomic classification, any virus normally transmitted via insects (arthropods) is commonly referred to as an arbovirus (from <u>ar</u>thropod-<u>bo</u>rne), and yellow fever and its close relatives are all arboviruses.

Most RNA viruses, including the *Flaviviridae*, have highly error-prone genomic replication, leading to a high mutation rate and rapid changes in their genome sequences over time. Such genomic changes can manifest in days, weeks, or months compared to the more stable DNA viruses where such changes accrue over years to decades. One of the puzzling aspects of the 17D vaccine strain of the yellow fever virus is its relative stability over decades of passage. As mentioned in the previous section, the parental 17D strain and a few derivative substrains have been in laboratory production for roughly 80 years with no major changes in its attenuated phenotype and suitability for vaccine production. Sequence comparison of the 17D strain and a low-passage stock of the original Asibi strain revealed 67 nucleotide changes resulting in 31 amino acid changes in the viral proteins (Beck et al. 2014). Subsequent work suggests that these mutations in 17D have improved the fidelity of its genomic replication and reduced its propensity to acquire mutations, an effect that likely accounts for its long-term genetic stability (Davis et al. 2019). While the specific mutational changes that increased the fidelity of replication are not yet identified, this chance acquisition of mutations during Theiler's passage of the Asibi strain produced the only flavivirus vaccine available throughout the twentieth century. The world was incredibly fortunate that Theiler stumbled upon a stable and attenuated version of the yellow fever virus, as no equivalent strain has ever been generated. Only recently has modern technology led to vaccines for some other flaviviruses: Japanese encephalitis virus (approved in 2009), dengue virus (approved in 2018), and tick-borne encephalitis virus (approved in 2021).

As successful and life-saving as the 17D vaccine has been, it is not without its faults. Like any live attenuated vaccine, 17D could potentially revert to virulence through mutations while replicating in a vaccinee. Fortunately, this has never been proven to occur, likely due to the low mutation rate of this strain. Live attenuated vaccines are also not recommended for pregnant women as they could potentially infect and harm the fetus. By accident, many pregnant women have received this vaccine and there is no evidence that fetal damage occurs from the 17D vaccine, but its use in pregnant women is still contraindicated and should only be considered when the risk of yellow fever is very high (Roukens and Visser 2008). More importantly, live vaccines are potentially dangerous to anyone with a weakened immune system. This can include the very young, the elderly, the sick, individuals on immunosuppressive drugs, or anyone with an underlying immune defect. Because of adverse events in infants, the vaccine is no longer given to individuals under 9 months old. Additionally, although the overall risk of serious adverse effects from the yellow fever vaccine remains very low (around 1/100,000 vaccinees), the risk increases significantly for those over 60 years of age. There have also been unexplained sporadic occurrences in localized areas where the rate of adverse effects was noticeably higher than the worldwide average (Roukens and Visser 2008; Montalvo Zurbia-Flores et al. 2022). These serious effects fall into two categories: yellow fever-associated viscerotropic disease (YEL-AVD) which resembles typical wild-type yellow fever and yellow fever-associated neurotropic disease (YEL-AND) resulting in encephalitis. All of these issues lobby for continued exploration of novel yellow fever vaccines that could replace 17D with similar efficacy and lower risk. At least 9 new yellow fever vaccines are currently in development with several already in phase I clinical trials (Montalvo Zurbia-Flores et al. 2022). Most of these are inactivated vaccines which would eliminate the safety concerns with the live 17D. Even more novel technologies such as DNA-based vaccines and recombinant viral vector vaccines are being explored, and given the success of mRNA vaccines for SARS-CoV2, that platform will likely be tested for yellow fever as well. Hopefully, in the next decade, the 17D vaccine can finally go into well-deserved retirement and be replaced with a safer, twenty-first-century yellow fever vaccine.

References

Adrian Stokes JHB, Hudson NP (1928) The transmission of yellow fever to Macacus rhesus. JAMA J Am Med Assoc 90(4):253–254

Ambrose CT (2019) An amended history of tissue culture: concerning Harrison, Burrows, Mall, and Carrel. J Med Biogr 27(2):95–102. https://doi.org/10.1177/0967772016685033

Bean WB (1983) Landmark perspective: Walter Reed and yellow fever. JAMA 250(5):659–662

Beck A, Tesh RB, Wood TG, Widen SG, Ryman KD, Barrett AD (2014) Comparison of the live attenuated yellow fever vaccine 17D-204 strain to its virulent parental strain Asibi by deep sequencing. J Infect Dis 209(3):334–344. https://doi.org/10.1093/infdis/jit546

Bos L (1999) Beijerinck's work on tobacco mosaic virus: historical context and legacy. Philos Trans R Soc Lond Ser B Biol Sci 354(1383):675–685. https://doi.org/10.1098/rstb.1999.0420

Brown F (2003) The history of research in foot-and-mouth disease. Virus Res 91(1):3–7. https://doi.org/10.1016/s0168-1702(02)00268-x

Bryan CS (1997) Discovery of the yellow fever virus. Int J Infect Dis 2(1):52–54

Bryant JE, Holmes EC, Barrett AD (2007) Out of Africa: a molecular perspective on the introduction of yellow fever virus into the Americas. PLoS Pathog 3(5):e75. https://doi.org/10.1371/journal.ppat.0030075

Carey M (1794) A short account of the malignant fever, lately prevalent in Philadelphia

Chambers TJ, Rice CM (1987) Molecular biology of the flaviviruses. Microbiol Sci 4(7):219–223

Chippaux JP, Chippaux A (2018) Yellow fever in Africa and the Americas: a historical and epidemiological perspective. J Venom Anim Toxins Incl Trop Dis 24:20. https://doi.org/10.1186/s40409-018-0162-y

Clements AN, Harbach RE (2017) History of the discovery of the mode of transmission of yellow fever virus. J Vector Ecol 42(2):208–222. https://doi.org/10.1111/jvec.12261

Crosby MC (2006) The American plague: the untold story of yellow fever, the epidemic that shaped our history, 1st edn. Berkley Books, New York

Cutter L (2016) Walter reed, yellow fever, and informed consent. Mil Med 181(1):90–91. https://doi.org/10.7205/MILMED-D-15-00430

Davis EH, Beck AS, Strother AE, Thompson JK, Widen SG, Higgs S, Wood TG, Barrett ADT (2019) Attenuation of live-attenuated yellow fever 17D vaccine virus is localized to a high-fidelity replication complex. mBio 10(5). https://doi.org/10.1128/mBio.02294-19

de Bernadi SA, Jacob Machado D, Guirales S, Janies DA (2020) FLAVi: an enhanced annotator for viral genomes of flaviviridae. Viruses 12(8). https://doi.org/10.3390/v12080892

Fontenille D, Diallo M, Mondo M, Ndiaye M, Thonnon J (1997) First evidence of natural vertical transmission of yellow fever virus in Aedes aegypti, its epidemic vector. Trans R Soc Trop Med Hyg 91(5):533–535. https://doi.org/10.1016/s0035-9203(97)90013-4

Haddow AJ (1967) X.-The natural history of yellow fever in Africa. Proc R Soc Edinburgh, Sect B: Biol Sci 70(3):191–227. https://doi.org/10.1017/S0080455X00001338

Hindle E (1928) A yellow fever vaccine. Br Med J 1(3518):976–977. https://doi.org/10.1136/bmj.1.3518.976

Marr JS, Cathey JT (2013) The 1802 Saint-Domingue yellow fever epidemic and the Louisiana Purchase. J Public Health Manag Pract 19(1):77–82. https://doi.org/10.1097/PHH.0b013e318252eea8

Mehra A (2009) Politics of participation: Walter Reed's yellow-fever experiments. Virtual Mentor 11(4):326–330. https://doi.org/10.1001/virtualmentor.2009.11.4.mhst1-0904

Monath TP (2010) Yellow fever. In: Artenstein AW (ed) Vaccines: a biography. Springer, pp 159–189

Monath TP, Barrett AD (2003) Pathogenesis and pathophysiology of yellow fever. Adv Virus Res 60:343–395. https://doi.org/10.1016/s0065-3527(03)60009-6

Montalvo Zurbia-Flores G, Rollier CS, Reyes-Sandoval A (2022) Re-thinking yellow fever vaccines: fighting old foes with new generation vaccines. Hum Vaccin Immunother 18(1):1895644. https://doi.org/10.1080/21645515.2021.1895644

Norrby E (2007) Yellow fever and Max Theiler: the only Nobel Prize for a virus vaccine. J Exp Med 204(12):2779–2784. https://doi.org/10.1084/jem.20072290

Oldstone MBA (1998) Viruses, plagues, and history. Oxford University Press, New York

Patterson KD (1992) Yellow fever epidemics and mortality in the United States, 1693-1905. Soc Sci Med 34(8):855–865. https://doi.org/10.1016/0277-9536(92)90255-o

Pierce JR, Writer J (2005) Yellow jack: how yellow fever ravaged America and Walter Reed discovered its deadly secrets. Wiley, Hoboken

Ploth D (2019) Walter reed at camp lazear: a paradigm for contemporary clinical research. Am J Med Sci 357(1):7–15. https://doi.org/10.1016/j.amjms.2018.06.024

Reagan RL, Brueckner AL (1953) Electron microscopy of yellow fever virus (17D strain). Am J Pathol 29(6):1157–1159

Reed W (1902) Recent researches concerning the etiology, propagation, and prevention of yellow fever, by the United States Army Commission. J Hyg (Lond) 2(2):101–119. https://doi.org/10.1017/s0022172400001856

Reed W, Carroll J (1901) The prevention of yellow fever. Public Health Pap Rep 27:113–129

Reed W, Carroll J, Agramonte A, Lazear JW (1900) The etiology of yellow fever-a preliminary note. Public Health Pap Rep 26:37–53

Rice CM, Lenches EM, Eddy SR, Shin SJ, Sheets RL, Strauss JH (1985) Nucleotide sequence of yellow fever virus: implications for flavivirus gene expression and evolution. Science 229(4715):726–733. https://doi.org/10.1126/science.4023707

Roukens AH, Visser LG (2008) Yellow fever vaccine: past, present and future. Expert Opin Biol Ther 8(11):1787–1795. https://doi.org/10.1517/14712598.8.11.1787

Theiler M (1930a) Studies on the action of yellow fever in mice. Ann Trop Med Parasitol 24(2):249–272. https://doi.org/10.1080/00034983.1930.11684639

Theiler M (1930b) Susceptibility of white mice to the virus of yellow fever. Science 71(1840):367. https://doi.org/10.1126/science.71.1840.367

5

Influenza: An Elusive and Evasive Foe

Keywords Genetic drift • Genetic shift • HN types • Reassortment • 1918 flu • Richard Shope

Abbreviations

CDC	Centers for Disease Control
cRNA	Complementary ribonucleic acid
EU	European Union
GISRS	Global Influenza Surveillance and Response System
HA	Hemagglutinin
IIV	Inactivated influenza vaccine
LAIV	Live attenuated influenza virus
MDCK	Madin-Darby canine kidney cells
NA	Neuraminidase
RBC	Red blood cells
USDA	United States Department of Agriculture
WHO	World Health Organization

An Ill Wind

Yellow fever was a horrid and feared disease in the Americas, but its impact was almost negligible compared to influenza. While thousands died from yellow fever, the global influenza pandemic of 1918 was a terrifying nightmare that killed 50–100 million people, including over 650,000 in the United

States (Nickol and Kindrachuk 2019; Oxford and Gill 2018). Some estimates suggest that by the pandemic's conclusion roughly one-third of the world's population had been infected. Sometimes still called the Spanish flu, this is a complete misnomer. No one knows exactly where this lethal influenza strain originated, but it wasn't Spain. Early outbreaks were observed in the United States, France, and China, suggesting that any one of these locations could have been ground zero for the subsequent pandemic (Brown 2018). Spain, unfortunately, took the blame for this illness merely because it was one of the few countries to widely report on the early cases. Because of the extensive coverage of the disease in the Spanish press, this flu initially became associated with Spain and acquired its inappropriate nickname (Kuszewski and Brydak 2000). Regardless of its origin, World War I in Europe provided a perfect setting for a highly contagious respiratory infection to spread among the troops living in crowded and often unsanitary conditions. With soldiers traveling back and forth from battlefields to their home countries, this virus was quickly and easily dispersed around the globe. As the virus began to sweep through cities and countries, one observer in Massachusetts reported that there were "several thousand cases in the city with a great shortage of nurses and doctors, … theatres, churches, gatherings of every kind stopped" (Solly 2020).

One of the remarkable aspects of the 1918 influenza was not only its rapid spread but also its high lethality. With a case fatality rate of greater than 2.5%, this 1918 influenza was an order of magnitude more deadly than any viral pandemic until the SARS-CoV-2 outbreak (Chap. 13) beginning in 2019. A second unusual feature was the age distribution of the fatal cases. Typically, seasonal influenza is most dangerous to the very young (under 5 years of age) and the elderly (over 65 years of age), with few fatalities in healthy individuals between those age ranges. Young children have less developed immune systems and likely haven't yet had a prior influenza infection to provide them with some residual protective immunity, so they are especially susceptible to the flu. They also are more likely to become severely dehydrated or develop serious pneumonia, both of which can be fatal if not treated promptly. Similarly, the elderly generally have declining immune functions which puts them at risk for more lethal responses to influenza infection. In contrast to these two typical risk groups, the 1918 flu was a horror for healthy young adults, the group most unaffected by typical seasonal flu. Twenty to forty-year-olds accounted for 50% of the fatalities, and less than 1% of deaths occurred in the over-65-year-old cohort (Taubenberger and Morens 2006). For the elderly population, we suspect that they may have had some partial immunity because of previous exposures to one or more somewhat related influenza strains earlier in their lives. Anyone born after these related strains left circulation would have been immunologically naïve with no immune memory and thus completely unprotected against the 1918 strain. Why the disease had such a high fatality in younger adults remains uncertain although there are clues and hypotheses to explain why the 1918 virus was such an atypical influenza strain. In 2005, scientists recreated the entire sequence of the 1918 influenza virus and began to examine its properties (Tumpey et al. 2005). The sequence and subsequent analysis revealed that the 1918 virus was mostly an avian influenza that crossed species into the human population (Worobey et al. 2014). Studies in animal models with reconstructed versions of the 1918 strain showed that it was highly virulent in mice, causing severe lung damage similar to what was observed in patient autopsy samples. Lung damage predisposes patients to secondary bacterial infections, and many researchers and historians speculate that in the pre-antibiotic era, it was these bacterial cases of pneumonia that proved fatal for so many young victims. Additionally, experiments with macaque monkeys revealed that the 1918 virus induced a severe aberrant immune response called a cytokine storm (Watanabe and Kawaoka 2011). Cytokines are small proteins released by many cell types in response to an infection. These proteins are modulators and regulators of our immune cells and are critical to mounting an effective immune response, including inflammatory responses. Unfortunately, when produced at excessive levels or for prolonged times, cytokines can trigger a life-threatening systemic inflammation. Many believe that cytokine storms were a significant factor in the dramatic fatality rate of the 1918 flu (and similarly may be contributing to SARS-CoV-2 deaths—Chap. 13). As intriguing as these speculations are, we may never know for certain just how all these possible factors combined to make the virus one of the deadliest pandemics in recorded history. Genetic analysis has yet to fully answer why this particular influenza virus was so lethal or exactly where and how it arose.

As bad as the 1918 pandemic was, it was neither the first nor last time that influenza has assaulted the world. Like the viruses discussed in Chaps. 2, 3 and 4, we believe that the influenza virus has a long association with human populations although the historical record for influenza is uncertain until around the sixteenth century. Older descriptions of diseases that could be influenza exist, such as the "cough of Perinthus" described in Hippocrates' Book of Epidemics from 412 BCE (Pappas et al. 2008). However, many other respiratory infections cause symptoms that overlap with those of true influenza. Because of similar symptoms, most ancient disease accounts lack sufficient distinctive information to be certain as to their causative agent. It wasn't until the second millennium CE that there began to be reports of disease outbreaks that are more reliably thought to be influenza (Barberis et al. 2016).

One presumptive epidemic in France in the early 1400s infected 100,000 people and was attributed to "smelly and cold wind", a reference to the medieval theory that diseases were caused by noxious fumes in the air. Around the same period, Italians coined the word influenza from the Latin *influentia*. *Influentia* means to influence and refers to the astrological belief that the stars affect or influence our susceptibility to diseases. Originally applied to any disease believed to be related to the stars, eventually, the term became specific to the respiratory infection we now know as viral influenza. Another name for influenza that became widely used was the French appellation "la grippe". Believed to have arisen in the 1700s, this word became anglicized as "the grip". In English-speaking countries, influenza was often referred to as the grip until well into the twentieth century when this term finally faded from common usage and was replaced simply by "the flu".

In addition to the sporadic European disease outbreaks in the Middle Ages that were likely influenza, what is considered to be the first recorded authentic influenza pandemic occurred in 1580 encompassing Asia, Russia, Europe, and parts of Africa. Over the next 600 years, human influenza established itself as an endemic virus that constantly circulates globally causing seasonal cases in the temperate zones, mostly in the winter months. Exactly why influenza cases peak in the winter is still debated, but lower temperatures and humidity seem to favor viral survival and transmission. The fact that people stay indoors more in the colder months likely also enhances the opportunities for a respiratory virus to spread from person to person via aerosols carrying the viral particles. Additionally, these viral particles can contaminate objects and surfaces where they can be picked up on our hands and inadvertently transferred to our mouths, eyes, and noses to start an infection. Once introduced into our bodies, seasonal influenza is characterized by a constellation of symptoms that include fever, cough, sore throat, runny or stuffy nose, muscle or body aches, headaches, and fatigue. Vomiting and diarrhea can occur in some people, though this is more common in children than adults. Unfortunately, these gastrointestinal symptoms in children have led to confusion in our terminology about this disease. Occurrences of diarrhea and vomiting in children are commonly referred to as the "stomach flu" when in fact these symptoms are seldom due to influenza. Instead, these upset stomachs are caused by any one of a large array of other viruses, completely unrelated to the influenza virus, that infect the gastrointestinal system. Using the term "flu" in connection with these gastrointestinal infections confounds the distinction between the influenza virus, a respiratory virus, and the distinctly different viruses that target our intestinal tract.

The symptoms of seasonal flu typically appear 1–4 days after infection and persist for 7–10 days. With much more intense symptoms than for a common cold, influenza patients may feel miserable for the duration of the illness. While most healthy individuals do recover fully with no lasting after-effects, influenza is definitely not a trivial viral infection. Each year in the United States roughly 10% of the population becomes infected and between 20,000 and 40,000 people die from this disease. For the vast majority of patients, the infection remains in the upper airway, and the immune system both contains and eliminates the virus resulting in complete recovery. In the more serious cases, mostly in the very young and the elderly, different scenarios occur. For some individuals, the immune system fails to squelch the infection, and rampant viral reproduction causes life-threatening damage to the lungs. In other cases, the seasonal flu acts more like the pandemic 1918 strain by causing cytokine storms and promoting secondary bacterial infections, both of which can be fatal. Current vaccines help to ameliorate these dangerous outcomes and reduce the overall clinical impact of influenza on the public but are far from perfect. As we will explore in subsequent sections, the influenza virus is a formidable adversary that we have only partially conquered even after a hundred years of battling this foe.

Influenza Hunters

In the aftermath of the devastating 1918 pandemic, influenza was recognized as an important and constant public health threat. Between the yearly burden of seasonal flu and the risk of future pandemic outbreaks, understanding this disease became a high priority for medical research. Additionally, unlike yellow fever which could be largely prevented by mosquito control, a respiratory disease like influenza was unstoppable by any simple public health measures. Face masks were used during the 1918 outbreak, but many people objected, and expecting the public to wear masks every winter for influenza protection was unsustainable (a grim foreshadowing of the COVID-19 pandemic). Antibiotics and other drugs to treat infections were largely nonexistent in that era, so the best path forward was the approach being vigorously pursued for other diseases, creating an effective vaccine. Regrettably, in 1918 the microorganism responsible for influenza was mistakenly acknowledged to be a bacterium. This incorrect belief greatly hindered research efforts as many scientists and physicians pursued this false notion to the detriment of finding the real culprit.

The story of influenza hunting began almost 20 years before the 1918 pandemic with Richard Pfeiffer, a German bacteriologist, who was a long-time assistant to and colleague of the distinguished Robert Koch (Fildes 1956). During his many years with Koch, Pfeiffer was exposed to the leading German bacteriologists of his time, and he became well-versed in the concepts of the bacterial basis of infectious diseases. While investigating an influenza outbreak in 1889, he turned his formidable bacteriological skills toward identifying the causative microbe. His studies revealed that influenza patients commonly had a small, rod-shaped bacterium in their nasal passages. The organism was fastidious and difficult to grow in culture making it problematic to study and characterize. Nonetheless, he named this organism *Bacillus influenzae*, and in an 1892 paper describing this finding he concluded that this bacterium was the etiological agent of influenza. This erroneous conclusion went mostly unchallenged for 40 years, perhaps due to his reputation as a thorough and meticulous researcher, as well as his close connection with Koch and the other eminent German bacteriologists. Years later, Pfeiffer's organism was renamed *Haemophilus influenzae* (the *H. influenzae* designation will be used henceforth), and we now know that this organism is part of the normal flora commonly found in our noses and throats. Pfeiffer's original finding of *H. influenzae* in flu patients just reflected the normal incidence of this bacterium in humans and not a causal relationship to influenza. However, *H. influenzae* is an opportunistic pathogen, meaning that it is typically harmless in healthy individuals, but it can sometimes spread into other parts of the body and cause life-threatening infections. For example, individuals with viral influenza can develop lung damage that allows bronchial colonization by *H. influenzae* leading to severe bacterial pneumonia. This organism was one of several pathogenic bacteria found in lung autopsy samples from victims of the 1918 influenza outbreak and likely contributed to the high death rate in that awful pandemic even though it wasn't the cause of influenza (Morens et al. 2008).

Pfeiffer's erroneous assertion about the bacterial etiology of influenza persisted well into the twentieth century, although by the 1918 pandemic, there were some suspicions that his bacterium was not the true agent of this disease. Some researchers had trouble repeating Pfeiffer's experiments, and his bacterium wasn't uniformly present in all influenza patients. Still, during the last decades of the 1800s, the notion that bacteria caused infectious diseases had become widely accepted, so Pfeiffer's pronouncement about his organism seemed eminently plausible. In contrast, the concept of viruses as alternative agents of disease was still new and mysterious. Without being about to grow or visualize viruses, the working definition of a virus was something that was very small, invisible to light microscopy, able to pass through filters that retained bacteria, and cause disease. However, to assess the disease-causing capability of the filtrate there had to be a suitable host for infection studies, and for human influenza, appropriate animal models were lacking. This constraint, coupled with the scarcity of influenza patients in many years, greatly limited influenza research until the 1918 pandemic provided cases in abundance, and for 25 years *H. influenzae* remained the leading candidate as the cause of influenza. Sadly, this misfocus wasn't just an ivory tower academic issue as much work was devoted to developing vaccines against *H. influenzae*, all of which proved useless in the 1918 pandemic.

The first scientist to present evidence to dispute Pfeiffer's claim was the French researcher, Charles Nicolle, who would eventually win a Nobel Prize in Medicine for his identification of lice as the vector of epidemic typhus (Schultz and Morens 2009). Nicolle was based at the Institut Pasteur in Tunisia and had been mentored by Emile Roux, Pasteur's highly accomplished assistant and colleague. Through his training and scientific connections, he was eminently familiar with Beijerinck's filtration method that distinguished between bacteria and the much smaller, invisible agents that passed through filters and which were becoming defined as viruses. Applying this technique to sputum samples from a few 1918 influenza patients, he showed that the filtrates caused disease in monkeys and two human volunteers, raising doubts about a bacterial cause. While his sample size was too small to be conclusive, other researchers were taking this same approach with similar findings. In Japan, Dr. T. Yamanouchi was doing human experiments on himself and 24 volunteers, also using sputum pooled from influenza patients (Taubenberger et al. 2007; Yamanouchi et al. 1919). Some volunteers received unfiltered sputum and others the filtered material. Both groups developed influenza symptoms, again supporting the belief that the disease-causing agent was not a bacterium as it wasn't trapped by the filter (In the twenty-first century, it's hard to imagine people volunteering for or being allowed to participate in such a potentially deadly experiment, but modern medicine owes much to the many brave citizens who willingly risked their lives to advance science). On the other side of the world at the Rockefeller Institute in New York City, American scientists were likewise pursuing the causative agent of influenza, albeit in a non-human system (Van Epps 2006). Peter Olitsky and Frederick Gates determined that nasal secretions from influenza patients would cause a brief, transitory illness when injected into the lungs of rabbits, along with visible lung pathology in sacrificed animals (Olitsky and Gates 1923). While influenza in rabbits didn't accurately mimic the disease in humans, this study indicated that rabbits could still be used to assess the presence or absence of the causative agent. Using this surrogate animal model rather than human volunteers, they too tested filtered and unfiltered samples for infectivity. Just like Nicolle and Yamanouchi, Olitsky and Gates showed that filtrates were still infectious. Unfortunately, they were actually able to culture a bacterium from their filtrate that they named *Bacterium pneumosintes* (now called *Dialister pneumosintes*) and were unable to conclusively rule out the role of this bacterium in human influenza. *D. pneumosintes* is an oral bacterium that is still not well-studied, but it is certainly not the causative agent of influenza. However, the presence of this bacterium in their samples confounded their study and helped maintain the confusion over whether or not influenza was a viral or bacterial infection.

While the above researchers each contributed to the growing body of evidence arguing for a viral cause of influenza, even collectively their work was not sufficient to convince everyone. Some researchers failed to replicate their studies in full or part and others simply denied the data and instead clung to the established belief that *H. influenzae* was the true culprit. As the 1918 pandemic wound down and cases diminished, finding patients to sample again became problematic and interest in influenza waned. It would be another decade before human influenza was convincingly proven to be caused by a virus, and influenza outbreaks in pigs would be instrumental in providing the groundwork for the human disease. As the human influenza pandemic was sweeping across Europe, Asia, and the Americas in 1918, another disease outbreak was noted by pig farmers in Iowa (Nelson and Worobey 2018; Zimmer and Burke 2009). This was a seemingly new swine disease that made hogs severely ill with a pronounced cough. Anecdotally it was observed that influenza outbreaks in farm families were closely followed by the appearance of this new disease in their swine herds. This temporal correlation suggested that these two illnesses might be the same disease although it would take another 10 years before Richard Shope performed the critical verifying experiments.

Richard Shope was an Iowan farm boy who became a physician and prominent virologist, focused mostly on diseases of animals. In 1929, while studying hog cholera he saw cases of the so-called "swine flu" and became intrigued with the disease. In his initial studies with sick animals, he isolated a bacterium that appeared identical to Pfeiffer's *B. influenzae* (Shope called it *B. influenzae suis* to indicate its swine origin, suis referring to a pig in Latin). However, he quickly proved that this organism did not elicit swine flu symptoms when injected into healthy pigs, so this microbe was not the disease-causing agent in pigs, and by analogy was unlikely to be producing influenza in humans. In a series of elegant papers, he carefully established that the real culprit was a filterable agent and that the agent could be serially passaged from pig to pig while maintaining its disease-causing capabilities (Lewis and Shope 1931; Shope 1931a, b, c, 1932). The successful serial passage of the disease is important as it indicates that the quantity of the causative agent is being maintained by replicating within each new host. In contrast, if the disease were caused by something nonreplicating, a poison or toxin for example, then that material would be diluted out with each passage through an animal, and eventually, there would be too little material remaining to elicit symptoms. These studies meticulously established that swine flu was an infectious disease and that the causative microbe met the definition of a virus. In subsequent work, Shope and others showed that the human and swine influenza viruses were antigenically related (Shope 1937; Smith et al. 1933). Decades later, genetic analysis of 1918 swine flu and human flu genomes confirmed their close relationship and suggested that the 1918 human virus was introduced into pig herds where it remains endemic to this day (Nelson and Worobey 2018).

Shope's thorough and convincing studies on swine flu severely weakened the case for a bacterial etiology of human influenza and provided a framework for other researchers to reexamine the human disease. Soon after his studies were published in 1931 and 1932, Great Britain was about to endure another influenza epidemic. By January 1933, there were more than 500,000 influenza cases reported across the British Isles, providing ample patient samples for analysis and testing. Following Shope's lead, a British research team consisting of Wilson Smith, Sir Christopher Andrewes, and Sir Patrick Laidlaw made a seminal discovery that finally established the viral nature of human influenza. One of the major hindrances in working with human influenza was the lack of a suitable animal model that recapitulated the human disease, so the British group embarked on a search for an appropriate research animal. Taking nasal samples from influenza patients they filtered them to remove bacteria and then tested the filtrates on various animal species. After repeated failures, they discovered that intranasal inoculation of the human patient filtrates into ferrets produced a constellation of symptoms resembling human influenza, while control experiments using nasal samples from healthy individuals did not affect the animals (Smith et al. 1933). Using ferrets, they were able to serially passage the disease, to show that infected ferrets transmitted the disease naturally to cage mates, and that swine influenza samples caused the same disease symptoms in ferrets as human samples. One of the researchers even developed influenza after being sneezed on by an infected ferret (Andrewes 1979). Additionally, they found that convalescent serum from individuals who had recovered from influenza had antibodies that neutralized the infectivity of the filtrates. Collectively, their results finally and convincingly argued that human influenza was a viral disease, and their work is acknowledged as the first isolation of the human influenza virus. This serendipitous discovery of the ferret's susceptibility to human influenza also established ferrets as the preferred animal model for mimicking human influenza infection, and ferrets remain a mainstay in modern influenza research.

Stumbling Towards A Vaccine

As important as ferrets were and still are for certain types of influenza investigations, these are moderate-sized and feisty animals that can be expensive to maintain and troublesome to work with. Subsequently, mouse strains were investigated and certain strains were found to be naturally susceptible to human influenza. Furthermore, seasonal influenza strains can usually be adapted to reproduce well in laboratory mice (Bouvier and Lowen 2010). While mice don't present with the same symptoms as humans (no fever, coughing, or sneezing), they are a useful system for propagating the influenza virus and conducting some types of research studies. Yet, as important as these animal models are for studying influenza, neither ferrets nor mice are optimal platforms for vaccine production. Another technical advance in the 1930s greatly facilitated subsequent influenza studies and eventually vaccine development. This advance was Ernest William Goodpasture's introduction in 1931 of fertilized chicken eggs as a medium for the growth of various microorganisms, including the emerging cadre of newly discovered viruses. Eggs were cheap, plentiful, easily handled (they didn't bite or scratch like ferrets), and convenient for laboratory work. One simply had to inject the egg with the test sample via a thin needle. Influenza (and many other viruses) will infect the cells of the embryo and replicate extensively. His methodology was quickly adopted by researchers around the world, and within a few years, both an American (Dochez et al. 1933) and a British (Smith 1935) group established that the human influenza virus could readily be cultured in eggs. This achievement provided a means for mass production of the influenza virus, and the egg growth system would soon become the mainstay of influenza vaccine production. Egg-grown influenza vaccines are still the most commonly available vaccines worldwide although cell culture-grown vaccines have been available in the United States since 2012 (Rajaram et al. 2020).

Once influenza was established as a viral disease and the non-human systems (ferrets, mice, and eggs) became available for the propagation of the virus, attention turned to the creation of a protective vaccine. By the 1930s, two successful approaches for vaccine development had been demonstrated, live attenuated viruses (vaccinia for smallpox and 17D for yellow fever) and inactivated viruses (rabies), and scientists were eager to transfer these approaches to influenza. Given the global impact of influenza and the world's sensitivity to this disease after the 1918 pandemic, it's not surprising that influenza vaccine researchers explored both options simultaneously with studies conducted in many countries. One of the first preliminary studies came from the USSR in 1937. Using the human influenza virus passaged in both ferrets and mice, the researchers created a live, attenuated influenza virus (LAIV) strain. After intranasal delivery to human subjects, these volunteers developed neutralizing antibodies against virulent influenza. In a similar experiment, a live influenza virus that had been propagated in chicken embryo tissue culture was tested for antibody induction in human volunteers after subcutaneous inoculation (Francis and Magill 1937). As with the USSR study, volunteers inoculated with this second LAIV again showed strong neutralizing antibody responses against the wild-type influenza virus. In both these studies, volunteers did not develop serious influenza, indicating that LAIVs were relatively safe yet still effective at generating the appropriate antibodies.

The collective animal and human studies strongly indicated that injection of LAIV induced neutralizing antibodies, however, it was still unproven that such antibodies protected the vaccine recipients against an actual influenza infection. The conundrum was how to test the protective effect since it needed to be done in humans. Community volunteers could be used but monitoring them and waiting for natural influenza cases to show up in a free-ranging group of adults could be difficult, time-consuming, and would require a large cohort to get a statistically significant difference between the vaccinees and the unvaccinated control group. A more efficient approach would be to use confined individuals where an influenza outbreak would spread readily among the entire group so that a much smaller number of vaccinated and control subjects could be used. In the 1930s, there was not yet any formal set of rules regulating the use of human subjects, and the use of convicts, soldiers, and wards of the state in research experiments was still an acceptable approach. Unlike Walter Reed whose ethical concerns for subject safety resulted in his creation of informed consent documents, most researchers of that era simply didn't believe that consent was necessary or even reasonable. In this context, Joseph Stokes from the Pediatrics Department at the University of Pennsylvania decided to test an LAIV vaccine on residents of the New Jersey State Home for the Feeble-Minded (Stokes et al. 1937a, b). There was no mention of informed consent or volunteers in his publications, and even if there were, in contemporary society, we maintain that individuals with diminished mental capacity cannot really give informed consent so are not suitable for such risky trials as testing a new vaccine. As abhorrent as Stokes' study would be considered today, his request to use these residents was quickly granted. In 1935 he initiated a small, controlled experiment by vaccinating some residents and leaving others unvaccinated. Over 2 years, he collected data on influenza cases and showed a significant reduction in cases among the patients injected with a vaccine derived from human influenza. However, even with this mostly positive outcome, there were troubling observations such as some vaccinated individuals who still became infected and concerns about how long the vaccine protection would persist. The authors speculated that there might be different influenza strains circulating with only some being neutralized by their vaccine, a harbinger of the confounding problem with influenza that is still with us today.

Over the next 6 years, Stokes and others conducted larger and more elaborate LAIV studies. Some included actual volunteers, but most used confined populations such as prison inmates and residents at state mental health facilities. Collectively, the data from these various trials were inconclusive as some studies showed no vaccine protection and others showed inconsistent results with vaccine protection in one tested cohort but not in another at a different location. More disappointingly, improved assays for measuring neutralizing antibodies revealed that antibody levels didn't correlate well with protection from influenza, i.e. some individuals with high levels of neutralizing antibodies still became infected and ill. One issue that became apparent and which contributed to erratic results with early LAIV vaccines was that there are two distinct types of influenza virus. There was the original type isolated by the British group in 1933 and an antigenically unrelated type first isolated in 1940. To distinguish them, the first isolate was named influenza A and the second influenza B. Both types are endemic in human populations around the world and contribute each year to the burden of seasonal influenza. Importantly, vaccines made with influenza A did not protect against influenza B infection. Consequently, for any influenza vaccine to be successful, it would need to protect against both types as well as any strain variants of either type.

In addition to the growing appreciation for the diversity of circulating influenza viruses that must be addressed in a vaccine, there was also growing dissatisfaction with live attenuated influenza as the vaccine platform. These crude attenuated viruses still caused disease in some individuals, might revert to more virulent forms, and could be very unstable during transport and storage. While these early LAIVs produced a strong immune response, their inherent problems caused some researchers to become disillusioned with LAIVs which led to a burgeoning interest in developing inactivated influenza vaccines. By the late 1930s, small initial studies by several groups (Andrewes

1937; Stock and Francis 1940; Salk et al. 1940) showed that the influenza virus could be partially purified and treated to produce an inactivated influenza vaccine (IIV). So now there were two options and a choice to be made about which type of influenza vaccine to pursue. Ultimately the United States Army chose the IIV type which would become the global standard.

As the world entered the 1940s, Europe was engaged in a brutal fight with Germany and the United States was growing closer to entering what would become World War II. The prior experiences with yellow fever during the Spanish-American war and influenza in World War I had sensitized the U.S. military command about the devastating effect disease could have on troops. With the yellow fever problem already solved by Theiler's vaccine, influenza suddenly became a top priority with the military. In 1941, the Army Epidemiological Board created the Commission on Influenza headed by Thomas Francis, a top influenza researcher at the University of Michigan. Francis brought in a young Jonas Salk (of subsequent poliovirus vaccine fame) to run the daily research activities, and by 1942 the Commission had an egg-grown, formalin-inactivated vaccine ready for testing. This was a trivalent IIV vaccine containing three different influenza viruses, two prevalent strains of type A and the type B virus. Field trials were conducted in 1943 at eight universities using students who were members of the Army Specialized Training Program. Over 6000 students were vaccinated with a similar number of unvaccinated controls. The trial was a clear success with the aggregate data showing a greater than three-fold reduction in influenza cases in the vaccinated cohort compared to the control group. Based on these positive results, the Army adopted the vaccine in 1944 and commenced universal vaccination of all soldiers in 1945. That same year, the first commercial version of this IIV vaccine was also licensed in the United States. The vaccine was safe and effective, and the production process was reasonable for generating large numbers of doses, so all that remained was to implement mass vaccination campaigns to protect the public. It looked like science had defeated influenza, just like rabies, smallpox, and yellow fever. Sadly, that optimism only lasted until 1947.

The influenza outbreak of 1946 was a B virus and the vaccine performed wonderfully, so the results of 1947 were completely unexpected. Among both the military and civilian vaccinees, there was no protection against the influenza virus that circulated in 1947. Studies quickly showed that the 1947 influenza strain was antigenically very different than the three strains present in the vaccine. Scientists were already aware that there were strain differences that could be distinguished by antibodies, but it was assumed that these antigenic differences were modest and wouldn't totally abrogate the vaccine's efficacy. The appearance of this new 1947 strain and the unforeseen failure of the vaccine highlighted an unanticipated feature of influenza that would become a chronic problem that has yet to be solved. Unlike many common viruses that are fairly stable, influenza is a highly protean virus whose surface antigen sequences can change so rapidly that antibodies against this year's strain may not recognize next year's strain. Because influenza is a constantly shifting antigenic target, neither prior infection nor a vaccine offers lasting immune protection against this virus. The only solution so far is to create a new vaccine each year matched to the strain or strains predicted to be predominant that season. Having a new vaccine each year requires that people be re-vaccinated annually to ensure that they have neutralizing antibodies against the coming strain or strains. Unfortunately, predicting the coming strains is not an exact science, and influenza vaccines are rarely perfectly matched with the upcoming strains. This mismatch results in a fairly low protective efficacy for yearly influenza vaccines, typically only 40–60%, compared to greater than 90% protection for most other viral vaccines.

Drifting and Shifting—The Vaccine-Confounding Biology of Influenza Viruses

Influenza viruses are classified in the family known as the Orthomyxoviridae. These viruses were originally termed myxoviruses from the Greek word myxa meaning mucous, an apt description for one of the classic symptoms of human influenza. The ortho prefix was added later to distinguish the influenza viruses from another viral family that will be covered in subsequent chapters, the Paramyxoviridae. In addition to influenza types A and B, the orthomyxovirus family contains two other influenza types that can infect humans. Known as types C and D, these two types were discovered in 1947 and 2011, respectively. Influenza C causes mild respiratory tract infections resembling a common cold. It has never been associated with epidemics or pandemics and instead seems to be a ubiquitous endemic virus that infects most people in childhood (Sederdahl and Williams 2020). Given its relatively innocuous clinical effects, no vaccine against type C has ever been developed. Influenza type D is primarily a disease in cattle although pigs and other domestic animals (sheep and goats) are also susceptible to this virus (Su et al. 2017). Some people have antibodies against influenza D indicating that they were infected, but there is no indication that this virus can cause serious illness in humans so again no vaccine has been pursued. Given the low clinical impact of types C and D, neither will be discussed further.

Orthomyxovirus aren't just pathogens for humans and this family contains numerous viruses that infect a wide variety of animal species, both invertebrates and vertebrates (Wille and Holmes 2020). Genetic analysis of widely diverged orthomyxoviruses suggests that this family has existed for hundreds of millions of years. These viruses likely originated in early invertebrates and have co-evolved with animals ever since. Current human type A influenzas all appear to be zoonotic diseases that initially arose in avian populations before spreading to humans, and birds remain the largest reservoir for these type A viruses. This is problematic for two reasons. First, the migratory nature of many birds enables viral spread over large geographic areas, so these avian viruses are widespread. Second, influenza in birds is a gastrointestinal disease. Infected birds shed the influenza virus in their feces which contaminates water sources and allows viral spread to different avian species that consume the water, including domestic birds such as chickens and ducks. Cross-species transmission spreads the avian viruses into humans, either directly through contact with infected birds or in some cases possibly through an intermediate host such as another domestic animal. In particular, pigs are susceptible to avian influenzas if they drink contaminated water and can then pass the virus on to their human handlers. This ubiquitous distribution of avian viruses in both wild and domestic animals creates an environment that is conducive to the development of dangerous new influenza viruses.

Like the yellow fever virus discussed in the last chapter, orthomyxoviruses have an RNA genome. During the reproduction of any RNA virus in an infected cell, the RNA genome must be repeatedly copied to generate new genomes for the progeny virions. The copying is performed by a viral enzyme called an RNA polymerase. RNA polymerases are proteins that catalyze the joining together of the individual molecular building blocks (called ribonucleotides) to form chains that are the new RNA molecules. To reproduce an RNA virus genome is a 2-step process. First, the RNA polymerase will use the original genome (let's call this gRNA for genomic RNA) as the template to make a copy (Fig. 5.1). This copy (let's call this the cRNA for complementary RNA) will have a sequence that is complementary to the original genome. To generate new gRNAs, the RNA polymerase must then repeatedly copy the cRNA and the resulting products are now identical to the gRNA. For influenza, its RNA polymerase is comprised of three different viral proteins called PA, PB1, and PB2 (Dou et al. 2018). These three proteins bind together to form a trimeric complex that is the functional RNA polymerase.

Across the different families of RNA viruses, their RNA polymerases vary widely in what is known as fidelity. Fidelity is a measure of how many mistakes a polymerase makes when it copies a template. A useful analogy is to

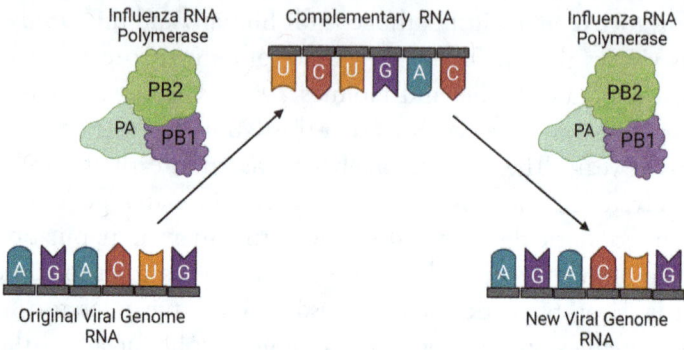

Fig. 5.1 RNA Replication. The genome for influenza consists of eight RNA segments. Each segment is a long, single-stranded molecule comprised of linked ribonucleotides. All RNAs use only four ribonucleotides designated A, C, G, and U. Each ribonucleotide can recognize and form an interaction only with a specific partner: A always pairs with U, and C always pairs with G. This base pairing function is essential for RNA replication. Shown is a small portion of RNA containing six ribonucleotides and the influenza trimeric polymerase complex comprised of the PA, PB1, and PB2 subunits. The polymerase uses the original RNA genome as the template to direct the assembly of a complementary RNA. In the figure, the polymerase will position a U with the first A on the template and position a C with the adjacent G. Using its enzymatic activity, the polymerase will then chemically link the U and C together. Next, it will position a U with the second A and link this U to the UC to generate UCU. The polymerase will continue down the template strand positioning complementary ribonucleotides and then joining them to the growing complementary chain. Once the complete complementary RNA is formed, the polymerase will use it as the new template and repeat the synthesis process. The complement to the complementary strand regenerates the original genomic sequence. Repeating this process using the complementary strands as the template produces multiple new genomic RNAs for incorporation into the new progeny virions

think of retyping a page of text over and over. In this scenario, the original text is the RNA genome template, the letters of the alphabet are the ribonucleotides, and your hands are the polymerase. As you type you may make mistakes by hitting the wrong key. This results in incorrect letters (ribonucleotides) being incorporated into the newly copied text (genome). You may be an excellent typist (high fidelity) who makes very few mistakes or a poor typist (low fidelity) with frequent errors. If you had no mechanism to correct your errors then these mistakes would randomly occur and remain every time you copied the text. Like the poor typist, it turns out that the influenza polymerase is a low-fidelity enzyme that makes frequent errors as it copies the viral gRNA and cRNAs. The influenza RNA polymerase lacks any mechanism to correct its errors, so whatever errors it makes remain in the newly synthesized genomes. These errors in the RNA are mutations, and if they occur within a gene then they may change the amino acid sequence of the protein encoded by that gene.

Mutations introduced by the low-fidelity influenza RNA polymerase can occur in any gene, but for immunity and vaccination the critical viral gene is called HA and it encodes the protein known as hemagglutinin. The HA protein is located on the surface of the virion where it projects like a stalk or a spike with a small globular head. Importantly, this is the molecule that the influenza virus uses to attach itself to cells. HA can bind to a human molecule called sialic acid which is found on the surface of cells in our respiratory system (as well as other places). When an influenza virus is inhaled, if it encounters a cell with sialic acid then the virus can attach itself to the cell through HA tightly binding to the sialic acid. Once bound and juxtaposed with the cell surface, the virus triggers an invagination of the cell membrane so that the virion is brought into the cell inside a small vesicle. Inside the cell, the HA protein is critical for releasing the virion from the vesicle so that the virus is now free inside the cell and can initiate its replication process. Without HA a virion could not attach to the cell, be taken up, or be released from the internalizing vesicle, so a functional HA is essential for infection. Neutralizing antibodies against the influenza virus work by binding to HA on the virion surface and physically interfering with its functions. Neutralizing antibodies coat the HA and can prevent its activity in three ways: antibodies may block HA's ability to attach to sialic acid so the virion can't attack a cell, can interfere with the virion invagination step, or may prevent virion release from the vesicle. Antibody interference with any of these steps prevents the cell from becoming infected. The seasonal influenza problem arises because the neutralizing antibody interaction with the HA protein is extremely specific for exact amino acid sequences in HA. Since the influenza virus accumulates mutations very rapidly, the amino acid sequence of its HA protein can change significantly from year to year, a phenomenon known as antigenic drift. This drift can greatly reduce or completely abrogate the ability of existing neutralizing antibodies to recognize the new version of the HA protein, leaving us susceptible to infection by the new strain. It is this extensive seasonal antigenic drift that necessitates new vaccines each year and continual re-vaccination to maintain adequate immunity. The virus never stops changing and we either have to tolerate annual vaccinations or accept the risk of disease. Fortunately, most other common RNA viruses have RNA polymerases with greater fidelity and/or error-correction mechanisms (known as proofreading) and don't drift this quickly. Consequently, the neutralizing antibodies induced by vaccinations against other RNA viruses such as yellow fever (Chap. 4), polio (Chap. 6), or measles and mumps (Chap. 7) retain their efficacy for decades or even whole lifetimes.

While antigenic drift is the major contributor to our constant battle with seasonal influenza, another feature of influenza viruses called antigenic shift is even more worrisome. Antigenic shift is an event that abruptly changes the properties of influenza viruses. Unlike antigenic drift which gradually changes the influenza virus each year, a shift is a sudden and often dramatic change in the virus' properties. The result of the shift may include heightened transmissibility, enhanced virulence, or a major change in viral surface proteins that makes the new virus resistant to our existing anti-influenza immunity. Such shifts have typically resulted in the global pandemics that occurred periodically over the last 100 years.

To understand antigenic shifts, we need to consider the influenza virus genome and its properties. Many RNA viruses have a single piece of RNA as their genome (called nonsegmented genomes), but others such as influenza have a genome consisting of multiple separate pieces of RNA (segmented genomes; Fig. 5.2). It might initially seem bizarre that a viral genome would be in pieces, but plants and animals also typically have segmented genomes. For example, humans have 23 DNA segments that we call our chromosomes, and for those viruses with segmented genomes, each piece of RNA is like a chromosome. Influenza types A and B both have genomes comprised of eight RNA segments, and each segment contains a gene that produces one or more viral proteins. Importantly, when new viruses are being assembled within an infected cell, each progeny virus must incorporate one copy of all eight segments, and influenza has mechanisms that ensure that the correct cohort of eight different RNA segments is packaged into each virion.

Because it contains multiple genome segments, influenza (or any other segmented genome virus) can undergo a genetic process known as reassortment (Fig. 5.2). Reassortment is somewhat akin to sexual reproduction in that there is a mixing of chromosomes from two parents to give an offspring with a unique genetic composition. Just like sexual reproduction, for reassortment to occur there must be two different parental influenza viruses, such as two different strains that I'll call X and Y. Secondly, these two viruses must coinfect the same cell. Each infecting virus will replicate its eight RNA genome segments (X1-8 and Y1-8) and all these RNA pieces will be within a single cell. When it comes time to assemble the replicated RNA segments into a new virus particle, each virion will incorporate 1 copy of segments 1–8, but whether it packages an X-type or Y-type segment is random. For example, one virion might end up with segments 1–4 of the X type and segments 5–8 of the Y type while another virion could have segments X1 and 2-7Y. Any resultant virion with a mixture of X and Y-type segments is called a reassortant. With 8 segments involved, 256 possible combinations of reassortants can be

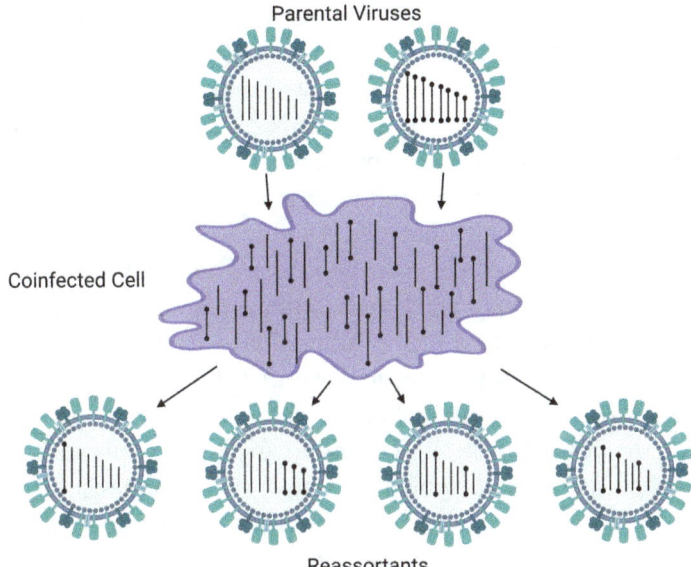

Fig. 5.2 Influenza Reassortment. Shown are two different influenza strains that are designated the parental viruses. Each parental virion contains eight genomic RNA segments that are drawn as different length lines to distinguish the segments. When a cell is coinfected with these two viruses, each parent will replicate many copies of its genomic segments. As new virions are assembled, each virion will receive one copy of segments 1–8, but which parent provides that segment is random. When a virion is generated that has genomic segments from two different parents it is called a reassortant. Shown below the coinfected cell are four possible reassortants (out of 256 total combinations). Reassortants may have properties significantly different than either parental virus

produced in any coinfection. And just as children resemble their parents but have different and unique characteristics, so can reassortants have new and novel properties that distinguish them from either parental virus.

To complicate the reassortment picture even further, within the influenza A-type viruses, there are not only different strains (viruses with small genetic differences) but also different subtypes (Dou et al. 2018). Subtypes are based on major differences in the two viral surface proteins, hemagglutinin (HA) and a second protein called neuraminidase (NA). While HA is critical for viral entry to start an infection, NA functions to release new virions from the infected cell. To date, there are 18 known HA subtypes (designated H1-18) and 11 NA subtypes (N1-11) present in bird populations. Every influenza A is designated by its HA (H for short) and NA (N for short) subtype, for instance, the types circulating in 2022 are an H1N1 and an H3N2. The H3N2 subtype has been around since 1968 when it appeared and caused a pandemic killing around one million people worldwide. Likewise, the H1N1 subtype emerged as a pandemic-causing virus in 2009 and has remained in circulation since then. New combinations of subtypes arise by reassortment during coinfections. If the H1N1 and H3N2 subtypes were to coinfect an individual then two new reassortment subtypes could develop, an H1N2 and H3N1. Theoretically, among the influenza A viruses in nature, every possible combination of HA and NA subtypes ($18 \times 11 = 198$) could exist although only around 130 combinations have been found so far. Anytime a new HN combination jumps species and enters human populations we call this an antigenic shift. Because these new combinations haven't been seen before by our immune system we lack protection against them, and they typically cause pandemics with case numbers well above seasonal averages.

Most of the reassortment is happening in animals, primarily birds, as human populations generally only have one or two subtypes circulating. And remember, it isn't just the HA and NA gene segments that can reassort during a coinfection. The six other genome segments reassort as well, so the total reassortant combinations among 198 possible H/N subtypes are enormous. Combine this huge pool of potential reassortants with the additional variation introduced by antigenic drift and the extent of influenza virus diversity is staggering. What really worries scientists and public health experts is a random antigenic shift that produces a novel virus with high transmissibility in humans along with high virulence like the 1918 pandemic virus. Although there is no way to prevent such a dangerous strain from arising in animal populations, there are protocols in place for early detection and elimination of potentially lethal new reassortants. In 1952, the World Health Organization (WHO) created the Global Influenza Surveillance Network (subsequently renamed the Global Influenza Surveillance and Response System [GISRS] in 2011). Member countries of the network work closely with animal surveillance agencies, such as the United States Department of Agriculture (USDA). The goal of this network is to detect new influenza strains and prevent their spread into human populations so that they don't become pandemics. Currently, over 150 research facilities in 113 countries collectively monitor influenza viruses around the world by constantly collecting and analyzing human and animal samples. The data amassed by the network serves two purposes. First, tracking human influenza virus cases helps identify genetic drifts and is used to inform vaccine development for each coming year. Vaccine development and production take about 6 months, so this is a continuous process that seeks to identify the most prevalent strains occurring each influenza season and incorporate those into the next year's vaccine. Second, monitoring domestic animal populations, primarily birds (chickens, ducks, turkeys, etc.) and swine, is a sentinel system. The biggest threat to humans is a highly pathogenic reassortant that infects domestic animal populations where it could easily spread to individuals who have contact with the infected animals. Early detection of such strains allows culling of the infected animals to prevent viral spread among the animal population and thwart cross-over into humans. This can be a costly action that may involve killing thousands of birds or pigs, but it is the only mechanism to ensure that potentially dangerous influenza strains don't become established in humans. The 2020 SARS-CoV-2 outbreak and the resulting COVID-19 pandemic is an instructive example of how one small animal virus can invade human populations with devastating consequences. The worldwide public health system works tirelessly to avoid future influenza pandemics, however, influenza surveillance is not a perfect system. It is always possible that a new and dangerous influenza virus could slip through into humans, so vaccines remain an essential tool for controlling seasonal and pandemic influenza. The entry of the highly pathogenic avian H5N1 type influenza into domestic poultry in 2020 followed by its spread to cattle in 2024 is just such a troubling event. The prevalent human strains are H1N1 and H3N2, so we might not have much immunity against the H5N1 type. Fortunately, this strain seems to have limited pathogenicity in humans and little if any human-to-human transmission capability. Still, this needs constant monitoring as genetic drift or shift could turn this currently innocuous virus into a raging killer.

Eighty Years of Vaccine Evolution

The 1940s successfully produced the first influenza vaccines, but there was ample room for improvement as the initial vaccines were quite crude by modern standards. The protective efficacy was limited and the adverse side effects were numerous. Localized reactions at the site of injection such as inflammation, soreness, and swelling were common. In addition, many individuals, especially young children, suffered systemic effects such as fever and flu-like symptoms. To make these first influenza vaccines, embryonated chicken eggs were inoculated with an appropriate virus strain and incubated to allow viral replication. The newly produced progeny virions accumulated in the allantoic fluid from the eggs and this fluid was harvested to provide the virus stock for the vaccine doses. It was discovered that the influenza virus binds to chicken red blood cells (RBCs) and this was used as a one-step purification to separate the virions from the allantoic fluid. After harvesting the fluid, RBCs were added to adsorb the virions followed by centrifugation to collect the RBCs with the bound virions attached. The virions were subsequently eluted from the RBCs, inactivated by formalin treatment, and were ready for use as the vaccine. However, this simple purification step did not effectively remove all the contaminating proteins and other biomolecules present in the allantoic fluid. These residual components contributed to unwanted reactions at the injection site, as well as causing allergic reactions in some individuals. Additionally, there were incidents where the formalin inactivation was incomplete resulting in live viruses present in the vaccine dose. Nonetheless, even with these caveats, both the military and the general public accepted this vaccine as there was no alternative for protecting people from influenza, and even a poor vaccine was considered better than no vaccine.

In the two decades following its introduction into general use, influenza vaccine manufacturing remained mostly unchanged. There were some modest tweaks in the production process but the vaccine was not significantly improved. It wasn't until the 1960s that two advances helped create safer and more effective versions of the vaccine. The first advance was the application of more sophisticated centrifugation protocols. These procedures improved the purity of the allantoic fluid-derived virions by more effectively removing the egg biomolecules compared to the original RBC purification method. The resultant vaccine material with fewer contaminants had decreased reactions at the injection site, making it more palatable to the recipients. The second improvement was the discovery that incubation of the vaccine with certain chemicals (originally either ether or a detergent called Tween-80, but other detergents are now used—see Table 5.1) greatly reduced the systemic effects (Davenport et al. 1964). These chemicals disrupt the virion membrane and facilitate the removal of the virion RNA and inner virion proteins whose presence in the vaccine contributed to the systemic issues. The remaining virion material is enriched for the antigenically important surface proteins, NA and HA, that elicit the neutralizing antibody response. This so-called "split" influenza vaccine was licensed for use in the United States in 1968 and this type of vaccine has remained in use ever since as it is well-tolerated by most individuals. There are four split vaccines produced by different manufacturers that are available in the United States: Afluria, Fluarix, FluLaval, and Fluzone (Table 5.1). All four are egg-based, quadrivalent vaccines containing two type A strains and two type B strains.

One downside of the split vaccines wasn't appreciated until years later when it was discovered that these vaccines were not as effective at inducing the immune response, particularly in the elderly population. To improve the vaccine efficacy in the older population there have been two strategies adopted: increasing the amount of viral protein in the vaccine or addition of an

5 Influenza: An Elusive and Evasive Foe

Table 5.1 Summary of current influenza vaccines

Vaccine Trade Name	Manufacturer	Production Platform	Type	Components	Inactivating Agent	Age
Fluarix	GlaxoSmithKline	Egg	Inactivated	Split with deoxycholate	Formaldehyde	≥6 months
FluLaval	GlaxoSmithKline	Egg	Inactivated	Split with deoxycholate	UV Light + Formaldehyde	≥6 months
Fluzone	Sanofi Pasteur	Egg	Inactivated	Split with Triton X-100	Formaldehyde	≥6 months
Fluzone High Dose	Sanofi Pasteur	Egg	Inactivated	Split with Triton X-101	Formaldehyde	≥65 years
Afluria	Seqirus	Egg	Inactivated	Split with sodium taurodeoxycholate	Beta-propiolactone	>5 years
Fluad	Seqirus	Egg	Subunit with adjuvant	HA/NA	Formaldehyde	≥65 years
FluMist	AstraZeneca	Egg	Live Attenuated	Whole virions	None	2–49 years
Flucelvax	Seqirus	Cell Culture	Subunit	Enriched for HA+NA	Beta-propiolactone	>4 years
Flublok	Sanofi Pasteur	Cell Culture	Subunit	HA	None	≥18 years

adjuvant to the vaccine. FluZone High-Dose (Table 5.1) has 4 times the amount of HA protein as the standard influenza vaccine. This increased amount of antigen generally triggers a more robust immune reaction and elicits a better protective response in older individuals than the standard dose vaccines. The other available approach is the Fluad vaccine (Table 5.1) which includes an adjuvant along with the standard dose of inactivated influenza viruses. Adjuvants are substances that boost the immune response to the target antigen(s), generally in a nonspecific fashion that isn't dependent on the antigen being used. Adding an adjuvant to a vaccine produces stronger and sometimes broader immunity without raising the amount of antigen in the vaccine dose. The commonly used adjuvants are generally inexpensive compounds so it can be more cost-effective to add an adjuvant rather than increasing the antigen content as the antigen is the more expensive component. Both Fluzone High-Dose and Fluad are now the recommended vaccines for individuals of age 65 or greater.

All of the above vaccines are still egg-based products generated from virus-infected embryonated eggs, a process that hasn't changed since its inception in the 1930s. Any egg-based vaccine still contains trace amounts of egg proteins so these vaccines can be problematic to anyone with an egg allergy. There are also manufacturing concerns since a huge number of eggs are needed to produce the millions of influenza vaccine doses distributed each year. Eggs need to be constantly replenished which involves significant costs for shipping, storage, and handling. Additionally, should there be a crisis in egg production, for example, an avian influenza outbreak requiring the slaughter of millions of chickens, then vaccine production could be impacted. These limitations with egg-based vaccines led to alternative production methods, two of which are now licensed in the United States. Flucelvax and Flublok (Table 5.1) are both quadrivalent subunit vaccines produced in cell culture rather than chicken eggs. Cell culture has the advantage that the cells can be grown and maintained indefinitely in the manufacturing facility so there are no supply or shipping issues (Rajaram et al. 2020). Cell production is also fairly easy to scale up for the large quantities needed for commercial vaccine production, and this method is routinely used for several common vaccines. For Flucelvax, the four influenza viruses for the vaccine are grown in a dog kidney cell line called MDCK (Madin-Darby Canine Kidney). The viruses are inactivated and partially purified to enrich the HA and NA proteins. In contrast, Flublok is a completely recombinant vaccine that doesn't involve any influenza viruses. Instead, the gene for HA is cloned from the four influenza types chosen for the vaccine. The cloned HA genes are expressed in an insect cell line and their proteins are purified to give a vaccine that consists solely of HA protein without any other influenza components. Since there is no egg step for either vaccine, both are specifically recommended for people with egg allergies, although anyone can use them.

While multiple inactivated vaccines for influenza are in wide use, inactivated vaccines generally produce less effective immune responses than live, attenuated virus vaccines. Inactivated viruses mostly invoke a B-cell immune response that produces antibodies that bind to and neutralize circulating virions so that they cannot infect new cells. These neutralizing antibodies can be quite effective in preventing or reducing the clinical outcome of subsequent infections, yet they represent only half of the acquired immune response. In contrast, attenuated viruses are still able to infect cells and replicate to a low level which can elicit both a B cell and T cell response. The T cells attack virus-infected cells and destroy these viral factories, thus preventing the production of new virions. This combined B and T cell attack provides a broader and often more long-lasting immunity. The initial attempts at influenza vaccines in the 1930s and 40s tried to develop and use attenuated strains, but technical issues, including reversion to virulence, limited their success and they were supplanted by the inactivated vaccines. In recent decades there has been renewed interest in developing live, attenuated influenza vaccines (Treanor 2020), and there is now one licensed version available in the United States called FluMist (Table 5.1). FluMist is also a quadrivalent vaccine containing cold-adapted influenza viruses. Human viruses typically reproduce best at 37 °C (human body temperature). For the vaccine, mutant forms of the viruses were developed that grew well at 25 °C but poorly at 37 °C. These viruses could be grown for production at the lower temperature, but when introduced into vaccinees would only be able to replicate slightly, just enough to stimulate the desired B- and T-cell responses but not enough to cause disease. This vaccine is delivered as a mist that is sprayed into the nasal passages. All the other influenza vaccines are delivered as injections, so FluMist is an alternative for those who don't like shots.

Except for FluMist, none of the influenza vaccines available in the United States contain live influenza and therefore they cannot cause the disease in vaccine recipients. Some people claim that the vaccine gave them the flu or have heard reports that this can happen, but it simply can't. Two explanations likely cover these supposed cases of vaccine-induced influenza. The first is that a very small number of individuals still develop significant systemic effects from the vaccine including fever and malaise. They may mistake these symptoms for a case of the flu when in fact it is merely an immunological response to the vaccine. The second explanation involves the timing of the vaccination and a community-acquired infection by the influenza virus. It takes roughly

1–2 weeks after vaccination to develop immunity, so during that period, you are still susceptible to infection if exposed. Given that the incubation period of influenza is 2–4 days, a natural exposure anytime from 4 days before the vaccination through the week after could result in a case of influenza before the vaccine had time to elicit protective antibodies. The coincidental timing might make it appear that the vaccination caused the disease, but it was really just a naturally acquired infection that happened to occur around the time of the vaccination. Since many people get their influenza vaccine when influenza is actively circulating in their community, it is not surprising that some individuals will contract an infection around the time of their vaccination. There are certain well-known side effects associated with influenza vaccines, but getting an influenza virus infection isn't one of them.

A Final Note

Scientists and manufacturers have explored influenza vaccine strategies for over 80 years and multiple commercial influenza vaccines of different types and characteristics exist. Sadly, none of them succeeds at the optimal level. Antigenic drift and antigen shift change the influenza virus surface proteins so rapidly that our acquired immunity, either through vaccination or previous infection, often quickly loses its effectiveness. This loss of immunity necessitates yearly revaccination to stay protected from disease. However, the constantly evolving nature of influenza viruses makes matching vaccines with next season's expected influenza strains a technical challenge that science has not yet mastered. The result is that in most years the efficacy of the available influenza vaccines is modest. The Centers for Disease Control (CDC) average influenza vaccine efficacy data for the 2015/2016 through 2019/2020 seasons were 48, 40, 38, 29, and 39%, all numbers way below the typical vaccine standard of ≥ 90 efficacy. The CDC does point out that even if they do become ill, vaccine recipients usually have less severe cases of influenza than nonvaccinated persons. Still, except for the high-risk populations (the very young and the elderly), how much benefit accrues from the recommendation in the United States that everyone should receive a yearly influenza vaccine? Production, distribution, and administration of the vaccine have a monetary cost and utilize a significant amount of healthcare resources, so is the public health benefit significant enough to justify this massive annual influenza vaccine campaign? Some other countries, including those in the European Union (EU), recommend a more targeted strategy that vaccinates only the most vulnerable populations. This includes the very young (under 5 years old), the elderly (65 years and up), pregnant women, those with underlying health issues, and healthcare workers. Which approach is most effective at reducing morbidity and mortality is still being evaluated and debated. And even if one approach is actually better, neither is as good as we would like it to be. What is really needed is an influenza vaccine that provides broad and lasting protection that is not diminished by the yearly changes in the viral surface proteins. Ideally, a single vaccination series would provide life-long immunity or would only require very infrequent booster shots to maintain adequate resistance to infection. This holy grail of influenza vaccinology has eluded scientists for 80 years, but researchers are creeping closer and closer to this so-called "universal" influenza vaccine (McMillan et al. 2021). The world would benefit enormously if influenza could finally be controlled as easily as we control other once-prevalent seasonal viruses like measles and mumps.

References

Andrewes CH (1937) Influenza: four years' progress. Br Med J 2(4001):513–515. https://doi.org/10.1136/bmj.2.4001.513

Andrewes C (1979) Growth of virus research 1928-1978. Postgrad Med J 55(640):73–77. https://doi.org/10.1136/pgmj.55.640.73

Barberis I, Myles P, Ault SK, Bragazzi NL, Martini M (2016) History and evolution of influenza control through vaccination: from the first monovalent vaccine to universal vaccines. J Prev Med Hyg 57(3):E115–E120

Bouvier NM, Lowen AC (2010) Animal models for influenza virus pathogenesis and transmission. Viruses 2(8):1530–1563. https://doi.org/10.3390/v20801530

Brown J (2018) Influenza: the hundred year hunt to cure the deadliest disease in history. First Touchstone hardcover edition. Touchstone, New York

Davenport FM, Hennessy AV, Brandon FM, Webster RG, Barrett CD Jr, Lease GO (1964) Comparisons of serologic and febrile responses in humans to vaccination with influenza A viruses or their hemagglutinins. J Lab Clin Med 63:5–13

Dochez AR, Mills KC, Kneeland Y (1933) Studies of the etiology of influenza. P Soc Exp Biol Med 30(8):1017–1022

Dou D, Revol R, Ostbye H, Wang H, Daniels R (2018) Influenza A virus cell entry, replication, virion assembly and movement. Front Immunol 9:1581. https://doi.org/10.3389/fimmu.2018.01581

Fildes PG (1956) Richard Friedrich Johannes Pfeiffer, 1858-1945. Biogr Mem Fellows R Soc 2:236–247. https://doi.org/10.1098/rsbm.1956.0016

Francis T, Magill TP (1937) The antibody response of human subjects vaccinated with the virus of human influenza. J Exp Med 65(2):251–259. https://doi.org/10.1084/jem.65.2.251

Kuszewski K, Brydak L (2000) The epidemiology and history of influenza. Biomed Pharmacother 54(4):188–195. https://doi.org/10.1016/S0753-3322(00)89025-3

Lewis PA, Shope RE (1931) Swine influenza: II. A hemophilic bacillus from the respiratory tract of infected swine. J Exp Med 54(3):361–371. https://doi.org/10.1084/jem.54.3.361

McMillan CLD, Young PR, Watterson D, Chappell KJ (2021) The next generation of influenza vaccines: towards a universal solution. Vaccines (Basel) 9(1). https://doi.org/10.3390/vaccines9010026

Morens DM, Taubenberger JK, Fauci AS (2008) Predominant role of bacterial pneumonia as a cause of death in pandemic influenza: implications for pandemic influenza preparedness. J Infect Dis 198(7):962–970. https://doi.org/10.1086/591708

Nelson MI, Worobey M (2018) Origins of the 1918 pandemic: revisiting the swine "mixing vessel" hypothesis. Am J Epidemiol 187(12):2498–2502. https://doi.org/10.1093/aje/kwy150

Nickol ME, Kindrachuk J (2019) A year of terror and a century of reflection: perspectives on the great influenza pandemic of 1918-1919. BMC Infect Dis 19(1):117. https://doi.org/10.1186/s12879-019-3750-8

Olitsky PK, Gates FL (1923) Investigations on the bacteriology of epidemic influenza. Science 57(1467):159–166. https://doi.org/10.1126/science.57.1467.159

Oxford JS, Gill D (2018) Unanswered questions about the 1918 influenza pandemic: origin, pathology, and the virus itself. Lancet Infect Dis 18(11):e348–e354. https://doi.org/10.1016/S1473-3099(18)30359-1

Pappas G, Kiriaze IJ, Falagas ME (2008) Insights into infectious disease in the era of hippocrates. Int J Infect Dis 12(4):347–350. https://doi.org/10.1016/j.ijid.2007.11.003

Rajaram S, Boikos C, Gelone DK, Gandhi A (2020) Influenza vaccines: the potential benefits of cell-culture isolation and manufacturing. Ther Adv Vaccines Immunother 8:2515135520908121. https://doi.org/10.1177/2515135520908121

Salk JE, Lavin GI, Francis T (1940) The antigenic potency of epidemic influenza virus following inactivation by ultraviolet radiation. J Exp Med 72(6):729–745. https://doi.org/10.1084/jem.72.6.729

Schultz MG, Morens DM (2009) Charles-Jules-Henri Nicolle. Emerg Infect Dis 15(9):1520–1522. https://doi.org/10.3201/eid1509.090891

Sederdahl BK, Williams JV (2020) Epidemiology and clinical characteristics of influenza C virus. Viruses 12(1). https://doi.org/10.3390/v12010089

Shope RE (1931a) The etiology of swine influenza. Science 73(1886):214–215. https://doi.org/10.1126/science.73.1886.214

Shope RE (1931b) Swine influenza: I. Experimental transmission and pathology. J Exp Med 54(3):349–359. https://doi.org/10.1084/jem.54.3.349

Shope RE (1931c) Swine influenza: III. Filtration experiments and etiology. J Exp Med 54(3):373–385. https://doi.org/10.1084/jem.54.3.373

Shope RE (1932) Studies on immunity to swine influenza. J Exp Med 56(4):575–585. https://doi.org/10.1084/jem.56.4.575

Shope RE (1937) Immunological relationship between the swine and human influenza viruses in swine. J Exp Med 66(2):151–168. https://doi.org/10.1084/jem.66.2.151

Smith W (1935) Cultivation of the virus of influenza. Brit J Exp Pathol 16(6):508–512

Smith W, Andrewes CH, Laidlaw PP (1933) A virus obtained from influenza patients. Lancet 2:66–68. https://doi.org/10.1016/S0140-6736(00)78541-2

Solly M (2020) What we can learn from 1918 influenza diaries. Smithsonian Magazine

Stock CC, Francis T (1940) The inactivation of the virus of epidemic influenza by soaps. J Exp Med 71(5):661–681. https://doi.org/10.1084/jem.71.5.661

Stokes J, Chenoweth AD, Waltz AD, Gladen RG, Shaw D (1937a) Results of immunization by means of active virus of human influenza. J Clin Invest 16(2):237–243. https://doi.org/10.1172/Jci100853

Stokes J, McGuinness AC, Langner PH, Shaw DR (1937b) Vaccination against epidemic influenza with active virus of human influenza - (A two-year study). Am J Med Sci 194(6):757–768. https://doi.org/10.1097/00000441-193712000-00002

Su S, Fu X, Li G, Kerlin F, Veit M (2017) Novel influenza D virus: epidemiology, pathology, evolution and biological characteristics. Virulence 8(8):1580–1591. https://doi.org/10.1080/21505594.2017.1365216

Taubenberger JK, Morens DM (2006) 1918 influenza: the mother of all pandemics. Emerg Infect Dis 12(1):15–22. https://doi.org/10.3201/eid1201.050979

Taubenberger JK, Hultin JV, Morens DM (2007) Discovery and characterization of the 1918 pandemic influenza virus in historical context. Antivir Ther 12(4 Pt B):581–591

Treanor J (2020) History of live, attenuated influenza vaccine. J Pediatr Infect Dis Soc 9(Supplement_1):S3–S9. https://doi.org/10.1093/jpids/piz086

Tumpey TM, Basler CF, Aguilar PV, Zeng H, Solorzano A, Swayne DE, Cox NJ, Katz JM, Taubenberger JK, Palese P, Garcia-Sastre A (2005) Characterization of the reconstructed 1918 Spanish influenza pandemic virus. Science 310(5745):77–80. https://doi.org/10.1126/science.1119392

Van Epps HL (2006) Influenza: exposing the true killer. J Exp Med 203(4):803. https://doi.org/10.1084/jem.2034fta

Watanabe T, Kawaoka Y (2011) Pathogenesis of the 1918 pandemic influenza virus. PLoS Pathog 7(1):e1001218. https://doi.org/10.1371/journal.ppat.1001218

Wille M, Holmes EC (2020) The ecology and evolution of influenza viruses. Cold Spring Harb Perspect Med 10(7). https://doi.org/10.1101/cshperspect.a038489

Worobey M, Han GZ, Rambaut A (2014) A synchronized global sweep of the internal genes of modern avian influenza virus. Nature 508(7495):254–257. https://doi.org/10.1038/nature13016

Yamanouchi T, Sakakami K, Iwashima S (1919) The infecting agent in influenza: an experimental research. Lancet 1:971–971

Zimmer SM, Burke DS (2009) Historical perspective--emergence of influenza A (H1N1) viruses. N Engl J Med 361(3):279–285. https://doi.org/10.1056/NEJMra0904322

6

Poliovirus: An Insidious Plague

Keywords Cytopathic effect • John Enders • Simon Flexner • Paralytic polio • Plaque assay • Reversion • Albert Sabin • Jonas Salk • Serotype • SV40 • Vaccine-associated paralytic polio

Abbreviations

CNS	Central nervous system
CPE	Cytopathic effect
GI	Gastrointestinal system
IPV	Inactivated poliovirus vaccine
OPV	Oral poliovirus vaccine
SV40	Simian virus 40
VAPP	Vaccine-associated paralytic polio
VDPV	Vaccine-derived polioviruses
WHA	World Health Assembly
WHO	World Health Organization

The Rise of Poliomyelitis

Poliomyelitis (from the Greek words polio = grey and myelos = marrow or of the spinal cord), polio for short, was one of the most feared diseases in the early twentieth century. Originally known as infantile paralysis, this disease terrorized families with the threat of death or paralysis to hapless children and adults. Creeping into towns and cities as the weather warmed each summer, it was never predictable where and when this insidious disease would strike. Once cases occurred, entire towns shut down as fearful families kept their children home and avoided public places. Fortunately, most infected people had either no symptoms or mild nondescript symptoms such as fever, headache, and gastrointestinal distress (nausea and vomiting) from which they fully recovered. Only about 1% of all infections progressed to severe disease that attacked the central nervous system (CNS), but these cases could be tragic. These individuals often suffered permanent damage leading to partial or full paralysis, with the most severe cases leading to death because of paralysis of the breathing muscles. As one frightened young patient stated, "I could talk, I could open and close my eyes, and I could turn my head from side to side on my pillow, but otherwise I could not move at all" (Kehret 1996). While only a small minority of the total infections, these paralytic cases were often devastating and left behind a legacy of permanently crippled victims. The 1952 outbreak was the worst in U.S. history with approximately 58,000 cases, 3145 deaths, and over 21,000 people left with some degree of paralysis. While the subsequent 3 years saw a modest decline in cases, the early 1950s still produced a total of over 100,000 cases and nearly 4000 deaths. Keep in mind that the number of cases reported was likely a vast underestimate. Inapparent infections would not have been detected and even mild gastrointestinal cases might not have been attributed to polio. With polio rampant, the introduction of the first polio vaccine in 1955 was greeted enthusiastically by the public. Mass vaccination events organized by fearful communities were widespread as parents desperately sought to protect their children. Because of these public health efforts, within 10 years polio was tamed and by 1965 there were only 72 total polio infections reported in the U.S. with 16 deaths. By 1980 the U.S. was polio-free and has remained free ever since. The few naturally acquired cases in this country since 1980 have been in individuals who became infected elsewhere in the world before returning home. This remarkable conquest of a devasting infection in one generation was all thanks to vaccines and our campaigns of mass inoculation against this disease. Similar campaigns in every country have nearly abolished this disease. In 2022, only three countries (Afghanistan, Mozambique, and Pakistan) reported endemic polio for a total of 30 confirmed cases (There were an additional 871 vaccine-derived cases from 24 countries and we'll return to this below). Like smallpox, the virus causing polio has no natural reservoirs other than humans. If we can stop all human cases we should one day soon see the complete elimination of poliovirus from the world.

Scientists suspect that poliovirus has circulated among humans for at least several millennia, but with only a low incidence of the paralytic disease. The oldest potential evidence for polio infection comes from ancient Egypt (Galassi et al. 2017; Friedrich Koch 1985). Two Egyptian carvings from the period 1500–1300 BCE each show a man with a crippled leg using a walking stick, a pathology consistent with a case of paralytic polio. Additionally, there are Egyptian mummies with leg shortenings and foot deformities that could also be due to polio. However, as with the viral diseases presented in previous chapters, none of these supposed cases of polio can be proven definitely as other clinical causes for the observed limb features are possible such as congenital malformations. Similar doubt exists about the few historical writings that contain descriptions of polio-like illnesses. Hippocrates of ancient Greece wrote of a seasonal illness occurring in the summer and autumn that could inflict paralysis. This description is reminiscent of modern polio outbreaks but is not sufficiently detailed to be definitive. Likewise, the few biblical passages that portray individuals with paralyzed limbs could reflect a polio infection or could be due to other diseases. And beyond these limited examples, written records of other ancient cultures don't reveal any major epidemics that resemble paralytic polio outbreaks. With scant written records and no physical evidence of the virus in historical samples, the presence of polio in BCE times is still speculative. A more convincing historical case of paralytic polio has been made for a male skeleton, designated Halbturn 59, from the fourth century CE (Berner et al. 2021). Extensive paleopathology performed on this skeleton, including X-rays, CT scans, histology, and cross-sectional morphology, was highly suggestive of paralytic polio while ruling out other likely causes. Still, this is only a single individual, and even if Halbturn 49 truly had paralytic polio, one fact is certain: the paucity of visual, written, and physical evidence for this illness strongly suggests that the paralytic sequela of polio was never rampant in human populations before modern times. While it is likely that the causative virus has been circulating among humans for many centuries, it apparently did so without causing much frank disease. Paralytic cases in children were so infrequent that polio was not recognized as a distinct disease until it was described in the late 1700s by Michael Underwood, an English physician (Dunn 2006). In 1784 he published a lengthy book based on his medical observations with children entitled "A Treatise on the Diseases of Children, With General Directions for the Management of Infants from Birth". In this work, he provided the first modern description of polio as a condition causing "debility of the lower extremities". Underwood asserted that the condition was "not a common disorder" and that it "usually attacks children previously reduced by fever; seldom those under one, or more than four or five years old". He also accurately described the clinical presentation and the range of patient outcomes although he had no explanations for the cause or prevention of this illness.

It would take another 50 years before Underwood's observations would be carefully extended by a German physician, Jacob Heine (Pietrzak et al. 2017). In an 1840 publication, he not only described the physical and clinical characteristics of patients afflicted with this disease but also shared treatment options and his use of orthopedic devices intended to help patients regain some mobility. To codify this disease as a distinct clinical entity and distinguish it from other paralytic conditions, he coined the term infantile spinal paralysis. His terminology of "infantile paralysis" would persist well into the twentieth century until the clinical name of poliomyelitis was adopted. Importantly, Heine presented some modest pathological evidence that led him to postulate that the disease involved damage to the anterior horns of the spinal cord, the region that transmits motor signals to the skeletal muscles. This hypothesis would be confirmed in 1870 by the French neurologists, Jean-Martin Charcot and Alix Joffroy (Kumar et al. 2011), leading us one step closer to understanding the paralytic mechanisms of this disease.

Although the medical literature of the late 1700s and early 1800s began to recognize and define the disease of infantile paralysis, this illness remained uncommon during that period. It wasn't until the late 1800s that the rare and sporadic nature of polio began to change into the more epidemic form that plagued the twentieth century. Rather than rare single cases, small, clustered outbreaks suddenly began to appear in Europe, initially in Norway and Sweden. First reported in 1890 by the Swedish pediatrician, Karl Oskar Medin, he prophetically noted the new epidemic nature of this disease, a harbinger of what was coming in the twentieth century. A similar situation was beginning in the United States with an outbreak in Boston of 26 cases in 1893 followed by a larger outbreak the next year in the state of Vermont. The Vermont epidemic in the summer of 1894 was carefully observed and reported by a local physician, Charles Caverly (1896). He noted 132 cases, mostly in children under 6 years of age, with 119 incidences of paralysis and ultimately 18 deaths. Such smallish outbreaks continued on both continents over the ensuing decade before polio exploded after the turn of the century.

The first so-called major polio outbreak occurred in Sweden in 1905 (Axelsson 2009). Unlike the previous small epidemics that were confined to local areas, this outbreak spread across the entire country and afflicted over 1000 people in that nation. The large case numbers and the widespread distribution of cases were intently studied by Dr. Ivar Wickman leading to several seminal observations about the disease. First, based on the pattern of case occurrences over both time and geography, he concluded that this was likely an infectious disease and not due to some noninfectious cause. This deduction was important because prior to this outbreak the etiology had been uncertain. Few physicians had ever noted the disease and there was no abundant folklore detailing this condition, likely because cases of paralytic polio were so scarce before this large outbreak. Since polio had never been a conspicuous problem, very few ideas had been espoused about possible causes and mechanisms for this disease. By postulating an infectious cause, Wickman helped start the scientific and medical communities on the pathway that would eventually lead to our modern understanding of this illness. Furthermore, since there was often no obvious physical or familial connection between affected individuals, he reasoned that there must be frequent inapparent cases that help spread the infection, another hypothesis that would one day be proven correct. He also defined the incubation period and identified two stages of the disease which he called the minor and major illness phases. The minor phase typically began 3–4 days following a known exposure and presented with mild systemic symptoms such as fever and gastrointestinal issues. Patients who failed to recover at the minor phase progressed to the major phase around days 8–10 and began to exhibit severe symptoms indicative of CNS involvement. His perceptive observations were well-received by the medical community and became the accepted dogma for polio in the early twentieth century.

Building on Wickman's pronouncement that polio was a contagious disease, the search for the causative microbe began. By this era, the concept of bacteria and viruses as disease-causing entities was becoming more widely accepted, and the hope was that a specific microorganism could be identified for every infectious disease. There were now many techniques for growing and characterizing bacteria, and the functional definition of viruses as invisible, filterable agents was well-known in scientific circles. What was lacking was an animal model for testing patient samples. Polio was too dangerous for experiments using human subjects, so an animal that could be infected and develop symptoms similar to the human disease was essential for proving Koch's postulates. The credit for first solving this problem belongs to Karl Landsteiner, an Austrian physician-scientist working at the University of Vienna (Schwarz and Dorner 2003). Landsteiner was a pathologist who performed thousands of autopsies and in 1908 he isolated spinal cord material from a 9-year-old child recently deceased from infantile paralysis. Unable to grow any bacteria from this material, he checked for an infectious agent in his sample by injecting ground-up spinal cord into several common laboratory animals, including guinea pigs, rabbits, and mice. When none developed any disease, he turned to two primates, a rhesus monkey and a baboon (Goldman and Schmalstieg 2019). The baboon died eight days after intraperitoneal injection and the rhesus developed paralysis of the legs. Upon autopsy, both animals had spinal cord pathology that resembled human polio cases, confirming that he had actually transferred the human disease. In subsequent work done in collaboration with the Pasteur Institute, he produced identical results after filtration of his human samples to remove all cells and bacteria, thus defining the infectious agent as a member of this mysterious new class of microorganisms called viruses. This marked the fourth human disease identified as a virus behind rabies, smallpox, and yellow fever. Landsteiner published these results in 1909 and soon thereafter showed that the agent could be neutralized by serum from patients recovering from polio, an indication that patients developed protective antibodies. He even tried to develop procedures both for chemically inactivating the virus and for creating attenuated versions for vaccine use. While he failed in his vaccine attempts, these results lead some researchers to predict optimistically that a vaccine might be soon available; sadly it would take nearly 50 years and thousands of deaths before a vaccine was finally produced.

Continuing his polio work at the Pasteur Institute, Landsteiner showed that the polio agent was not confined to spinal tissue and could be isolated from numerous other bodily tissues of infected monkeys. In particular, Landsteiner showed that the oral region was rich in the virus where it could be found in the pharynx, salivary glands, and tonsils. This observation would ultimately become important for understanding the transmission of poliovirus but was mostly unappreciated for decades. As seminal as Landsteiner's work with poliovirus was, it was not his major career achievement and infectious disease was only a portion of Landsteiner's research focus. After leaving the Pasteur Institute and returning to Vienna in 1911, he left polio behind and turned to other research problems. Now he is remembered primarily as a preeminent immunologist who won the Nobel Prize in 1930 for discovering the three human blood group types, A, B, and O, a finding that was essential for the development of blood transfusions and organ transplants. Even though Landsteiner did not continue his polio work, his discovery of the viral nature of this disease greatly influenced polio research. At last, the causative agent was isolated and classified, convincingly eliminating competing theories about the disease's etiology. Science now had a solid foundation for further work to understand and defeat this illness even though the public was slow to understand and accept this information. Finding the virus early on was a critical advance as these initial small polio outbreaks were just a harbinger of what was coming. The next 40 years would see continual outbreaks of polio across Europe and the United States. No longer just a sporadic disease with only a few isolated cases, polio exploded in the twentieth century to become a frightening yearly occurrence that killed and crippled children and terrorized parents. In the United States alone there were nearly 400,000 cases and 50,000 deaths from 1910 to 1950. While these overall numbers are modest in comparison to some other viral diseases, for example, the COVID-19 pandemic, the crippling consequences of polio and its preference for targeting children made it a feared and hated disease until its conquest through vaccination.

False Starts and False Hopes

Simon Flexner was a giant of American medicine in the 19th and 20th centuries. He was born in 1863 in Louisville, Kentucky to immigrant parents and began his studies there. With an educational background in pharmacy and medicine coupled with postgraduate work in pathology at Johns Hopkins University Medical School, he became a broadly trained clinical researcher. By the early 1900s, he was already a prominent pathologist and bacteriologist working on bacterial meningitis when he was recruited to the Rockefeller Institute for Medical Research where he later became the director. As the director, he was instrumental in promoting experimental science, rather than mere observation, as a tool to address medical issues, particularly infectious diseases (Rous 1948). Early in his tenure during the first decade of the 1900s, the United States was beginning to suffer increasing polio outbreaks like the European nations. By 1907 there had been growing pockets of polio cases throughout the Northeast and Midwest, with well over 2000 cases reported in New York City alone, Rockefeller's home. With this burgeoning public health threat on his doorstep, Flexner shifted his bacteriological studies to polio research. Seizing on Landsteiner's work, Flexner quickly confirmed the filtrable nature of the polio agent and embarked on a 30-year quest to understand this disease and its cause. His first major contribution was establishing that the polio agent could be serially passaged by intracerebral inoculation of successive Rhesus monkeys, a protocol that obviated the need to continually obtain fresh isolates from recently deceased polio victims. As with influenza and the yellow fever virus, this was a critical requirement for effective study in the laboratory setting. Now any human polio isolate could be maintained, propagated, amplified, and studied perpetually in the lab.

Flexner came from a research background studying bacterial meningitis and believed that polio would have similar mechanisms. In particular, he was convinced that polio entered its victims through the nasal portal where it would travel the olfactory nerves to the brain and then spread to the spinal cord. Pursuing this hypothesis, in 1910 he conducted nasal inoculation experiments on monkeys and was able to transmit the disease and produce brain and spinal pathology (Flexner and Lewis 1910). These results solidified his belief that the nasal route via the olfactory nerves was the authentic entry point for the virus and that poliovirus was strictly neurotropic, a belief that became his dogma. Between 1910 and 1940, Flexner published over 30 papers on polio infections, fiercely maintaining throughout that the neural route was the only effective mode of poliovirus transmission (Flexner 1936). Tragically, he was completely wrong but it would take nearly 30 years for the field to fully repudiate his error. His stature in the scientific community, his directorship of America's most prominent research institute, and his position as editor of the prestigious Journal of Experimental Medicine made challenging his opinion difficult. As a result, the field was unnecessarily delayed in discovering that the real route of poliovirus transmission was oral-fecal. This was a sad but not unique example in the history of science where personal prestige coerced a field and prevented the critical evaluation and acceptance of conflicting data (Grimshaw 1995).

Other aspects of Flexner's work also unintentionally contributed to the mistaken conclusion that poliovirus was exclusively a neurotropic organism. Like Landsteiner before him, Flexner established Rhesus monkeys as his animal model system for polio studies, and this species was widely adopted and became the standard in the field. However, Rhesus monkeys turn out to be fairly unique among primates in that they are naturally resistant to poliovirus and can only be infected by unnatural protocols such as abrasive nasal challenge or direct inoculation into the peritoneal cavity, bloodstream, or brain (Flexner and Lewis 1909). Consequently, attempts by Flexner and others to cause disease by oral administration failed even though this is the authentic route for polio to spread among humans. An additional confounding problem arose as Flexner passaged his polio isolate from monkey brain to monkey brain over several years. Through this process, he eventually created a stable strain that was designated MV. He generously provided this strain to other investigators and it became widely used in polio studies by many labs. Unknowingly, however, during repeated passaging of this strain in simian brain tissue, it lost its natural properties and evolved into a strictly neurotropic virus. This outcome shouldn't have been unexpected or even unanticipated since viral adaptation during passaging had already been observed for other viruses. However, neither Flexner nor other investigators appeared to have considered this eventuality, and MV was simply treated as a wild-type strain. Like an oddly self-fulfilling prophecy, Flexner's belief that poliovirus infection was restricted to nerve cells was transmuted into reality with the MV strain. As he and other researchers experimented with MV, they failed to detect significant infection of non-nervous tissue because this strain now lacked that ability (Sabin and Olitsky 1936). This misconception that poliovirus infection was confined to nerve cells would persist well into the 1930s even as conflicting data were accumulating for the poliovirus infection in other tissue types. Even the renowned Albert Sabin, who would go on to develop the oral polio vaccine, would contribute to this false dogma by reporting that the MV strain would only grow in human embryonic brain tissue and not non-nervous tissue (Samuel L. Katz 2011).

Throughout the 1910s into the 1930s, as Flexner was both directly and inadvertently shaping the field of polio research with his dogma and his MV strain, other missteps about polio were occurring in the areas of public health and treatment. The increasing regularity of polio outbreaks in the United States, along with the growing numbers of deaths and paralytic sequelae, was causing panic and extreme responses (Rogers 1992). The general public knew little about the scientific studies on poliovirus and likely understood even less. To the public, it was immaterial whether polio was a virus, a bacteria, or something else altogether. Their only concern was how to avoid and prevent polio infections each year as the temperature warmed during the spring and summer months. With the true mechanism of transmission unknown, different communities grasped at any likely source of the disease, often with misguided as well as worthless results. In some cities, swimming was prohibited (beaches or pools were closed), playgrounds were barred, and theaters were shuttered to prevent children from congregating in large groups. Other communities blamed polio spread on insects (flies especially), vermin, or pets. While we now know that none of these creatures can carry or spread poliovirus, it didn't stop New York City from killing roughly 80,000 cats and dogs during the large polio epidemic of 1916. Much wasted and futile effort went into ineffective prevention strategies that had no actual impact on polio's spread. Even more disheartening was the prejudice that polio invoked against the poor, the homeless, and the immigrant populations. Americans had a long-standing belief that immigrants were inherently dirty and somehow associated with disease spread in general. Many people also believed that polio transmission was related to dirty conditions such as filthy streets, unsanitary foods, and poor personal hygiene. This combination of beliefs exacerbated the prejudice against immigrants and set them up as easy targets to blame for polio outbreaks. Although polio cases weren't confined to the poorer areas of cities where immigrants typically lived, the middle and upper classes still often shunned these areas and continued to ascribe polio outbreaks to the lower socioeconomic classes. If there was anything positive about this prejudice, it was that many cities tried to improve sanitation in poor neighborhoods as a polio-preventative strategy. While sorely needed and useful to control other diseases, the effect on polio eradication was negligible. It would be decades after the discovery of poliovirus before these hurtful and unfounded assumptions about the poor and immigrant populations would finally be dismissed and replaced with a scientifically accurate understanding of polio transmission.

Along with the public health failures, there were numerous mistakes made in polio treatment. As case numbers climbed, physicians lacked any proven remedy to treat the acute illness or the paralytic aftereffects. With no standard of care, they responded based on their own biases and their own ingenuity. Some treatment approaches were benign while others were almost certainly harmful to the patients, but all were utterly useless (Wyatt 2014). Conventional therapies consisted of tonics, anti-irritants, and the topical application of various substances touted to improve health and remove disease-causing substances. More invasive treatments included the administration of tetanus or diphtheria antitoxins (Johnston 1916), convalescent serum from recovered polio patients, or strychnine (rat poison). Even more extreme approaches often directly targeted the spinal column as this was seen as the prime organ for attack by polio. There are reports of spinal injections using substances such as adrenaline, quinine, or urea hydrochloride. It is highly unlikely that any of these treatments had a beneficial effect, and patients often died after such treatments although it was never established if death was from the injections or from polio itself. As an alternative to injecting supposedly helpful compounds into the spine, some physicians promoted the removal of spinal fluid via a lumbar puncture. This approach was touted to relieve "pressure" in the spine that they believed contributed to spinal damage. In a variation on this approach, Dr. George Retan at the Syracuse University School of Medicine was funded by the National Foundation for Infantile Paralysis to develop a procedure for removing spinal fluid and replacing it with a salt solution. He hypothesized that there were toxic products in the spinal fluid that could be removed and diluted with his salt solution (Friedrich Koch 1985). Fortunately, this strategy never made it to human patients as many of the test monkeys died and the project was eventually dropped.

A more egregious episode of failed treatment involved a Stanford University researcher, Edwin Schultz (Schultz and Gebhardt 1938). Dr. Schultz was a proponent of chemoprophylaxis, the use of chemical substances to prevent viral infection. In the 1930s, most researchers were still operating under Flexner's dogma that poliovirus entered via the nasal route and infected the olfactory nerves. Schultz, along with many other researchers, believed that a nasal spray could be developed that would block poliovirus entry to the olfactory nerves in the nasal passages. Different investigators tried sprays of tannic acid (Sabin et al. 1936) or picric acid (Armstrong 1936), both compounds that had been used as topical treatments for burns and infections. Experiments in monkeys showed that these treatments had at least some protective efficacy, likely because tests were typically conducted with Rhesus monkeys using the MV poliovirus strain administered nasally. Under these conditions, the sprays were able to block some infections for this adapted virus delivered by this unnatural route. Based on the animal studies, many desperate physicians, private individuals, and even whole public health departments elected to implement these unverified approaches during polio outbreaks. Rather than wait for controlled trial results, they simply went ahead and treated at-risk individuals and communities. Unsurprisingly, attempts in humans showed little or no success for these nasal sprays since natural infections with the wild-type virus occurred via the oral route, not the nasal route. Additionally, even when actual human trials were conducted they were often small and poorly controlled. Human application of the sprays was also plagued by technical issues as children were less amenable to the procedure than restrained test monkeys, and it was difficult to ensure delivery of consistent quantities of the test material. Schultz's contribution to this failed research area was the firm belief that a nasal spray containing zinc sulfate would protect against polio (Schultz and Gebhardt 1937). Buoyed by his own animal studies, Schultz conducted a large human field trial during the 1937 polio outbreak in Toronto (Rutty 1996). Approved by the Ontario government, the plan was to administer the zinc nasal spray to 5000 children with another 5000 children as unsprayed controls. There was widespread community enthusiasm for this study and they easily enrolled 5233 children as test subjects. However, as with the other types of nasal sprays, the final analysis of the trial showed no significant difference in the number of polio cases in the sprayed cohort versus the control children. More unfortunate was the damage to the sense of smell in some spray recipients. It was known and expected that zinc sulfate would cause a temporary loss of smell (anosmia), but many children still hadn't regained this sense even months after the treatment, and it is unknown how many were permanently affected. Mercifully, without any significant successes, nasal prophylaxis faded away before too much damage was done as researchers, physicians, and the general public finally realized the futility of this approach.

The 1930s didn't just see failed treatment attempts but also failed vaccination trials. Drs. John Kolmer (1936) and Maurice Brodie (1936) independently developed and tested unsuccessful poliovirus vaccines. Kolmer, from Temple University in Philadelphia, chemically attenuated the virus for use as a vaccine while Brodie created a formalin-inactivated virus preparation for his vaccine. Both men rushed their vaccines into human trials prematurely after minimal animal studies with disastrous outcomes. Brodie's first trial in 1935 involved 3000 children in California, but it was hastily organized and poorly constructed so the results were inconclusive. A second trial a year later in North Carolina with about 1500 children was more properly controlled, but not a single case of polio occurred in either the vaccinated or the control group so nothing could be concluded about the vaccine's efficacy. At the same time, Kolmer was testing his attenuated vaccine with tragic results. After testing his vaccine on himself and his own children, Kolmer embarked on a large, but uncontrolled trial with over 10,000 children. Among his vaccine recipients, there were both deaths and cases of paralytic polio, although whether from the vaccine or natural infection could not be determined. Kolmer maintained that his vaccine was a success and that the number of deaths and cases would have been higher without his treatment. However, follow-up trials conducted independently by the Commission for Infantile Paralysis showed that neither the Kolmer nor Brodie vaccines were effective in preventing polio and that Kolmer's supposedly attenuated vaccine was potentially dangerous. Outrage from both the public and the scientific/medical communities about these reckless trials with primitive vaccines halted any further vaccine development attempts and ended the careers of both investigators. After the vaccine scandal, Brodie was fired and moved to Detroit to become the laboratory director of a hospital where he died suddenly at the age of 36, possibly by suicide. Kolmer soon retired and the first era of polio vaccines came to an ignominious end.

The above examples illustrate that scientific discovery does not always follow the correct or most straightforward path. Like an explorer entering an unknown region, scientists have no precise map to guide their research choices. There can be many possible avenues to explore with no certainty about which route or routes will lead to the correct understanding of the research problem. Polio research had many earnest and well-intentioned researchers who were devoted to the public good yet ultimately contributed little to the final conquest of this disease, and in some cases hindered progress. It was highly unfortunate that so much effort and so many resources were devoted to unproductive research in the fight against polio. For decades after Flexner's pronouncement that poliovirus was a neurotropic agent that initiated infection in the nasal passages, the scientific community pursued this false dogma. Researchers were hampered by primarily using Rhesus monkeys along with the MV virus strain, which was an inadequate experimental system. Without the correct tools, developing accurate knowledge about the properties and real transmission mode of this virus was elusive, leaving physicians, public health officials, and the public handicapped in their efforts to prevent and treat this illness. However, science is not a static enterprise that remains perpetually stuck on incorrect information. Instead, science is collectively a dynamic and protean activity that relentlessly, albeit sometimes slowly, moves toward a greater and more accurate understanding of our world. While many polio investigators pursued unproductive research projects others were patiently discovering the true nature of this virus and its infectious process.

The Path Forward At Last

The early 1900s weren't just years of polio failures and fiascos as several seminal findings slowly began to illuminate the true nature of the disease, its transmission, and its prevention. Drs. Sabin and Salk eventually received the lion's share of the credit for conquering polio based on their vaccines, yet in the decades prior to their work, many other scientists made critical discoveries that were essential for the eventual vaccine successes. Even Flexner's incorrect contention in 1910 (Flexner and Lewis 1910) that poliovirus was strictly neurotropic and transmitted via nasal infection was quickly challenged, albeit unsuccessfully, by Dr. Wade Frost, an epidemiologist in the Public Health and Marine Hospital Service (Grimshaw 1995). Based on his own observations of polio cases, as well as studies from European outbreaks, Frost argued for the digestive tract as the more general route of infection. He noted that the epidemiological data suggested that there were likely many inapparent cases for each acute case and that these "silent" infections promoted transmission. This idea of an individual who could remain healthy while continuing to spread the virus, a so-called "passive carrier", was a novel concept that would prove to be quite common in viral diseases as we will see in subsequent chapters (Winkelstein 2009). Frost also criticized early primate experiments as being too artificial for human relevance as the test monkeys weren't naturally susceptible to poliovirus and could only be infected via direct injections. Unfortunately, although correct in most of his conclusions, Frost's position never gained widespread acceptance due to Flexner's prominence and active rejection of Frost's claims. Nonetheless, throughout the 1910s and 1920s, there was a slow accretion of data that conflicted with Flexner's model eventually leading to John Paul and James Trask revisiting the long-held beliefs about poliovirus.

Paul and Trask founded the Yale Poliomyelitis Study Unit in 1931 and quickly developed a research program to collect data on the polio outbreak of 1931/1932. Their detailed examination of affected families combined with the collection and testing of oral samples produced important results. Initially, their isolation of the virus from throat washes of children with only minor symptoms (headaches, vomiting, and sore throats) confirmed both the oral presence of poliovirus and supported Frost's assertion that many infections did not produce serious paralytic disease (Paul and Trask 1932; Paul et al. 1935). In extensive and precise studies throughout the 1930s, they repeatedly demonstrated that poliovirus could be isolated from oral samples in individuals with little or no apparent disease. Their subsequent detection of poliovirus in fecal samples and raw sewage further substantiated the notion that polio is primarily a gastrointestinal disease with a fecal-oral mode of transmission (Trask et al. 1938a, b; Paul et al. 1939) and finally disabused the field of Flexner's nasal-neurotropic dogma. They also did groundbreaking work comparing different human isolates and various laboratory-passaged polioviruses. Using sera from convalescent patients, they found that each serum could neutralize some isolates and not others, indicating that there were significant antigenic differences between isolates (Paul and Trask 1935, 1933), a result also reported earlier by the Australian group of Frank Burnet and Jean MacNamara (Burnet and MacNamara 1931). Collectively, these serologic studies established that there were distinct strains of polioviruses and it would subsequently be shown that there are three strains now designated types 1, 2, and 3. The existence of multiple poliovirus strains was critical information for future vaccine development. Since antibodies against one type would not neutralize the other two strains, any effective vaccine would have to generate neutralizing antibodies against all three types. The lack of this knowledge likely contributed to the failures of the Kolmer and Brody vaccine trials. Since their vaccine preps would have only protected against one poliovirus type, their vaccinated cohort would have been susceptible to infection by the other two types, rendering their vaccines useless in the real world. Ultimately, both Sabin and Salk would develop trivalent vaccines that protected against polioviruses types 1–3.

Along with the new epidemiological and clinical findings, important laboratory advances were being made on several fronts. One of the limitations of poliovirus work was the lack of a convenient animal model system. Monkeys were the primary animal used for poliovirus studies, however, monkeys were expensive, difficult to work with, and often in short supply. Plus, many research facilities simply weren't set up to house and handle primates. All these factors limited the number and type of experiments that could be performed and made many laboratory studies unfeasible. Attempts to transfer polio isolates from monkeys to other test animals were mostly unsuccessful until the work of Charles Armstrong in 1939. He managed to transfer polio from monkeys into the cotton rat and create a rat-adapted strain. Reasoning that this rat-adapted strain might be infectious for other rodents, he transferred the rat stain into the canonical laboratory white mouse by intracerebral injection. Recipient mice developed paralytic polio and the virus could be transferred from mouse to mouse (Armstrong 1939). By creating a mouse-adapted polio strain he gave scientists a cheap and plentiful animal system where many previously impossible experiments could be easily conducted. For example, he showed that the mouse system was suitable for the poliovirus neutralization assay. Before Armstrong's work, poliovirus neutralization tests were conducted with monkeys. Sera from patients or experimentally infected monkeys were mixed with poliovirus and then injected into the brains of test monkeys. If a serum contained neutralizing antibodies then the test monkey would survive and not develop paralytic symptoms. Armstrong's contributions enabled this assay to be performed with small, abundant mice instead of scarce and costly monkeys. This simple advance enabled scientists worldwide to conduct thousands of times more neutralization assays than they could when using primates. Now serology could be done to screen for anti-poliovirus antibodies in large populations rather than in just a few select patients. These large epidemiologic studies would further corroborate the silent spread of asymptomatic polio infections, especially among infants and very young children.

While Armstrong's production of the mouse model greatly facilitated some aspects of poliovirus research, primates would continue to play an essential role. One important advance in the 1940s was the burgeoning awareness that different types of primates varied in their sensitivity to poliovirus. Rhesus monkeys had been the standard animal model for poliovirus work, but this species could only be infected by massive nasal exposures, direct intracerebral injection, or other direct internal injections, none of which reflected the natural transmission of poliovirus through the oral-fecal route. This changed when David Bodian and Howard Howe fortuitously noted polio infection in two uninoculated chimpanzees (Howe and Bodian 1944). The two chimps were housed in cages adjacent to poliovirus-infected rhesus monkeys and apparently became infected through the natural transmission of viruses shed from the rhesus monkeys. Further studies demonstrated that chimps, as well as cynomolgus monkeys, could be infected by simple oral ingestion as was the case with humans. This was especially true with freshly isolated human poliovirus samples rather than the neuro-adapted strains like MV. Having primate models that more closely resembled natural human infection was invaluable for studying transmission, pathology, and immune responses to the virus. Now laboratory experiments could be conducted and compared in parallel with data from human cases.

Another highly significant contribution to the polio field was made by Dr. Dorothy Horstmann (Carleton 2011). With an M.D. from the University of California at San Francisco, Dr. Horstmann was hired by John Paul in 1942 to work on polio at the Yale Poliomyelitis Study Unit. She conducted one of the first comprehensive clinical evaluations of polio cases by sampling individual patients multiple times at multiple anatomical sites over many days, something that simply had not been done before. Her detailed results confirmed that oral shedding of the virus was very limited but that fecal samples remained positive for weeks, confirming the importance of fecal contamination for spreading the infection. Even more importantly, she solved a puzzle that was mystifying the poliovirus research community. By the early 1940s, the field had largely accepted that the oral-fecal route was the primary mode of poliovirus infection and transmission, not the nasal route. However, if polio was a gastrointestinal illness, how did the virus reach the brain to cause the paralytic damage? The bloodstream was an obvious potential route, yet many researchers had examined the blood of polio victims for traces of the virus without success. To assay for the virus, blood from patients was injected into test animals such as mice. If the test animal got sick then the patient had viremia (virus in the bloodstream), but this approach was almost uniformly negative. Using the chimpanzee model, Horstmann performed a simple yet elegant experiment that finally resolved this conundrum. She orally infected the animals, took daily blood samples, and tested for the presence of poliovirus in the blood (Horstmann 1952). Within a few days, the virus was readily detectable in the bloodstream proving that gastrointestinal infection did produce a viremia that could seed the virus to the central nervous system. Surprisingly, the viremia disappeared before the animals developed nervous system symptoms. Her correct interpretation of these results was that as the infection progressed, neutralizing antibodies accrued which neutralized the virus in the blood and prevented its further detection, i.e. neutralized virus wouldn't cause infection in the test animals so the patients appeared to have no viremia. However, by the time neutralizing antibodies arose the virus was already in the nervous system and causing the virus-induced pathology in the brain. The failure of many clinicians and investigators to detect viremia in human cases was purely a matter of timing. By the time patients presented with symptoms and were tested for viremia, the virus in the bloodstream was already neutralized and undetectable by the infection assay. Checking people early after suspected poliovirus exposure quickly confirmed that humans had viremia just as in the animal model, adding an important element to our understanding of the disease process.

Down the East Coast from Horstmann, Isabel Morgan and David Bodian of Johns Hopkins University were adding further critical knowledge to the poliovirus story. Morgan followed in the footsteps of her parents who were both biologists and obtained a Ph.D. in bacteriology from the University of Pennsylvania. After a stint at the Rockefeller Institute for Medical Research, she joined David Bodian at Johns Hopkins in 1944. Focusing on immunity to poliovirus she published several solo papers exploring the nature and role of antibodies that developed during experimental polio infection of monkeys. Her careful studies led to a method (serotyping) for distinguishing poliovirus isolates based on their neutralization by antisera (Morgan 1949). Working with David Bodian (Nathanson 2005), in 1949 they showed that all tested poliovirus isolates fell into one of three neutralization groups (called serotypes), thus finally defining that three and only three poliovirus strain types existed (Bodian et al. 1949). As mentioned above, this was critical knowledge for future vaccine development as any successful vaccine would need to produce neutralizing antibodies against all three circulating serotypes. Now investigators knew what the target really was and vaccine research could begin in earnest. Morgan also made contributions to vaccine development by showing that poliovirus could be safely inactivated with formalin and that the resulting preparation induced protective antibodies in monkeys without eliciting any disease (Morgan 1948). She also noted that this inactivated vaccine material needed repeated injections over time to produce effective levels of neutralizing antibodies, a "booster" concept that would be needed for the eventual Salk vaccine as well as other future inactivated type vaccines. While leaving behind an impressive legacy, including being the only woman acknowledged in the Polio Hall of Fame (Warm Springs, Georgia), Morgan left poliovirus research after 1949 to explore other areas and never returned to the polio field.

In the late 1940s and early 1950s, two additional groups made seminal contributions that were critical for the subsequent production of both the live oral (Sabin) and inactivated (Salk) polio vaccines. In the judgment of the Nobel Prize Committee, the most important advance was the development of a cell culture system for growing poliovirus, an achievement that garnered the Nobel Prize for John Ender and his colleagues, Thomas Weller and Frederick Robbins. As we saw in previous chapters, the growth and production of viral stocks in animals are adequate for research purposes but are highly problematic for vaccine manufacturing. Large-scale vaccine production requires massive amounts of viruses, a high degree of homogeneity and reproducibility in the viral batches, and stringent quality control at each step in the process, all features that are hard to achieve using animals as the source for generating the virus. Animals, even small laboratory mice, require voluminous caging and are expensive to house, feed, and maintain at the scale needed for millions of vaccine doses. Even inbred strains exhibit considerable genetic heterogeneity as well as harboring different endogenous viruses and other microorganisms that can affect the quality and quantity of the vaccine virus produced. Trying to meet vaccine production standards with animal-based virus stocks ranges from challenging to impossible. Likewise, large-scale production of poliovirus in eggs, which were the mainstay for the production of other viruses, failed because poliovirus doesn't grow well in eggs. Fortuitously, the era of intense poliovirus research coincided with the birth of tissue and cell culture technologies, and the former would become critical for the latter. Tissue culture began in the early 1900s with Ross Harrison at Yale who placed bits of embryonic frog tissue in lymph fluid and was able to maintain the living cells for several weeks. Over the next two decades, similar results were obtained with numerous tissue types from animals and humans. While a marvelous advance that allowed tissues and cells to be more readily visualized and studied outside the confines of the animal host, these cultured tissues inevitably became contaminated and died within a few weeks. Subsequent advances such as aseptic techniques (sterilization of all materials and supplies as well as handling protocols that minimized contamination), special culture flasks that promoted growth, defined synthetic media to nourish the tissues, and the introduction of antibiotics to prevent bacterial growth greatly extended the lifespan and usefulness of these cultures. With the ability to maintain uncontaminated tissue samples for prolonged periods, this technology was adopted by virologists who perceived its advantages over animals for producing consistent viral stocks. Tissue culture also made it possible to observe the effects of viruses on individual cells from the start of infection to cell death, something that was impossible in infected animals where cellular effects could only be seen late in infection when diseased or deceased animals were autopsied. Tissue culture, and later cell lines (immortalized cells that never senesce and can be propagated indefinitely), would become the mainstays of many subfields in biology including virology.

Into this burgeoning field of tissue culture came Yale Associate Professor John F. Enders. Enders hailed from a prestigious banking family and was by all accounts a modest and unassuming man. Perhaps feeling an obligation towards business, he had a peripatetic education and career before finally settling into the biological sciences (Bendiner 1982; Robbins 1994). With a Bachelor of Arts degree from Yale and 3 years as a Naval Reserve pilot, he first embarked on a career in real estate. Quickly finding that field unsatisfactory, he next considered teaching so he entered Harvard to study English, Celtic, and Teutonic literature. He received a Master of Arts from Harvard and was working on his Ph.D. in literature, hardly a career path destined for scientific success. But this is one of those tales where the vagaries of life conspire to change the course of science. While Enders was pursuing his literature Ph.D., his medical student roommate introduced him to the charismatic Hans Zinsser, the noted bacteriologist and scholar. Quickly becoming enthralled with scientific research, Enders abandoned his literature Ph.D. to pursue microbiology, receiving his doctorate in this field in 1930. Over the next 15 years, Enders' career at Harvard was steady but unremarkable. His interests expanded from bacteriology to include immunology and eventually virology, and he rose to the rank of Associate Professor. Early on he developed an interest in the fledgling field of tissue culture, primarily as a vehicle to cultivate the mumps virus (Chap. 7). However, early failures with this technique and the interruption of World War II conspired to delay these studies until the postwar era. In 1947, Enders moved his laboratory to Boston Children's Hospital where he was joined by two young physicians, Thomas Weller and Frederick Robbins, both interested in establishing tissue culture systems for viral growth. Poliovirus was not their focus as mumps, chickenpox (herpes varicella-zoster virus. Chap. 10), and a suspected diarrheal virus were their initial targets. With Weller and Robbins leading the laboratory work, new techniques were developed to culture embryonic tissues, both chicken and human, in flasks. By changing the nutrient medium in the flasks every 3 to 4 days they observed that the cells could be maintained for prolonged periods, providing a suitable environment for viral infection and propagation.

After successfully propagating the mumps virus in their culture system, they explored growing the chickenpox virus in cultures of embryonic human skin and muscle tissue. When setting up for their first attempt with chickenpox there were extra flasks of human cells. Rather than waste these leftover flasks, on a whim, they infected these cells with the Lansing strain (type 2) of polio that was languishing in their laboratory freezer. Previous investigators were unable to grow poliovirus in culture except in neural tissue, but neural tissue was problematic for vaccines. Purifying viruses away from cellular material was difficult and nerve cell components were already known to have toxic effects in rabies and yellow fever vaccines, so using this cell type for polio vaccine production was not ideal. However, these prior cell culture experiments were typically done using the MV strain of polio which unknowingly had become stringently neuro-adapted. One can only speculate that Enders and his colleagues hoped that the Lansing strain might exhibit different properties from MV and would possibly replicate in the non-neural tissue, hence their humble initial experiment. With a result that would change the field, they observed that the Lansing strain multiplied in their non-neural cultures and retained its paralytic capacity in mice (Enders et al. 1949). To ensure that this wasn't a property unique to the Lansing strain they went on to show that 13 different fresh isolates of both types 1 and 2 all grew in human non-neural tissues. Their careful and compelling results finally convinced the scientific community that wild-type poliovirus could be readily cultured in many tissue types, including nonhuman cells such as monkey kidney cells. The ability to grow poliovirus in cell types that were abundant and less problematic than neural tissue removed one of the last impediments to vaccine development. Their tissue culture techniques were quickly adopted by other labs and were instrumental in the development of protocols to propagate other medically important viruses including the agents of measles, German measles, and chickenpox.

The Enders group also made another seminal contribution to virology by observing that poliovirus-infected cells changed their morphological appearance distinctively within a few days post-infection and could be identified as infected simply by visual inspection under a microscope. Rather than testing samples for the presence of poliovirus by the slow and expensive process of awaiting disease symptoms after animal inoculation, now you could rapidly determine positive samples just by examining the effect on cultured cells. Enders coined the term "cytopathic effect" (CPE) for this virus-induced cellular damage. Subsequently, many other viruses also exhibit distinctive CPE on cultured cells and this term has become standard nomenclature in the lexicon of virology. For these ground-breaking contributions, Enders was awarded the 1954 Nobel Prize in Physiology or Medicine, initially as the sole recipient. Upon receiving the news of this award, he declined unless Drs. Weller and Robbins were added and the three of them ultimately received the prize and the fame. Two years after receiving the Nobel Prize, Harvard finally promoted him to full Professor. Anticlimactically, none of the three continued in polio research although all went on to successful careers in other areas.

One last piece needed to enable the development of the oral poliovirus vaccine was the plaque assay pioneered by Marguerite Vogt and Renato Dulbecco. A plaque assay is a method to isolate and quantitate individual infectious viral particles. Originally developed by researchers studying viruses of bacteria (called bacteriophages or phages for short), the procedure involves spreading a highly diluted suspension of the phages on a nutrient agar plate whose surface was entirely covered by bacterial growth (a so-called "lawn" of bacteria). Each of the few phages in the suspension would infect a different bacterial cell. As each phage reproduced in its target cell, it would eventually kill and lyze the cell releasing hundreds to thousands of new phages. These progeny phages would infect the cells surrounding the initially infected cells and repeat the process of reproduction, cell lysis, and further infection. As more and more bacterial cells are lysed it creates a hole or plaque in the bacterial lawn that is visible (Fig. 6.1). Each plaque results from one initial phage that started the infection, so counting the plaques tells you the number of infectious phages in the suspension. Additionally, since all the resultant progeny phages in each plaque arose from one original phage, the plaque phage population is genetically homogenous (aside from any mutations acquired during reproduction). By reaching into the plaque with a pipet the researcher can harvest a relatively pure culture of that particular phage type. In this way, stocks of phages with different genetic properties can be isolated and purified.

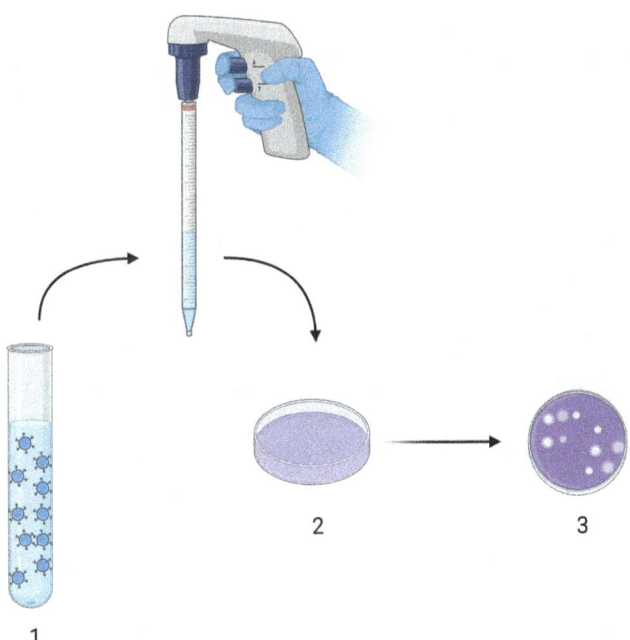

Fig. 6.1. The viral plaque assay. A fixed quantity of a dilute suspension of virions (1) is pipetted onto a solid monolayer of cells in a plate (2). After allowing the virions to attach to the cells for a short period, the plate is covered with a layer of soft agar. As each virion infects a cell it will replicate and eventually lyse that cell to release the progeny virions. Released virions are trapped by the agar and can only infect new cells immediately surrounding the initially infected cell. As the initial infection proceeds, each infected cell lyses to release new virions that infect the surrounding cells. Through successive rounds of infection, lysis, and release, the cumulative lysis of adjacent cells causes small holes to form in the cell monolayer (3). As multiple rounds of viral replication occur the holes (the plaques) spread outward to become larger and easily visible. Each plaque represents an initial infection by one virion, so the number of plaques equals the number of infectious virions in the solution

Before Renato Dulbecco, the plaque assay had not been applied to animal or human viruses as cell culture provided some unique challenges compared to working with bacteria. In the early 1950s, Dulbecco was able to adapt the phage plaque assay to the Western Equine Encephalitis virus (a mosquito-transmitted virus of horses) grown on chicken fibroblast cells, establishing that this technique was feasible for animal viruses (Dulbecco 1952). Within 2 years, he and Marguerite Vogt successfully applied his method to poliovirus, finally providing a means to quickly and accurately quantify the number of infectious viral particles in any stock preparation (Dulbecco and Vogt 1954). Importantly, plaquing could be used to isolate individual poliovirus mutants with different growth and pathogenic properties. This assay was critical for developing the Sabin oral vaccine which required attenuated viral mutants that lacked neurovirulence yet still replicated effectively in the gastrointestinal (GI) system. Now individual plaque isolates could be tested and characterized to find virus strains with exactly the needed properties. Let's use a deck of cards as an analogy to further examine the plaque concept. Suppose each card in the deck is a poliovirus with a slight variation in sequence and properties. You only want the two of clubs which is the variant with no neurovirulence and excellent growth in the GI tract. If you always use the whole deck to infect animals or cells in culture, you will always produce a viral stock that contains all 52 variants. Plaquing allows you to spread the 52 cards out, pick only the two of clubs, and use that single variant to infect new cells. Now your viral stock will be a pure and homogenous stock of just the two of clubs and you can maintain this pure stock indefinitely. The work of Vogt and Dulbecco provided poliovirus researchers with this final tool needed to do precise genetic dissection of poliovirus stains and also had a far-reaching impact well beyond poliovirus. Plaquing became a widely used staple of virology and has been applied to numerous different viruses since its inception in the 1950s. Much of modern virology would not have been possible without this relatively simple yet enormously powerful technique to separate and isolate different viral variants.

Picornaviruses Revealed

The virological advances in the 1940s and 1950s not only enabled poliovirus vaccine development but also placed polioviruses in a broader context. The knowledge that poliovirus was not exclusively neurotropic, but instead was primarily an infectious agent of the intestinal tract, fostered the identification of related viruses affecting this organ system. Investigators began to examine fecal samples from both healthy and ill humans for new poliovirus strains or related viruses. During a polio outbreak in Coxsackie, New York, fecal samples from two suspected patients revealed a novel virus distinct from poliovirus yet with many similar features. These 1948 isolates acquired their name from their town of origin and the eponymous coxsackievirus group was formed. Within a few years, additional related viruses had been isolated from the feces of healthy individuals. These newly isolated viruses caused cytopathic damage in cultured cells yet did not seem to be causing human disease so were termed **e**nteric (of the gut) **c**ytopathic **h**uman **o**rphan (no disease association) viruses or echoviruses for short. Over the following decades, numerous subtypes of both coxsackie and echoviruses were discovered that all infected the human gastrointestinal system. Many are relatively harmless and their infections are either asymptomatic or present with typical gastrointestinal symptoms such as vomiting and diarrhea. Others can cause a wide range of significant symptoms, including meningitis, encephalitis, and heart issues, though none are as routinely devastating as poliovirus. Coxsackieviruses and echoviruses, together with poliovirus and a group of respiratory viruses called rhinoviruses, are all related and were ultimately classified as the *Enterovirus* genus within the picornavirus family. Another clinically important picornavirus that we will encounter in Chap. 9 is the hepatitis A virus, one of the agents of viral hepatitis.

The word picornavirus itself is descriptive of the viruses in this family as "pico" means small and "rna" signifies the type of nucleic acid used for the viral genome. Members of this family have positive-sense RNA genomes of around 7500 bases, a size below average for RNA viruses. These small genomes are packaged into icosahedral capsids of 15–30 nanometers in diameter. Lacking an outer membrane envelope, these naked capsids are quite resistant to harsh conditions. This allows the virions to survive passage through the stomach (high acidity and digestive enzymes) to infect the mucosa of the small intestines where the virus reproduces extensively. Newly assembled virions are similarly sturdy and when shed in the feces can survive for long periods in the environment. Poliovirus can remain infectious for several weeks on surfaces, in sewage, and in contaminated soil or water where it can be ingested directly or picked up on the hands and introduced into the nose or mouth. This lengthy survival outside the host facilitates the transmission of the virus to new victims and explains why poliovirus and other picornaviruses spread so readily in families, schools, daycare centers, and any environment where people are in close contact.

Breaking this train of transmission through public health measures alone was unlikely to ever be completely successful and is actually believed to be responsible for the dramatic rise in paralytic polio cases in the twentieth century (Rogers 1992). Prior to modern sanitation developments in the late 1800s and early 1900s, poliovirus was likely an endemic and ubiquitous infection. Without consistent access to pure water and food, and with primitive sewage disposal, the virus would spread easily among families and communities. Most people typically were exposed to the virus early in infancy when infections tended to be mild or asymptomatic and rarely progressed to the paralytic sequela but still produced life-long immunity. Additionally, since most adult women were infected in infancy, they produced anti-poliovirus antibodies that were present in their breast milk. These maternal antibodies were passed to their infants and further protected the babies from any serious neurological consequences of poliovirus infection acquired early in life. This cycle was repeated generation after generation to maintain poliovirus in human populations with only very infrequent occurrences of paralytic disease. However, as cities and towns began to focus on improved public sanitation there were advances in sewage removal and treatment coupled with an emphasis on food and water cleanliness. Access to indoor plumbing facilitated personal cleanliness (think hand washing) and safe, sanitary removal of bodily wastes. These changes in public and personal hygiene were critical for preventing diseases such as cholera and typhoid fever yet created a new problem with poliovirus. Now infants were less likely to become exposed to and infected by poliovirus, and they would become older children or even adults with no immunity to poliovirus. This effect also created fewer immune adult women who could pass on protective antibodies to their infants. Over time, as the percentage of older poliovirus-susceptible individuals in the population increased, the disease began to take its modern form. If these older children or adults became infected, they were now at a much higher risk for developing neurological complications leading to paralysis or death. What was once a mostly silent infection became the scourge of the early 1900s due to changing sociological factors rather than any change in the virus itself. Additional public health improvements were unlikely to totally eliminate the virus and its pernicious spread, so a vaccine to protect the susceptible population was sorely needed.

The Salk Vaccine—Immediate Success and Unforeseen Problems

By the early 1950s, there was enough understanding of poliovirus biology and the host immune response to this virus to resume serious vaccine development. The fiascos of Kolmer and Brodie in the 1930s were long past and there was sufficient knowledge to avoid their grievous mistakes. A variety of other viral and bacterial vaccines were in clinical use, and producing an actual poliovirus vaccine finally seemed inevitable. However, the race for the vaccine was becoming a highly competitive enterprise with two ardent camps, one touting an inactivated vaccine strategy and one adamant that a live, attenuated vaccine was the best solution. There were clear precedents for effective vaccines of both types, but ultimately the National Foundation for Infantile Paralysis favored the inactivated vaccine type as the quicker and easier route to success. Pathogen inactivation with chemicals such as formalin was a well-established procedure while trying to develop appropriate attenuated versions of all three poliovirus types might be problematic. Jonas Salk of the University of Pittsburgh proposed producing poliovirus stocks in monkey cell culture using Ender's techniques. Cultured cells would be cheaper than animals, more consistent to maintain, and the virions could be more easily purified away from contaminating materials before being inactivated. The Foundation favored this approach over attenuation and funded Salk to develop a protocol for the purification and inactivation of culture-grown virions. His resulting trivalent candidate vaccine containing all three poliovirus types produced no disease in monkeys and induced protective levels of anti-poliovirus antibodies. With no contraindications from these preclinical studies, human subjects were the next step.

Just as for earlier vaccines seen in previous chapters, the 1950s still lacked stringent ethical standards for human trials. The post-World War II Nuremberg Code of Ethics for medical experimentation was drafted in 1947 but was never formally adopted by any of the Allied nations. Many nations in the 1950s, including the United States, had no laws or regulations dictating standards for conducting human subjects research. Consequently, the phase I subjects typically were chosen from among the most vulnerable and defenseless populations, individuals who weren't given a choice to participate and who often were incapable of informed consent. Salk's first human inoculations went into 78 residents of the D. T. Watson Home for Crippled Children in Pennsylvania (Salk 1953). Since most of these children were already crippled from polio, the intent was to confirm that the vaccine would raise antibody levels and to ensure that the vaccine caused no adverse reactions. The trials were continued with an additional 68 children from the Polk State School, an institution for individuals with mental disabilities. These Polk children had no history of polio infections so the effectiveness of the vaccine at inducing high levels of anti-poliovirus antibodies could be easily evaluated. As horrible and unacceptable as these studies were by modern standards, they were not atypical for medical research in that era. The resulting data were deemed sufficient to establish that Salk's vaccine had no major safety concerns and was capable of stimulating the production of protective antibodies. While these studies were ongoing in 1952, the United States was experiencing its single worst year for poliovirus with over 3000 deaths and more than 21,000 cases of paralysis. This public health crisis demanded an urgent response, and the Foundation rapidly implemented one of the largest clinical trials in US history.

To oversee the inactivated vaccine trial, the Foundation created the Vaccine Advisory Committee and appointed Thomas Rivers, a noted virologist affiliated with the Rockefeller Institute, to direct the operation. Rivers has experience with vaccinia virus, influenza, and chickenpox virus, but hadn't worked on poliovirus so was considered an unbiased expert who would impartially conduct and analyze the trial. He implemented a stringent trial design consisting of three groups of elementary school-aged children, the vaccine group plus two control groups (Monto 1999). One control group received a placebo vaccine while the other control was unvaccinated and served as an observational control. Importantly, this entire study was conducted as a double-blind trial where neither the vaccine deliverers nor the vaccine recipients knew who was receiving the true vaccine versus the placebo, a design that helps prevent any bias in the analysis. Ultimately, over 1.5 million children in the United States, Canada, and Finland participated in this trial, and the results were unequivocal, the vaccine worked, and vaccinated individuals didn't get paralytic polio. Children were now safe and parents could stop fearing the summer months. Almost immediately after announcing the result on April 12, 1955, the inactivated poliovirus vaccine (IPV) was approved for public use. In anticipation of this approval, six companies (Eli Lilly, Park Davis, Wyeth, Pitman-Moore, Cutter Laboratories, and Connaught) had already produced and distributed millions of doses of vaccine so public vaccination began immediately.

Although the Salk vaccine was and is highly effective, it was not without problems in the years immediately following massive public vaccination. One tragic issue was an outbreak of polio cases among a subset of vaccine recipients where there were 192 paralytic cases and 10 deaths (Offit 2005). All the infections were traced back to Cutter Laboratories where a manufacturing issue caused incomplete inactivation of the virulent virus in some vaccine doses. This horrific mistake led to revised and improved procedures for viral inactivation along with the implementation of more rigorous and stringent safety assessments for each vaccine batch, a manufacturing criterion that has become the standard for vaccine production. A second and more insidious problem was the discovery in 1959 that millions of doses of the inactivated poliovirus vaccine were inadvertently contaminated by an unknown virus that was subsequently named Simian virus 40 (SV40) (Garcea and Imperiale 2003). SV40 was a silent infection in the monkey cells used to grow poliovirus so SV40 was propagated and amplified along with poliovirus during the preparation of vaccine stocks. Unfortunately, SV40 was more resistant to formalin than poliovirus, resulting in many vaccine doses containing infectious SV40 even though the poliovirus was completely inactivated. Horrifyingly, it was quickly demonstrated that this new simian virus could transform normal cultured human cells into cancerous cells, raising the concern that millions of people, mostly children, had been inoculated with a cancer-causing virus. Steps were taken to change production and screening protocols to eliminate SV40 contamination and since 1964 there has no longer been any SV40 in IPVs produced in most countries. However, there was nearly a decade of IPV delivery using the SV40-contaminated vaccines and the potential effect of SV40 infection has been studied ever since (Wilson 2022). Nearly 60 years on there is still debate about whether or not SV40 caused any adverse health effects in those individuals receiving SV40 through their poliovirus vaccines. The good news is that most reliable studies have not seen any health consequences of SV40 infection, although it is still possible that certain rare cancers in some individuals may be attributable to their childhood exposure to this vaccine contaminant.

In addition to the solvable manufacturing problems that arose in the early days of IPV vaccination, there is an inherent biological limitation with this vaccine. Like all injected inactivated vaccines, IPV primarily induces B cell-derived circulating antibodies in the bloodstream. When poliovirus that is replicated in the gastrointestinal tract enters the bloodstream, these antibodies will bind to and neutralize the virions to prevent spread to the central nervous system. Blockage at the viremic stage provides excellent protection from paralysis or any other neurological complications from poliovirus infection and eliminates the major clinical concern with these infections. What the vaccine does not do is prevent poliovirus infection of the gut. People fully vaccinated with IPV are still susceptible to poliovirus infection; if introduced orally, the virus will reproduce in their small intestines and be spread in their feces. Thus, IPV protects the individual from neurological damage but does not eliminate the virus or protect the public as the virus is still free to circulate regardless of how many people in the community are vaccinated with IPV. Any unvaccinated individuals in the community will then be at risk for infection and possible paralytic consequences. While this limitation is not ideal, it didn't invalidate IPV as an important tool to combat poliovirus. After its introduction in the late 1950s, IPV dramatically reduced poliovirus deaths and paralytic cases in every country where it was used, incontrovertibly demonstrating its effectiveness. Still, many researchers of that era thought that a live, attenuated poliovirus vaccine would be superior to IPV and the race was on.

The Sabin Vaccine—Greater Success but a New Problem

Jonas Salk's vaccine success and his prominent support from the National Foundation for Infantile Paralysis did not dissuade other researchers from continuing the development of attenuated versions of all three poliovirus types. The main competitors were Hillary Koprowski of the Wistar Institute in Philadelphia, Herald Cox of Lederle Laboratories, and Albert Sabin at the University of Cincinnati. Each group used the standard approach of passaging human prototype poliovirus strains through animals and/or cultured cells. Mutations accumulate during passaging and the progeny viruses whose mutations favor reproduction in the new host eventually predominate. With luck, a variant may arise that just happens to have reduced neurovirulence in humans yet is still capable of enough replication in the human gut to induce a protective immune response. This is a stochastic approach that requires many rounds of passage and extensive animal testing to evaluate the properties of the resultant mutant strains. After passaging, Vogt and Dulbecco's plaque assay was important for isolating viral populations derived from a single mutant virion that could be further screened and propagated as a relatively pure viral stock. Ultimately, all three groups succeeded in generating attenuated versions of polioviruses types 1, 2, and 3, and their respective trivalent oral poliovirus vaccines (OPVs) were ready for human testing. For Sabin's strains, the first tests were on young adult volunteers from the Federal Penitentiary in Chillocothe, Ohio (Sabin 1965). Convinced by these preliminary studies that his strains were safe, he further tested them on his wife, his children, and several of their playmates, another dramatic example of a practice that would be anathema to today's researchers. Progressing through these early studies, by the late 1950s, all three groups had decreed their candidate vaccines ready for large-scale human testing. However, by this time the Salk vaccine was already widely in use throughout the United States. Because so many American children were already vaccinated, most clinical trials of the OPVs had to be conducted elsewhere. Koprowski chose Northern Ireland and the Congo, Cox tested in Latin America, and Sabin conducted his trials in South America. Sabin also provided his strains to a Soviet colleague and they were tested on millions of Soviet Union children (Horstmann 1991), an extraordinary example of East-West collaboration during the burgeoning Cold War. A comparison of the various trial results indicated that Sabin's strains were superior in conferring rapid and robust protection with fewer adverse effects. This led to the speedy licensing of Sabin's OPV by the United States in 1961 while the strains of Koprowski and Cox were relegated to the also-ran category.

The Sabin OPV had several immediate advantages over the Salk IPV. First, it was cheaper to produce as there was no need for the complex formalin inactivation process. Second, it was given as an oral dose rather than an injection. Typically the vaccine dose was soaked into a sugar cube that was simply eaten by the recipient. Not only did children greatly prefer this route to a shot, but it also made delivery of the vaccine much simpler and less expensive. It didn't take trained personnel to give injections and there was no need for syringes, needles, or any of the other paraphernalia needed for administering an injectable vaccine, thus making mass community vaccination campaigns a much simpler task. And lastly and most importantly, the OPV was simply a better vaccine than the IPV. Compared to inactivated vaccines, attenuated vaccines generally provide a broader immunity by invoking both humoral (antibody production by B cells) and cellular (T cells) immunity. By activating both arms of the adaptive immune system, attenuated vaccines tend to provide stronger protection and more long-lasting immunity, and this was true for the OPV. Additionally, since the attenuated poliovirus strains in the OPV replicated in the gastrointestinal tract, there were two important consequences. First, OPV induced mucosal immunity in the gut, something the IPV did very poorly. Mucosal immunity protects the gastrointestinal tract against any future poliovirus infection. This meant the OPV recipients who were later exposed to wild-type poliovirus did not have viral replication in the intestines and did not shed viruses in their feces. Thus the OPV not only protected the individual but also protected the community by preventing continual transmission of the wild-type virus, something the IPV was unable to do. Secondly, because the attenuated viruses in the OPV transiently replicated in the gut of the vaccine recipients, these vaccine strains were shed in their feces for several weeks. It turned out that the shed vaccine strains could spread to other individuals just as wild-type poliovirus would. Anyone contracting these attenuated strains would become immunized as if they had received the vaccine. This secondary spread of the vaccine strains was highly beneficial as it helped spread vaccination beyond just the originally vaccinated person. For example, vaccinating one child in a household might spread the attenuated strains to everyone else in the family, thus protecting the entire family and amplifying the community's resistance to poliovirus.

The ease and superiority of the OPV led to it rapidly supplanting the IPV. Within 3 years of its licensing, nearly 100 million doses were administered in both the United States and the Soviet Union, with similar massive distribution in developed nations around the globe. Poliovirus infections, both paralytic cases and deaths, plummeted in every country with an effective vaccination campaign. Summer was no longer a dreaded season when parents feared letting their children roam. Swimming pools, movie theaters, fairs, sporting events, and public gatherings were no longer dangerous venues to be avoided. OPV was so successful that by 1968 the IPV was no longer being given in the U.S., and by the early 1970s OPV was the predominant vaccine in use around the globe. The last naturally occurring case of poliovirus in the U.S. was detected in 1979, ending nearly a century of death and disability attributed to this small virus. The virtual elimination of this once horrendous disease from nearly every country is a magnificent tribute to the power and effectiveness of both the IPV and OPV. Neither Salk nor Sabin ever made any attempt to patent or commercialize their vaccines as they both wanted to ensure unfettered access to their life-saving inventions; any company or country was free to use their strains and protocols to produce these vaccines, and their largesse remains to this day.

As simple and effective as the OPV was, one tragic issue arose with this vaccine, the rare reversion of the vaccine strains to neurovirulence. Remember that all three poliovirus types in the vaccine were derived from wild-type viruses and contain multiple mutations that render them unable to invade and damage the nervous system. These mutations are just small changes in the nucleotide sequence of the poliovirus genome. For example, within the wild-type genome, there could be a portion of the RNA with the sequence AUACAGUGCCA. The attenuated strain might have a single nucleotide change in this sequence converting the underlined G to an A to give the sequence AUACAAUGCCA. However, such mutations aren't necessarily permanent changes. If a second replication error should happen to occur at that underlined A, the A could by chance be replaced with a G. This second mutation would restore the RNA to its original wild-type sequence, a genetic event referred to as reversion. As the attenuated strains replicate in the gut of vaccine recipients (and the guts of anyone who contracts these strains from the vaccinee) the viruses will continue to accumulate random replication errors in their genomes. When these mutations accrue this generates new variant strains of poliovirus collectively called vaccine-derived polioviruses (VDPVs). Most mutations are harmless or detrimental to the virus. However, on very infrequent occasions, a VDPV will arise where all the mutations critical for attenuation are reverted to the wild-type sequence and this virus regains its virulence once again. The unfortunate individual harboring this virulent virus is then a risk for paralytic polio, and the disease in these rare cases is known as vaccine-associated paralytic polio (VAPP). Estimates vary from different studies, but the incidence of VAPP is roughly 1 case per every two million vaccine doses delivered.

The appearance of potential VAPP cases was recognized within a year after the massive distribution of the OPV began in 1961, and by 1962 the U.S. Surgeon General instigated a review of the vaccine (Henderson et al. 1964). Over the next 2 years, several hundred polio cases were reviewed, with indeterminate results. While several dozen cases were adjudged as likely due to the vaccine, there was no definitive test in that era that could distinguish community-acquired polio from vaccine-acquired polio. Given the low incidence rate of possible VAPP cases and the huge benefit of the OPV for eliminating wild-type polio from the country, the risk associated with the attenuated vaccine was deemed acceptable, and usage of the OPV continued unabated for the next three decades. The last endemic case of polio in the U.S. occurred in 1979, and in 1994 the World Health Organization (WHO) declared the entire Americas (North, Central, and South) free of polio. With the complete absence of wild-type polio in the 1990s, the issue of VAPP resurfaced in the U.S. and the risk-benefit analysis changed. If there was no risk to the population from wild-type polio, was even the small risk of VAPP tolerable? Ultimately, the Advisory Committee on Immunization Practices, the American Academy of Pediatrics, and the American Academy of Family Practice all decided that using the OPV was no longer acceptable and they recommended returning to the safer IPV. Since the year 2000, the U.S. no longer uses the OPV vaccine and all new poliovirus vaccinations use the IPV. From 1995 to 2010, several other countries, notably Canada, Mexico, Australia, and much of Europe also returned to solely using the IPV to vaccinate their populations. The rest of the world persisted longer with the OPV although most countries have now switched to a two-vaccine regimen. Vaccinees first receive the IPV to create the neutralizing antibody immunity that protects them from paralytic disease. Subsequently, they receive the OPV to induce a broader and more long-lasting immunity. Now if any of the three attenuated viruses in the oral vaccine revert to virulence, the vaccinees are already protected by the circulating antibodies generated by the IPV, thus negating the possibility of VAPP. This strategy combines the strengths of each vaccine and virtually eliminates the small risk associated with the OPV, helping the world move closer and closer to zero cases of poliovirus.

The Poliovirus Endgame in the Post-Vaccine World

Like smallpox, polio is a virus restricted to human hosts, and shed virions can only survive in the environment for a few months. If all poliovirus cases could be eliminated then any residual virions in soil, water, or sewage should rapidly become inactive and this virus would disappear from the Earth. With the dramatic success of the poliovirus vaccines, by the 1980s the notion of total poliovirus eradication was gaining momentum in scientific and political circles. The elimination of the virus from whole countries and regions predicted that global eradication was an achievable feat if enough resources were committed to vaccination and public health. Most developed nations were already poliovirus-free and the remaining cases were concentrated in poorer countries that lacked sufficient healthcare infrastructures to test for cases and conduct massive immunization campaigns. In 1988 at the 41st World Health Assembly (the WHA is the decision-making body of the WHO that is comprised of representatives from the member nations), a resolution was passed committing the WHO to the task of polio eradication globally. The resolution set an eradication date of the year 2000, a seemingly realistic target with the world already poised on the verge of conquering poliovirus.

Regrettably, we are already more than 20 years past the year 2000 deadline, and the virus is still lurking in a small number of countries. The final endgame has been much slower than anticipated mostly due to secondary factors rather than vaccine issues. Poliovirus eradication requires extensive vaccine coverage in every community, something that is very challenging in some countries. Certain populations live in isolated and remote areas lacking modern amenities which makes providing vaccines and medical personnel to these areas difficult. Warfare, regional conflicts, and political unrest or instability compound the geographical obstacles making it difficult if not impossible to deliver and administer vaccines as needed. Religious objections, distrust of outsiders, and fear of modern medicine further collude to limit widespread vaccination in some areas even when vaccines are available. Notwithstanding all these factors, national and international vaccine programs slowly penetrated more and more communities and year by year reduced the wild-type poliovirus cases down to only six reported cases in 2021. As of 2022, only Pakistan and Afghanistan reported cases of naturally occurring wild-type poliovirus infections, and while the case numbers are small they remain a dangerous source of poliovirus that must be eliminated to achieve the eradication goal. Several other countries, mostly in Africa, are still struggling with VAPP cases. These cases are troubling as virulent VDPVs can spread in the same fashion as wild-type poliovirus and they remain a source of possible reintroduction of the virulent virus into other countries. Hopefully, the introduction of the dual IPV-OPV vaccination strategy will eventually prevent any new VAPP cases. Additionally, the attenuated type 2 poliovirus strain used in the OPV (type 2 is the one most likely to revert) has been improved by making it genetically more stable and thus less likely to acquire mutations that return it to virulence (De Coster et al. 2021). These advances will enhance the fight in the remaining endemic areas and should finally push wild-type poliovirus into extinction.

With case numbers so small and confined to relatively obscure locations, some may wonder why the fight against poliovirus remains urgent. The answer is simply that unless a virus is 100% gone it's not gone at all and the entire world needs to continue vaccinating to ensure our safety (Chumakov et al. 2021). Travelers from the remaining regions with natural or vaccine-associated polio cases can carry the disease to any country in the world and potentially reintroduce the virus as happened in London, New York, and Jerusalem in 2022. Genetic analysis confirms that this outbreak is a VDPV strain that likely originated in Afghanistan or Pakistan. The unknown VDPV carrier apparently traveled to the United Kingdom and viruses shed from this individual gained a foothold in the Orthodox Jewish community of London, a susceptible group due to a low poliovirus vaccination rate. Silently infected travelers from this community likely carried the virus both to Jerusalem and to Rockdale County in New York, another Orthodox Jewish community with a high percentage of unvaccinated individuals. It was in Rockdale where a young man developed polio-induced leg weakness leading him to be identified as the first polio victim in the U.S. in over a decade. Sadly, analysis of sewage in both New York City and London found the VDPV, indicating that the virus had already spread widely through these cities. While not yet virulent, the detected VDPV strains could mutate further in any infected person to become dangerous as happened to the unfortunate young man in Rockdale. This cautionary tale regrettably demonstrates just how simple it is for poliovirus to reemerge unless vaccination rates in every city, county, state, and country of the world remain high. After 60 years of struggle, we've nearly conquered this horrid virus and have it largely confined to a few last enclaves. It would be foolish and tragic if indifference, complacency, and vaccine hesitancy allowed this scourge to once again spread throughout the world. Every nation needs to recommit to maintaining widespread polio vaccination so that we can finally drive poliovirus extinct and prevent even a single additional death or paralytic case.

References

Armstrong C (1936) Prevention of intravenously inoculated poliomyelitis of monkeys by intranasal instillation of picric acid. Public Health Reports (1896–1970) 51(10):241–243. https://doi.org/10.2307/4581773

Armstrong C (1939) Successful transfer of the lansing strain of poliomyelitis virus from the cotton rat to the white mouse. Public Health Reports (1896–1970) 54(52):2302–2305. https://doi.org/10.2307/4583135

Axelsson P (2009) "Do not eat those apples; they've been on the ground!": polio epidemics and preventive measures, Sweden 1880s-1940s. Asclepio 61(1):23–38. https://doi.org/10.3989/asclepio.2009.v61.i1.270

Bendiner E (1982) Enders, Weller, and Robbins: the trio that 'fished in troubled waters'. Hosp Pract (Off Ed) 17(1):163, 169, 174–165 passim. https://doi.org/10.1080/21548331.1982.11698030

Berner M, Pany-Kucera D, Doneus N, Sladek V, Gamble M, Eggers S (2021) Challenging definitions and diagnostic approaches for ancient rare diseases: the case of poliomyelitis. Int J Paleopathol 33:113–127. https://doi.org/10.1016/j.ijpp.2021.04.003

Bodian D, Morgan IM, Howe HA (1949) Differentiation of types of poliomyelitis viruses; the grouping of 14 strains into three basic immunological types. Am J Hyg 49(2):234–245

Brodie M, Park WH (1936) Active immunization against poliomyelitis. Am J Public Health Nations Health 26(2):119–125. https://doi.org/10.2105/ajph.26.2.119

Burnet FM, MacNamara J (1931) Immunological differences between strains of poliomyelitic virus. Brit J Exp Pathol 12(2):57–61

Carleton HA (2011) Putting together the pieces of polio: how Dorothy Horstmann helped solve the puzzle. Yale J Biol Med 84(2):83–89

Caverly CS (1896) Notes of an epidemic of acute anterior poliomyelitis. J Am Med Assoc 26(1):1–5

Chumakov K, Ehrenfeld E, Agol VI, Wimmer E (2021) Polio eradication at the crossroads. Lancet Glob Health 9(8):e1172–e1175. https://doi.org/10.1016/S2214-109X(21)00205-9

De Coster I, Leroux-Roels I, Bandyopadhyay AS, Gast C, Withanage K, Steenackers K, De Smedt P, Aerssens A, Leroux-Roels G, Oberste MS, Konopka-Anstadt JL, Weldon WC, Fix A, Konz J, Wahid R, Modlin J, Clemens R, Costa Clemens SA, Bachtiar NS, Van Damme P (2021) Safety and immunogenicity of two novel type 2 oral poliovirus vaccine candidates compared with a monovalent type 2 oral poliovirus vaccine in healthy adults: two clinical trials. Lancet 397(10268):39–50. https://doi.org/10.1016/S0140-6736(20)32541-1

Dulbecco R (1952) Production of plaques in monolayer tissue cultures by single particles of an animal virus. Proc Natl Acad Sci USA 38(8):747–752. https://doi.org/10.1073/pnas.38.8.747

Dulbecco R, Vogt M (1954) Plaque formation and isolation of pure lines with poliomyelitis viruses. J Exp Med 99(2):167–182. https://doi.org/10.1084/jem.99.2.167

Dunn PM (2006) Michael underwood, MD (1737-1820): physician-accoucheur of London. Arch Dis Child Fetal Neonatal Ed 91(2):F150–F152. https://doi.org/10.1136/adc.2005.074526

Enders JF, Weller TH, Robbins FC (1949) Cultivation of the lansing strain of poliomyelitis virus in cultures of various human embryonic tissues. Science 109(2822):85–87. https://doi.org/10.1126/science.109.2822.85

Flexner S (1936) Respiratory versus gastro-intestinal infection in poliomyelitis. J Exp Med 63(2):209–226. https://doi.org/10.1084/jem.63.2.209

Flexner S, Lewis PA (1909) The transmission of epidemic poliomyelitis to monkeys: a further note. JAMA LIII 23:1913–1913. https://doi.org/10.1001/jama.1909.92550230039003a

Flexner S, Lewis PA (1910) Experimental epidemic poliomyelitis in monkeys. J Exp Med 12(2):227–255. https://doi.org/10.1084/jem.12.2.227

Friedrich Koch GK (1985) History. In: The molecular biology of poliovirus. Springer Vienna, pp 3–14

Galassi FM, Habicht ME, Ruhli FJ (2017) Poliomyelitis in Ancient Egypt? Neurol Sci 38(2):375. https://doi.org/10.1007/s10072-016-2720-9

Garcea RL, Imperiale MJ (2003) Simian virus 40 infection of humans. J Virol 77(9):5039–5045. https://doi.org/10.1128/jvi.77.9.5039-5045.2003

Goldman AS, Schmalstieg FC (2019) Karl Otto Landsteiner (1868-1943). Physician-biochemist-immunologist. J Med Biogr 27(2):67–75. https://doi.org/10.1177/0967772016670558

Grimshaw ML (1995) Scientific specialization and the poliovirus controversy in the years before World War II. Bull Hist Med 69(1):44–65

Henderson DA, Witte JJ, Morris L, Langmuir AD (1964) Paralytic disease associated with oral polio vaccines. JAMA 190:41–48. https://doi.org/10.1001/jama.1964.03070140047006

Horstmann DM (1952) Poliomyelitis virus in blood of orally infected monkeys and chimpanzees. Proc Soc Exp Biol Med 79(3):417–419. https://doi.org/10.3181/00379727-79-19398

Horstmann DM (1991) The Sabin live poliovirus vaccination trials in the USSR, 1959. Yale J Biol Med 64(5):499–512

Howe HA, Bodian D (1944) Poliomyelitis by accidental contagion in the chimpanzee. J Exp Med 80(5):383–390. https://doi.org/10.1084/jem.80.5.383

Johnston H (1916) Tetanus antitoxin in poliomyelitis. Med Rec 90(7):292–293

Kehret P (1996) Small steps: the year I got polio. Albert Whitman, Morton Grove, Ill

Kolmer JA (1936) Vaccination against acute anterior poliomyelitis. Am J Public Health Nations Health 26(2):126–135. https://doi.org/10.2105/ajph.26.2.126

Kumar DR, Aslinia F, Yale SH, Mazza JJ (2011) Jean-Martin Charcot: the father of neurology. Clin Med Res 9(1):46–49. https://doi.org/10.3121/cmr.2009.883

Monto AS (1999) Francis field trial of inactivated poliomyelitis vaccine: background and lessons for today. Epidemiol Rev 21(1):7–23. https://doi.org/10.1093/oxfordjournals.epirev.a017989

Morgan IM (1948) Immunization of monkeys with formalin-inactivated poliomyelitis viruses. Am J Hyg 48(3):394–406. https://doi.org/10.1093/oxfordjournals.aje.a119251

Morgan IM (1949) Differentiation of types of poliomyelitis viruses; by reciprocal vaccination-immunity experiments. Am J Hyg 49(2):225–233. https://doi.org/10.1093/oxfordjournals.aje.a119272

Nathanson N (2005) David Bodian's contribution to the development of poliovirus vaccine. Am J Epidemiol 161(3):207–212. https://doi.org/10.1093/aje/kwi033

Offit PA (2005) The Cutter incident: how America's first polio vaccine led to the growing vaccine crisis. Yale University Press, New Haven

Paul JR, Trask JD (1932) The detection of poliomyelitis virus in so called abortive types of the disease. J Exp Med 56(3):319–343. https://doi.org/10.1084/jem.56.3.319

Paul JR, Trask JD (1933) A comparative study of recently isolated human strains and a passage strain of poliomyelitis virus. J Exp Med 58(5):513–529. https://doi.org/10.1084/jem.58.5.513

Paul JR, Trask JD (1935) The neutralization test in poliomyelitis: comparative results with four strains of the virus. J Exp Med 61(4):447–464. https://doi.org/10.1084/jem.61.4.447

Paul JR, Trask JD, Webster LT (1935) Isolation of Poliomyelitis Virus from the Nasopharynx. J Exp Med 62(2):245–257. https://doi.org/10.1084/jem.62.2.245

Paul JR, Trask JD, Culotta CS (1939) Poliomyelitic virus in sewage. Science 90(2333):258–259. https://doi.org/10.1126/science.90.2333.258

Pietrzak K, Grzybowski A, Kaczmarczyk J (2017) Jacob Heine (1800-1879). J Neurol 264(7):1545–1546. https://doi.org/10.1007/s00415-017-8454-7

Robbins FC (1994) John F. Enders. J Pediatr 124(1):155–157. https://doi.org/10.1016/s0022-3476(94)70276-4

Rogers N (1992) Dirt and disease: polio before FDR. Health and medicine in American society. Rutgers University Press, New Brunswick

Rous P (1948) Simon Flexner and medical discovery. Science 107(2789):611–613. https://doi.org/10.1126/science.107.2789.611

Rutty CJ (1996) The middle-class plague: epidemic polio and the Canadian state, 1936-37. Can Bull Med Hist 13(2):277–314. https://doi.org/10.3138/cbmh.13.2.277

Sabin AB (1965) Oral poliovirus vaccine. History of its development and prospects for eradication of poliomyelitis. JAMA 194(8):872–876. https://doi.org/10.1001/jama.194.8.872

Sabin AB, Olitsky PK (1936) Cultivation of poliomyelitis virus in vitro in human embryonic nervous tissue. Proc Soc Exp Biol Med 34(3):357–359

Sabin AB, Olitsky PK, Cox HR (1936) Protective action of certain chemicals against infection of monkeys with nasally instilled poliomyelitis virus. J Exp Med 63(6):877–892. https://doi.org/10.1084/jem.63.6.877

Salk JE (1953) Studies in human subjects on active immunization against poliomyelitis. I A preliminary report of experiments in progress. J Am Med Assoc 151(13):1081–1098

Samuel L, Katz CMW, Robbins FC (2011) The role of tissue culture in vaccine development. In: Plotkin SA (ed) History of vaccine development. Springer, pp 145–149

Schultz EW, Gebhardt LP (1937) Zinc sulphate as a chemoprophylactic agent in experimental poliomyelitis. Proc Soc Exp Biol Med 35(4):524–526

Schultz EW, Gebhardt LP (1938) The use of zinc sulfate solution for the prevention of poliomyelitis in man. JAMA J Am Med Assoc 110:2024. https://doi.org/10.1001/jama.1938.02790240048022

Schwarz HP, Dorner F (2003) Karl Landsteiner and his major contributions to haematology. Br J Haematol 121(4):556–565. https://doi.org/10.1046/j.1365-2141.2003.04295.x

Trask JD, Vignec AJ, Paul JR (1938a) Isolation of poliomyelitic virus from human stools. Proc Soc Exp Biol Med 38(1):147–149

Trask JD, Vignec AJ, Paul JR (1938b) Poliomyelitis virus in human stools. JAMA J Am Med Assoc 111:6–11. https://doi.org/10.1001/jama.1938.02790270008002

Wilson VG (2022) Viruses: intimate invaders. Springer. https://doi.org/10.1007/978-3-030-85487-4

Winkelstein W Jr (2009) The epidemiology of polio: Wade Hampton Frost and Bulletin Number 90. Epidemiology 20(3):460. https://doi.org/10.1097/EDE.0b013e3181989a5d

Wyatt HV (2014) Before the vaccines: medical treatments of acute paralysis in the 1916 new york epidemic of poliomyelitis. Open Microbiol 12(8):144–147

7

Measles/Mumps/Rubella (MMR): The Childhood Trifecta

Keywords Congenital rubella syndrome • Edmonston strain • German measles • Maurice Hilleman • Jeryl Lynn strain • Memory cells • Rubeola • Andrew Wakefield

Abbreviations

AGMK	African green monkey kidney cells
ALV	Avian leukosis virus
CDC	Centers for Disease Control
CPE	Cytopathic effect
CRS	Congenital rubella syndrome
EZ	Edmonston-Zagreb strain
FDA	Food and Drug Administration
HDCS	Human diploid cell strains
HPV-77	High passage virus 77
M&RI	Measles & Rubella Initiative
MMR	Measles, mumps, rubella vaccine
RA27/3	Rubella aborted fetus sample 27/3
SSPE	Subacute sclerosing panencephalitis
Ts	Temperature sensitive mutants
WHO	World Health Organization
WI-38	Wistar Institute cell line 38

A Fabricated Controversy

For much of the twentieth century, contracting measles, mumps, and rubella (German measles) was a common occurrence with most youngsters falling ill with each virus during childhood. While generally milder than the viral infections in the preceding chapters, all three of these diseases could have serious complications, including death. With millions of cases occurring around the world each year these diseases collectively produce significant morbidity and mortality, primarily in children. Fortunately, the second half of the twentieth century saw the development of effective vaccines against each of these viruses with a dramatic reduction in cases, serious clinical complications, and fatalities. Ultimately the three separate vaccines were combined into a single dose known as the trivalent MMR (measles/mumps/rubella) vaccine to facilitate distribution and reduce the number of shots needed. However, before detailing the science and history behind the development of each individual vaccine in the MMR injection, it is necessary to review an ugly episode in vaccine science. In 1998, Dr. Andrew Wakefield and colleagues published a fraudulent study implying that the MMR vaccine was associated with an increased risk for autism (Wakefield et al. 1998). Perhaps no vaccine has suffered more unjustly than the MMR vaccine due to Wakefield's spurious accusation. While convincingly disproven and discredited over the subsequent decade, Wakefield's premise created enormous and undeserved public distrust in vaccines that still resonates today, not only for the MMR vaccine but for vaccines in general.

Our current understanding suggests that autism results from a combination of genetics and environmental influences that occur mostly during pregnancy or birth. In the 1990s, these biological risk factors for autism were less defined, leaving both parents and physicians of that era desperately searching for causative agents. The 1990s also saw an alarming trend of rapidly increasing numbers of children identified with autism, fueling speculation that some new environmental factor was contributing to the rising case numbers. While we now know that much of this increase in case numbers was due to better awareness of this condition, improved diagnostic methods, and reclassification of what constitutes autism versus other atypical intellectual abilities, at the time it appeared that some new exposure was predisposing young children to develop autism. Into this emotionally charged knowledge vacuum came the British physician, Andrew Wakefield, and his MMR case study published in the prominent journal *The Lancet* (Wakefield et al. 1998). He and his colleagues reported on 12 children, ages 3–10, with inflammation of the digestive tract (enterocolitis) and regressive development disorders. The parents of 8 of the 12 stated that these symptoms arose after their MMR vaccine. While the article stopped short of declaring a causal connection between the MMR vaccine and the development issues, the implication was clear. Rather than addressing the severe limitations of this small case study without any control group, the media hyped this speculation and stoked an already present anti-vaccine sentiment. Wakefield subsequently became an ardent and vocal opponent of the MMR vaccine and has remained a leading anti-vaxxer to this day, further contributing to the controversy (Wakefield 1999, 2010). The combined effect of the media reports and Wakefield's crusade was a widespread public belief that the MMR vaccine caused autism, leading to declining compliance with MMR vaccination in Britain, much of Europe, and the United States. Sadly, Wakefield's study was not only scientifically flawed but also fraudulent. Various investigations concluded that there were significant violations of scientific and ethical standards. First, Wakefield was paid to perform his study by lawyers representing parents who believed that the vaccine injured their children. This serious conflict of interest was not revealed when his study was published and represents an egregious breach of standards that immediately taints the entire study. Conflicts of interest are allowed in scientific research, but such conflicts must be publicly acknowledged so that the work presented can be evaluated with full knowledge of the circumstances. Additionally, Wakefield filed a patent for his own version of an MMR vaccine before his publication, so discrediting the current version would have had potential financial benefits for him. Even more importantly, a subsequent review of the original medical records of the study's subjects found major discrepancies between their actual conditions and what was presented in Wakefield's paper (Godlee et al. 2011). These discrepancies completely invalidated the paper's finding and can only be ascribed to incredibly shoddy work or outright fabrication (Deer 2011). As the validity of Wakefield's paper became compromised, 10 of his co-authors withdrew their names from the publication, and the journal *Lancet* retracted the article (Horton 2004). Ultimately, in 2010 the British General Medicine Council stripped Dr. Wakefield of his medical license for unprofessional conduct.

As Wakefield and his MMR work were losing credibility for the numerous flaws in his study, the scientific community launched more reliable studies to explore any possible connection between the MMR vaccine and autism. Unlike Wakefield's very small study of only 12 children, scientists and major scientific organizations in different countries initiated large-scale investigations searching for any adverse effects from the MMR vaccine, including autism (DeStefano and Shimabukuro 2019). The first such reports simply looked at population correlations to see if the overall incidence of autism increased after the first introduction of the MMR vaccine into a country. Studies in the United Kingdom, the United States, Japan, and Canada all failed to find any associated increase in autism rates as the MMR vaccine became available and widely used in these countries. While negative, these large population studies may miss weak correlations or be confounded by other factors that can't be accounted for in the analysis. Stronger negative data were obtained from both case-control studies and cohort studies. In case-control studies, the MMR vaccination history was compared for children with autism versus matched controls without autism. Again, studies from the United States, Europe, and Japan found that neither the MMR vaccine nor the measles vaccine alone were associated with an increased risk of autism. Similarly, cohort studies used medical records to identify children at birth and divide them into two groups, those who received the MMR vaccine and those who did not receive the vaccine. Then those two groups are analyzed years later to determine the rate of autism in each group. One of the largest cohort studies was conducted in Denmark and included over 500,000 children born between 1991 and 1998 (Madsen et al. 2002). No MMR vaccine risk for autism was seen and subsequent cohort studies in other nations were likewise negative. As all these larger and more valid studies accumulated, Wakefield's claim of an MMR-autism connection became thoroughly disproved by legitimate scientific studies. An extensive review of all the available studies by the United States National Academy of Medicine in 2001 concluded that there was no evidence to support the claim that the MMR vaccine was a cause of autism (2001). The subsequent 20-plus years have confirmed this conclusion with additional well-conducted studies all failing to find any risk for autism associated with the MMR vaccine. Following the scientific method, Wakefield's original assertion has undergone rigorous testing and has been proven untrue. Unfortunately, that seed he planted sprouted and continues to spread like an aggressive and difficult-to-eradicate weed. Vaccine hesitancy, and especially resistance to the MMR vaccine, has become more commonplace. As measles remains endemic in several areas of the world, it can be easily reintroduced to any community. Pockets of unvaccinated individuals remain at risk for this disease as exemplified by several recent measles outbreaks in the United States and other developed nations (Flores and Immergluck 2023). Continued effort is needed to reassure the public and maintain high vaccination levels so that measles never again establishes itself as a common-place disease threatening our children.

Measles, The "First Disease"

In the nineteenth and twentieth centuries, scientists and physicians began to characterize and classify childhood diseases that produced skin rashes. Without knowing the actual causative agents, they distinguished each disease based on clinical features and simply listed them numerically. First on the list was measles and it became known as the "first disease". Measles is a respiratory disease caused by one of the most infectious human viruses, the rubeola virus. Most estimates give rubeola an R_0 value of around 12–18 (Hubschen et al. 2022), meaning that each person with a case of measles is likely to infect 12–18 other persons. The infection typically initiates in the respiratory system when a person breathes in viral particles shed in the coughs or sneezes from an infected individual (Perry and Halsey 2004). The virus is also present in the saliva and nasal secretions of an infected person and can be spread via these fluids, for example by sharing utensils or drinks. Virus survival on surfaces is generally limited to a few hours, but touching contaminated surfaces can transfer viruses to the hands from which viral particles can then be introduced into the eyes, nose, or mouth to start the infectious process. The virus slowly spreads within the infected individual, replicating in a variety of tissue types and incubating roughly 1–2 weeks before symptoms begin. The initial disease stage is fairly nondescript and can include high fever (>101 °F), cold-like symptoms (coughing, sneezing, runny nose), and conjunctivitis (red, watery eyes). While less obvious than the physical symptoms, patients at this stage also develop small, whitish spots on the inner cheeks of the mouth known as Koplik spots. These spots are a useful diagnostic indicator of measles before the characteristic rash appears. The infected individual begins to shed viruses during this preliminary period and is contagious even before realizing that they have measles. Within a few days after the appearance of these early symptoms, a blotchy rash consisting of flat, irregular, red spots begins on the face and spreads downward to the trunk and extremities. Unlike smallpox where the skin lesions are sites of virus replication and are teeming with viral particles, the measles skin rash is due to an immune response and is not a significant source of infectious virus. Instead, viruses infecting the capillaries in the skin stimulate an inflammatory immune response that causes cell damage leading to the itchy, reddish, and discolored skin lesions that are the hallmark of this disease.

Uncomplicated measles usually resolves within a week of the rash's appearance, and patients are infectious for most of this period. While the obvious presentations of measles are the rash and the clinical symptoms, it is critical to note that the virus is widely disseminated throughout the internal organs where more serious manifestations of the infection occur, especially in young children. These complications range from secondary infections such as ear infections to life-threatening encephalitis. The encephalitis was particularly problematic because survivors were often left with mental disabilities, seizures, or deafness. Tragically, some patients even develop a rare but inevitably fatal sequela known as subacute sclerosing panencephalitis (SSPE) that can arise months to years after a measles infection. In the years preceding the vaccine release in 1963, the United States had an estimated 500,000 cases per year with 500 deaths (Hinman et al. 1983). While this yearly death rate seems modest, the preventable death of anyone, especially a child, is a tragedy. More surprisingly, the measles yearly death rate was often comparable to the yearly deaths from polio which elicited much more public fear. Perhaps this was due to measles having a regular seasonal occurrence with consistent yearly case numbers compared to the more sporadic nature of poliovirus outbreaks. The frequency and familiarity of measles seemed to evoke less fear and more acceptance that this disease was just a normal part of childhood illnesses. What was likely less appreciated by the public was that even routine measles was associated with a high incidence of secondary infections, both viral and bacterial. Although the actual numbers remain uncertain as these secondary infections were not reportable, measles was certainly responsible for significant illness during childhood with considerable healthcare costs and lost school time for children. Even to this day, measles kills over 100,000 people per year worldwide, mostly young children in countries where vaccination is not universal. So, far from being an innocuous childhood disease, measles is a viral infection with a high rate of morbidity and a significant rate of mortality, all reasons for the priority given to vaccine development against this virus.

Enders' Game

Like smallpox, measles originated as an animal disease that spread to humans and adapted to become a strictly human disease. The closest relative of measles is a cattle disease known as bovine rinderpest, an RNA virus of the Paramyxovirus family (Dux et al. 2020). We suspect that a rinderpest-like virus spread into humans sometime after the domestication of cattle roughly 3000 years ago. As cattle husbandry became more common and human settlements grew in size, this facilitated the spread of the precursor virus. Molecular studies indicate that by the sixth century CE, the presumptive cattle virus had evolved into human measles and was now strictly limited to causing human infections. By the tenth century CE, the disease was already widespread in the Middle East and was clearly described by the Persian physician Rhazes. Spreading across Eurasia, measles became endemic in the Old World, and as human populations were infected and survived they slowly developed a measure of resistance to the virus. Denizens of the New World were not as fortunate, and the sudden introduction of measles via European explorers caused massive epidemics in these naïve populations. Only smallpox caused more death and destruction in the New World, and there are well-documented accounts of massive devastation when measles entered susceptible populations. For example, measles descended upon Hawaii in 1848 and killed one-third of the population of that island nation. Anyone who survived these tragic epidemics would never describe measles as an innocuous virus. Unfortunately, once introduced, measles readily spread in the Western hemisphere, eventually becoming endemic throughout the New World just as it had in the Old World. While vaccination ultimately reduced measles incidence and medical impact, even into the twenty-first century, the virus remains a worldwide concern and is far from eliminated.

After generations of global misery from measles, the control of the rubeola virus began in the mid-1900s as multiple scientific advances were finally marshaled to attack the pervasive infection. In the 1800s and early 1900s, several primitive attempts used patient blood or nasal secretions to show that the disease could be transmitted to other people, indicating that there was an infectious agent present in these fluids. While the recipients developed immunity, heralding the possibility of vaccine development, these crude initial studies were dangerous. Using human fluids containing virulent wild-type viruses as the source material meant that there was an inherent risk of serious disease in the recipients, forestalling this approach as a viable vaccine strategy. Importantly though, some of these early investigators used the filter sterilization approach to investigate the nature of the measles organism. By the 1920s, the functional definition of viruses as invisible agents that passed through bacteria-retaining filters was well established and many human and animal diseases, including measles, fell into the virus category (Rivers 1927). A second necessary step for eventual vaccine development was a system for propagating the viral agent and producing large quantities of material for vaccine use. While certain monkeys could be infected with human measles, using animals for vaccine production is problematic as discussed in previous chapters. Ideally, eggs, tissues, or cultured cells are the preferred growth platform for viral vaccine production. These materials are relatively inexpensive, are convenient to manipulate in research labs, and are scalable for the generation of large quantities of viral particles. Unfortunately, years of studies using chicken embryos or chick embryonic tissues gave inconsistent and unreliable results with measles, stymying further progress toward a vaccine. The breakthrough once again came from the laboratory of John Enders.

Enders and his Harvard colleagues were pioneers in using cultured human cells to grow and propagate numerous viruses, including the recalcitrant measles virus (Enders et al. 1957; Enders 1957). Their work in the early 1950s showed that the measles virus could be isolated from multiple infected patients and that it would reproduce in cultured human cells, both nasal tissue and kidney tissue. Viral infection in these cultures was confirmed by an unusual property of the measles virus, syncytial formation. Syncytia is the fancy term for cell-to-cell fusion. When measles infects a cell, it produces changes in the cell membrane that cause the infected cell to fuse to a neighboring uninfected cell. Now instead of two independent cells, there is one large cell with two nuclei. This clever strategy allows the virus inside cell number one to spread to cell number two without ever venturing outside the cells where it could be recognized and destroyed by the immune system. Additionally, this fusion doesn't stop with two cells and can continue to spread by fusing more and more uninfected cells to give rise to a giant, multinucleated cell full of virus particles. These giant cells are easily visualized under the microscope which is a convenient marker to determine if a culture has been successfully infected with rubeola.

Needing clinical samples to test in their culture systems, the researchers in the Enders lab anxiously waited for measles outbreaks. In 1954 when measles struck the Fay School, a boarding school in the Boston area, Thomas Peebles from the Enders lab rushed there to collect samples from student volunteers (Griffin and Oldstone 2009). One of the young boarders who donated a throat swab was 11-year-old David Edmonston. Back in the lab, Edmonston's sample quickly generated syncytia in a human kidney cell culture, was easily passaged in cultured cells, and became the lab's preferred measles strain. As was by then a fairly standard approach, the Edmonston strain was repeatedly passaged in kidney cell culture in hopes of generating a milder, attenuated version of the virus. In what can only be described as serendipity, as this passaging process slowly progressed, the lab lost its source of kidney cells. These kidney cells were primary cultures meaning that they derived directly from kidney cells taken from surgical patients. Such cultures have a finite life span and can only divide a limited number of times before the cells die off and have to be replaced with newly harvested kidney cells. When their surgeon supplier had no more kidney patients to donate cells, Enders turned to an alternative source, placental cells. Boston Lying-In Hospital (now Brigham and Women's Hospital), a maternity hospital for indigent women, was nearby and could routinely supply placentas from daily births. Not only was this tissue more consistently available, but Enders reasoned that infecting tissue that was not a normal target for measles could help select for more highly attenuated variants. To their delight, the Edmonston virus readily infected this type of human cell, and from there was passed into chicken eggs and finally into chick tissue. By this point, this successive passaging in different cell types resulted in a virus stock that caused no symptoms in monkeys yet induced anti-measles antibodies, key features for a potential vaccine strain. As was typical of that era, the next step was for the lab members to test the stock on themselves to see if it was safe. When none of the lab volunteers suffered any ill effects from their injections, it was time for clinical trials to test the actual efficacy of this potential vaccine. With clinical testing not being Enders' interest, he freely distributed his attenuated Edmonston stock to other researchers so that more labs could investigate the strain and conduct the studies that would lead to implementing a successful vaccine for measles. Among those receiving the strain was Maurice Hilleman, the head of viral research at Merck Laboratories, who would go on to become a giant in the vaccine field.

Handoff to Hilleman for a Touchdown

It is surprising that Maurice Hilleman is not a household name because his contributions to improving human health in the twentieth century are almost incalculable (Hilleman 1999). In previous chapters, we've seen a few persistent and dedicated individuals who have each created a life-saving vaccine. In contrast, during his career, Hilleman was responsible for over 40 vaccines against both viral and bacterial diseases. His extraordinary accomplishments have saved more lives than any other single person and by some estimates, his vaccines collectively save eight million lives per year, year after year. Born in rural Montana and growing up during the Depression, a life of scholarship and scientific achievement was not a likely outcome of his impoverished early environment. However, his drive and intellectual abilities ultimately led him to college and then to a Ph.D. in microbiology and virology awarded in 1944 from the University of Chicago. To the dismay of his faculty teachers and mentors, he eschewed academia and insisted that industry was where he could translate scientific findings into actual products to improve human health. Working first at E.R. Squibb and Sons, a pharmaceutical company, and then at Walter Reed Army Institute of Research, he spent a decade studying diseases relevant to the military and honing his skills in vaccine development. While this early work was important and successful, his contributions to vaccinology would flourish when he joined Merck & Company in 1957 as the Director of the newly established Department of Virus & Cell Biology Research.

Merck was working on a killed measles vaccine when Hilleman arrived, but he soon grew skeptical of their ability to create a safe and effective vaccine from an inactivated measles virus, and his intuition proved prescient. Several companies, including Merck, marketed killed measles vaccines in the early 1960s that ultimately were taken off the market. Upon public release, it became clear that these killed vaccines failed to provide adequate protection, had significant adverse reactions, and caused a severe atypical infection in some recipients who were subsequently exposed to wild-type measles. By the end of 1967, all of the killed measles vaccines were no longer available in the United States. Fortunately, Hilleman believed that Ender's live, attenuated Edmonston strain could be developed into an effective measles vaccine even if it was initially far from perfect as a vaccine candidate. In initial small clinical tests, the Edmonston strain also had too many side effects from fevers to significant skin rashes, yet it was still licensed as a measles vaccine in 1963. Regardless of its problems, Hilleman remained convinced that there was a path forward for the Edmonston strain.

In Hilleman's era, all the other attenuated human virus vaccines had been created using the process of repeated passage through cells that were not the normal target tissue of the virus, typically either animal cells or human cells from an organ the virus didn't naturally infect. As discussed in previous chapters, viruses frequently make replication errors while reproducing, and this results in changes, i.e. mutations, in their genomes. Viral stocks are typically heterogeneous and contain a low level of many mutant variants co-existing along with the original wild-type genome. Forcing a virus to reproduce in foreign cells causes it to adapt by selecting mutants that function well in the new cellular environment. Mutants that reproduce more effectively in the new cell type eventually predominate and the original wild-type virus is lost. If the adapted mutant now has less capacity to replicate in its original target cells then it has become attenuated and may be suitable as a vaccine. Of course, to be clinically useful the attenuated strain must still invoke an immune response that produces protection from the wild-type virus. As you might imagine, there is no guarantee that passaging will ever generate a strain that is both appropriately attenuated and sufficiently immunogenic. Choosing the cell types and the number of passages to use often relied on intuition or convenience (e.g. what cells and tissues a lab had easy access to) rather than any rational basis, and the process was as much art as science in those early decades of vaccine development. Nonetheless, Hilleman was confident that additional passaging could improve the Edmonston strain and generate a variant that had the needed properties of an effective measles vaccine.

Throughout the early and mid-1960s, Hilleman and others tinkered with the Edmonston strain in hopes of creating an improved vaccine strain through further passaging in nonhuman cells. Ultimately, three derivative versions of the Edmonston strain were developed and marketed by different companies: the Schwarz strain, the Edmonston-Zagreb (EZ) strain, and the Moraten ("More attenuated Enders") strain developed by Dr. Hilleman at Merck. Genetic analysis decades later showed that all three strains had numerous mutations compared to wildtype measles virus, although the precise molecular basis for their attenuation remains unclear (Griffin 2018). Surprisingly, the Schwarz and Moraten strains have identical sequences even though they were supposedly derived independently. While each strain possibly acquired identical mutations during passaging, some inadvertent cross-contamination along the way in a lab handling both strains cannot be ruled out. Nonetheless, all three strains were successful at protecting against measles with minimal adverse effects and all have been widely used globally as measles vaccines. With an effective measles vaccine in hand, Dr. Hilleman's focus turned to vaccines for rubella and mumps.

The Measles Endgame

The introduction of the measles vaccine in the 1960s produced an immediate and dramatic reduction in measles cases in every country that mandated this vaccine. Parents and healthcare workers were eager to eliminate this pervasive and potentially dangerous childhood illness. For example, in the United States, there were over 1.3 million measles cases reported from 1960 to 1962. This reported number likely represented only a small fraction of the true cases as most estimates suggest that there were 6–8 times more cases that were not reported. This huge burden of sick children each year, along with hundreds of tragic and needless deaths, created a strong demand for the vaccine. To ensure access to the vaccine, the United States passed the Vaccination Assistance Act in 1965 which provided funding for measles vaccination thus eliminating the financial burden for parents. After the release of the first vaccine in 1963, there was a steady decline in cases each year of the decade. With government funding and improved vaccines becoming available over the subsequent years, the measles immunization campaign was wildly successful. By 1969 there were only around 25,000 measles cases reported, a 95% decline in cases from the peak year of 1962. In a mere six years, measles went from a yearly epidemic affecting hundreds of thousands of children to a rarity. There was an upsurge in cases in the 1970s when the Vaccination Assistance Act ended, but by then many states were beginning to recommend measles vaccination prior to school entry. The vaccine dosage schedule has evolved over the years, but currently, the Centers for Disease Control (CDC) suggests a first dose between 12 and 15 months of age with a second dose between ages 4 and 6. The widespread acceptance of measles vaccination has kept measles immunization rates high, and this herd immunity helps protect even unvaccinated individuals. Over the last decades of the twentieth century measles cases dwindled, this illness faded into obscurity, and in the year 2000 measles was declared officially eliminated in the United States. While there are still occasional measles outbreaks in the United States, there is no longer endemic measles in this country. In every instance, the outbreak can be traced to the importation of measles by someone, often an unvaccinated susceptible person, who has traveled to another country where measles still exists. If this returning infected person encounters other unvaccinated people, measles can quickly spread causing mini epidemics in the susceptible community. It is this inherent threat of measles reintroduction that requires continued vigilance for measles vaccination even though most Americans have never seen and will never see a case of measles. The global reservoir of measles remains a constant source of danger that necessitates continued measles vaccination even in the absence of domestic disease. Should vaccination rates decline significantly, herd immunity would be compromised, giving measles the potential to reestablish itself as an endemic disease in the United States once again.

Measles vaccination has been ongoing for 50 years, but with a highly effective vaccine (97% protection rate with 2 doses and life-long immunity) and a virus with no non-human reservoirs, shouldn't it be possible to eradicate measles globally as was done for smallpox and has nearly been achieved for polio? The answer is unequivocally yes although in practice progress is hindered by many barriers, both scientific and nonscientific (Gastanaduy et al. 2021). Currently, all 195 countries in the world have a childhood measles immunization program, but the effectiveness and extent of those programs vary. The scientific barrier is the extremely high contagiousness of measles (R_0 = 12–18). This immutable high R_0 necessitates a very high level of population immunity, typically 92% or greater, to generate herd immunity. Without herd immunity, measles can remain endemic with continual spread among unprotected individuals. Consequently, very high immunization coverage must be obtained to eliminate measles from a country or region. As we saw for poliovirus, immunization coverage can be limited by social, political, and geographic constraints. Countries without an extensive healthcare infrastructure can have difficulty in providing the vaccine to areas that are remote, impoverished, or involved in armed conflicts. Social or religious concerns about vaccines in general can limit acceptance and political indifference may prevent adequate funding for vaccine programs. Even in countries that have been declared measles-free, declines in vaccine coverage due to public concerns or apathy can allow measles reintroduction and establishment once again as an endemic disease. Nonetheless, there is optimism that we may someday drive the rubeola virus into extinction. Over the last two decades, the global burden of measles has been substantially reduced. Since 2001 there has been a coordinated effort to eradicate measles by the World Health Organization (WHO), the American Red Cross, the United Nations Foundation, the United Nations Children's Fund, and the Centers for Disease Control (CDC). Their Measles & Rubella Initiative (M&RI) has seen measles vaccination rates increase steadily around the world in the twenty-first century and there are now 122 countries with greater than or equal to 90% vaccination coverage, a level that is close to ensuring the elimination of measles from those nations. Still, estimates suggest the annual number of measles infections worldwide is in the millions, so the human race is far from free of this disease. There is no target date yet for measles eradication and we can only hope that more money, effort, and vigilance will prevail over this virus in the not-too-distant future. Sadly though, the trend is not all positive. The COVID pandemic disrupted routine vaccination programs in many countries leading to decreased measles vaccine coverage. On top of the COVID-imposed issues, some countries are facing increasing resistance to vaccination in general which is resulting in lower measles protection. The World Health Organization reported in 2022 that there were a combined nine million measles cases in the world leading to 136,000 needless deaths, so this battle is far from over.

A Measles Puzzle

One lingering mystery about the rubeola virus has been the high rate of secondary infections in measles patients. Many viruses can cause temporary immunosuppression which makes the patient more susceptible to other pathogens, but this property is especially pronounced with measles infection. Unlike many virulent viruses, deaths from rubeola are typically caused by secondary infections rather than directly by the virus. Even more surprisingly, this increased susceptibility persists years after recovery from the measles infection which implies some nefarious effect of rubeola on the immune system. Recent studies shed some mechanistic light on this phenomenon although much remains to be learned (Laksono et al. 2018; Petrova et al. 2019). A key feature of prolonged immunosuppression is the ability of the rubeola virus to bind directly to memory B and T cells of the immune system and infect them. Memory immune cells are generated during any infection with a pathogen and are specific for that pathogen. Memory B cells produce anti-pathogen antibodies while memory T cells recognize and destroy pathogen-infected cells in our bodies. These memory cells persist in the circulation for years afterward and constantly surveil the body searching for any repeat infection by their target pathogen. If their pathogen is encountered, the memory cells proliferate and activate rapidly to attack the pathogen and provide immediate protection that prevents a recurrence of the disease. If you lose the memory cells that recognize a particular pathogen then you will no longer be protected against the pathogen, and it will be as if you had never had that infection before. During measles, the infection of existing memory cells by rubeola triggers the immune system to attack and destroy these virus-containing cells. The elimination of the infected memory cells is critical to recovery from measles, but it permanently removes whatever pathogen protection was supplied by the killed memory cells. Infection with measles literally robs the patient of their immune memory. Years of acquired immunity against numerous pathogens can be destroyed this way leaving the patient highly vulnerable to secondary infections, a condition that persists until new memory cells are slowly generated. Fortunately, this immunosuppression effect does not occur with the measles vaccine strain, so vaccination has no adverse effect on pre-existing immunity. Moreover, since the vaccine prevents measles infection it also protects against this measles-induced loss of acquired immunity, another key incentive for getting the measles vaccine.

Lumps and Mumps

Another once-common childhood disease that has largely been eliminated by vaccination is mumps (scientific name *Mumps orthorubulavirus*). Mumps is an odd word and its origin as a name for this disease is uncertain. It may derive from the old English word mump which meant grimace or sulk, but other origins from similar sounding words in Dutch and Icelandic have been proposed. Like measles, mumps is an RNA virus in the Paramyxovirus family that likely entered human populations from an animal source. Paramyxoviruses infect a large range of animals, and multiple animal species have been proposed as potential hosts of the primordial mumps virus. Perhaps the strongest case is for bats (Drexler et al. 2012) although there is no definitive answer to how mumps originated. Whatever the original animal host, mumps is now strictly a human disease with no animal reservoirs. We do know that both Chinese and Greek writings from around 2500 years ago describe a disease with the characteristic features of mumps, so it has been a human illness for at least that long.

While mumps is generally a much milder disease in children compared to measles and rarely causes any deaths (1 death per 10,000 cases), it is not completely innocuous. The virus is present in the saliva and nasal secretions of infected individuals and is released in aerosol droplets by coughing and sneezing. Inhalation or oral contact with these virus-laden droplets can result in an infection. Mumps has a long incubation period of 1–3 weeks with an average time to symptom development of 16–18 days post-exposure. Around 20% of infected individuals are asymptomatic though they can still be infectious to others. Another 40–50% of patients just have mild respiratory symptoms (coughing, sneezing, congestion) and nonspecific symptoms such as fever, headache, and malaise. Only about 30–40% of patients develop the classic mumps presentation of swollen parotid glands. The parotid glands are salivary glands located between the ears and the jaw. Mumps infection can cause noticeable swelling in these glands resulting in painful, bulging cheeks on one (unilateral) or both sides (bilateral) of the face. These swollen lumps gave children's faces an appearance that was sometimes referred to as chipmunk or hamster cheeks. Swelling in the glands persists from a few days to a week and may cause painful or difficult swallowing. Importantly, patients are infectious to others from 7 to 8 days before symptoms to roughly a week after symptoms begin so transmission can easily occur before the disease is recognized. There is no specific antiviral treatment but children generally only need minimal supportive care before they recover completely. Still, in the pre-vaccine era, nearly all children suffered mumps infections during childhood with the loss of 1–2 weeks of school time and the attendant parental work time. So even without a high mortality rate, the disease in its heyday had a significant societal impact.

Out of the Mouths of Babes

From 1930 to 1960, mumps cases in the United States varied from 50,000 to over 200,000 cases per year with an average of about 150,000 cases each year. With a 10–14 day duration of illness, this amounts to a cumulative average of 1.5–2.1 million sick days per year for Americans, mostly children. This number of sick days alone was justification for vaccine development, but there were also more serious complications that could arise from mumps infection. Mumps wasn't just a localized infection of the parotid glands and respiratory tract. Like many other respiratory viruses, mumps entered the bloodstream, and this viremia disseminated the virus widely throughout the body resulting in clinical manifestations in diverse organs. Surprisingly, the pathogenic mechanisms of this virus at these systemic sites are still poorly understood because no reliable animal experimental model exists. While certain primates and rodents can be artificially inoculated by direct injection with mumps, the resulting infection doesn't adequately reproduce what natural mumps causes in humans. Consequently, the relevance of mumps pathology in these animal models to actual human disease is questionable. Because of this limitation, most of what we know just comes from observing the effects of the infection in humans where the constellation of possible mumps symptoms is well documented even if the underlying pathogenic mechanisms haven't been defined. While less common than the classical presentation of swollen parotid glands, there are a variety of human sequelae associated with mumps infections (Rubin et al. 2015). The most common additional symptoms are painful orchitis (inflammation of the testes) in males and mastitis (inflammation of the breasts) in females. Each of these conditions occurs in about 30% of patients, primarily in post-pubertal individuals and only rarely in younger children. In males with orchitis, typically only one testicle is affected, and this can result in lasting testicular atrophy. Sperm counts may be permanently lowered although this is rarely significant enough to cause male sterility. Less common sequelae of mumps infections include inflammation of the ovaries (oophoritis—5–10% incidence), pancreas (pancreatitis—4% incidence), and central nervous system (meningitis and encephalitis—<1% incidence). The CNS infections are accompanied by headache, fever, vomiting, seizures, and other symptoms typical of brain infections. Additional but even less commonly observed inflammatory responses occur in the joints, kidneys, liver, gallbladder, or eyes. Fortunately, all of these inflammatory manifestations of a mumps infection tend to resolve completely upon recovery without lasting consequences. Sadly, the one complication that does persist is unilateral deafness which occurs in about 4% of patients. The pathogenic mechanism for this virus-induced deafness is unclear, but regardless of the biology, this is a significant permanent dysfunction. When there were 150,000 mumps cases per year in the United States this produced around 6000 cases of partial deafness year after year, a strong incentive for eliminating this disease through vaccine development.

By the 1960s, considerable work had been done on mumps although no effective vaccine had been developed. Thirty years before, Johnson and

Goodpasture had first established the viral nature of the mumps agent by the standard filtration method using saliva from mumps patients (Johnson and Goodpasture 1934). They also showed that the samples collected after filtration caused parotitis when injected into the parotid glands of rhesus monkeys, an observation supporting that they had the authentic agent of mumps. With the onset of World War II, the United States government viewed mumps as a disease likely to spread among young soldiers in close quarters and impede their combat readiness, making mumps a national security issue. As with yellow fever and influenza, government funding poured into mumps research and by the mid-1940s it was possible to isolate and grow the virus in embryonated chicken eggs. Once again, Enders' lab made invaluable contributions by improving the culture methods for this virus, developing a skin test to identify individuals who had already been exposed to mumps, and showing that mumps could be attenuated in culture while still retaining its ability to induce an immune response. There was even an inactivated mumps vaccine developed and licensed in 1948, though it failed to provide long-term protection and was eventually taken off the market. In the post-war decade, interest in a mumps vaccine waned as the disease just wasn't considered a priority public health concern compared to other more serious childhood illnesses like measles and polio. Diseases that caused significant mortality and/or severe disability were the main focus while mumps was considered by the public to be mostly a nuisance. Many parents, along with many healthcare providers, believed that the best way to handle mumps was to get children exposed and over it at an early age. This way they would become immune and would not suffer orchitis or mastitis which were much more common in cases acquired after puberty. However, even with minimal interest in mumps by the public, funding agencies, and private companies, some researchers continued to explore the virus and improve the methods for growing and testing possible vaccine candidates. One of those researchers was Maurice Hilleman whose wide-ranging interests and passionate dedication to vaccine development led him to experiment with mumps alongside numerous other viruses and bacterial agents of disease that his laboratory was studying.

By the late 1950s, Hilleman's group was well into measles vaccine development. Coincident with the measles studies, they realized that a mumps vaccine would be a useful adjunct for reducing cases of this second ubiquitous childhood illness and could possibly be combined into a dual measles/mumps vaccine. Chicken cells, either embryos (i.e. fertilized eggs) or chick embryo cell cultures, were the preferred culture media for virus production as they were cheap, convenient, and scalable for the large quantities needed for commercial vaccine production. However, a new barrier to vaccine development was the discovery that chickens and chicken cell cultures were frequently infected with an endogenous virus that became known as avian leukosis virus (ALV). Thus, any vaccine produced in these cultures would be contaminated with ALV. ALV is a retrovirus, and this family of RNA viruses has a unique lifecycle where the viral RNA is converted to DNA inside the infected cell. After the viral DNA is produced this foreign DNA is integrated into the cell's DNA to become a permanent part of the cell's genetic makeup. This integrated viral DNA (known as a provirus) directs the production of new virus particles that are released and go on to infect other cells. This type of infection is difficult to control as the only way to stop continual viral reproduction and spread is to kill every cell containing a provirus, something that the immune system is rarely able to do. Consequently, once an animal or human is infected with a retrovirus these infections tend to be chronic and life-long. For example, HIV, the cause of AIDS, is a type of retrovirus that causes chronic human infections.

In addition to causing chronic infections in chickens, ALV is an oncogenic virus that can cause cancer in infected birds. Although ALV doesn't infect humans, a cancer-causing virus is worrisome and not an acceptable contaminant in any vaccine even if the known risk is small. Coincidently, a risk assessment experiment had already inadvertently been done with the yellow fever vaccine. Retrospective testing discovered that yellow fever vaccines produced in chicken cells were often contaminated with ALV. These vaccines had been used in humans for years with no adverse effects and were still on the market. While the yellow fever vaccine experience suggested that ALV was not harmful to humans, it still seemed prudent to eliminate this contaminant in future vaccines, if possible, and to remove any risk, no matter how small. The solution to this conundrum came from work at the University of California at Berkeley in the early 1960s where Harry Rubin and collaborators developed a flock of ALV-free chickens (Hughes et al. 1963). Using these birds, Hilleman was able to establish chicken cell cultures free of ALV that were then used as the production cells for multiple different viral vaccines, including measles and mumps.

With ALV-free cultures finally available it was time to attack the mumps vaccine issue directly. The first step for any vaccine is to isolate the virus from a clinical sample and to use this isolate as the starting stock for virus production and attenuation. As a prime example of serendipity in science, in March of 1963 Hilleman's 5-year-old daughter, Jeryl Lynn, developed mumps with the common one-sided parotitis (Hilleman 2011). It's probably not standard parenting, but Hilleman immediately took throat samples and rushed them back to his Merck laboratory where he quickly established the virus in laboratory cultures. Using the customary attenuation approach of culturing human viruses in nonhuman hosts, Hilleman tested different passage strategies for Jeryl Lynn's mumps virus using ALV-free chick embryos and chick embryo cell cultures. By passing the virus stocks from embryo to embryo followed by multiple rounds in cell culture, the hope was that the virus would adapt to the chicken cell environment and lose its pathogenicity for humans while retaining its ability to induce neutralizing antibodies. But now there was a second barrier, the absence of an appropriate animal model of human mumps to test the safety and efficacy of the potential vaccine candidates. In a solution that would be absolutely unethical today, Hilleman did what many of his predecessors and contemporaries in vaccine research did, test his candidate vaccines on mentally handicapped children in state institutions (Stokes et al. 1967). Using the specious argument that he would potentially be protecting them from mumps, candidate vaccines were injected into hapless children to evaluate side effects and induction of an immune response. One candidate generated by a combination of 12 passages in fertilized eggs followed by 5–6 passages in chick embryo cell culture produced a viral stock (Jeryl Lynn B strain) that was highly immunogenic yet caused no adverse reactions. Encouraged by these small-scale early studies, Hilleman performed a series of larger studies using community volunteer families that included thousands of children (including his own younger daughter Kirsten). Typically, one child in the family was vaccinated and the others used as controls to see who got mumps and who didn't over the subsequent 1–2 years. The collective results confirmed that the Jeryl Lynn B strain was not only highly immunogenic but also provided a 95% protection rate from community-acquired mumps with minimal adverse effects, making it an ideal vaccine strain. In March of 1967, almost exactly 4 years after the mumps isolate was obtained from Jeryl Lynn's mouth, the Food and Drug Administration approved Jeryl Lynn B as Mumpsvax. Mumpsvax held the record as the fastest vaccine developed to market until the COVID-19 vaccine which is covered in Chap. 13.

Close But Not Quite for Mumps

The United States and Dr. Hilleman were not alone in their efforts to eliminate mumps. Similar research in other countries developed alternative vaccine strains such as the Urabe strain, the Rubini strain, and the Leningrad Zagreb strain. Many of these other strains are used around the world for mumps vaccination although some are less effective or have more adverse effects than the Jeryl Lynn strain. For example, the incidence of vaccine-associated aseptic meningitis is as high as 1 case per 1000 inoculations with some strains while it is less than 1 in 500,000 doses for Jeryl Lynn (Bonnet et al. 2006). Based on its safety and efficacy, Jeryl Lynn remains a standard component of the MMR vaccine used in the United States.

As with measles, the introduction of the mumps vaccine dramatically reduced cases worldwide. Nationwide vaccination campaigns transformed mumps from a ubiquitous childhood disease to a rarely seen illness in many countries. As Mumpsvax was being rolled out in 1968 the United States had over 150,000 mumps cases that year. A decade later the annual case numbers were below 15,000, an over 90% reduction due to the vaccine. Continued widespread mumps vaccination in the United States as part of the MMR vaccine further reduced the case numbers to only a few hundred per year by the end of the twentieth century, a level that has remained fairly consistent ever since. An illness once so common that it was considered a normal childhood occurrence has become so rare that generations of Americans have never seen a case or suffered through the symptoms. Similar success in decreasing mumps cases has occurred in every country adopting widespread mumps vaccination, greatly reducing the global burden of this virus and preventing much needless childhood morbidity and mortality.

The success of mumps reduction in many nations is remarkable and laudatory, but we are in no way close to eradicating this disease. Without a natural animal host reservoir, complete eradication is theoretically possible as was done for smallpox and is in progress for polio and measles. However, there has never been a sustained global initiative to eradicate mumps, likely due to its perception as a milder and less threatening illness. Since the start of the twenty-first century, the world has averaged nearly 500,000 cases of mumps per year. These case numbers illustrate that this virus is still widespread and persistent in most countries, including the United States which typically has several hundred cases each year. This persistence, even if at a very low level, means that the virus remains endemic and will spread rapidly if vaccination levels decline and herd immunity is lost (herd immunity for mumps requires a population immunization rate of around 90%). Compounding the problem of under-vaccination is the growing evidence of vaccine failure (Gokhale et al. 2023). In most of the mini-outbreaks in the United States over the last 20 years, the cases involved vaccinated people, often college-age individuals (note that some vaccinated individuals had only one dose rather than the recommended two doses). Available evidence suggests that the mumps protection from their childhood MMR immunization has waned enough in some people to render them susceptible again to mumps by the time they reach young adulthood. The close crowding of young people in schools and on college campuses is ideal for a respiratory virus to spread among susceptible individuals which may be why the mumps outbreaks have been primarily in these populations. However, the broader implication is that a significant fraction of adults may be poorly protected from mumps and could be at risk if exposed to the virus in the community. This unfortunate limitation of the vaccine may warrant additional booster doses to restore and maintain immunity. Recommendations haven't been changed yet and more studies will be needed to devise the best public health strategy going forward. Importantly, a vaccine that requires two initial doses plus booster shots downstream makes it much more difficult to ensure vaccine compliance across large populations. Compliance is difficult enough with one-and-done vaccines like polio, and a multi-dose schedule for mumps vaccination will complicate any future efforts to eradicate mumps nationally or globally. It may be that with current vaccines we cannot defeat mumps and the best we can do is keep this disease in check.

Rubella, The German Measles

Following the convention that measles is the "first disease" on the list of childhood rashes with fever, rubella is the "third disease" (the "second disease" is scarlet fever, a bacterial illness). The word rubella is derived from Latin and means "little red", a reference to the minor rash associated with this infection. Historically, rubella infections were misdiagnosed as mild cases of measles (rubeola) until the early 1800s when German physicians began to study it and clarified that it was a separate disease from true measles. Because it was first described as a new disease in the German medical literature it became known as German measles, and that name has persisted even though this is a worldwide disease with no specific clinical or biological connection to Germany. Like the other components in the MMR vaccine, rubella is also an RNA virus, but in a different family than measles and mumps. The scientific name for rubella is *Rubivirus rubella*, formerly assigned in the Togavirus family but reclassified in 2018 into a new family called *Matonaviridae*. Rubella was an orphan with no close relatives until the discovery in 2020 of two related animal viruses that are now included in the Matonavirus family (Bennett et al. 2020). Neither of rubella's relatives is known to infect humans, although one relative does infect a variety of animal species, indicating that cross-species transmission may be a feature of some members of this family. Almost nothing is known about rubella's origin or when it arose in human populations, but the existence of related viruses in animals is consistent with a possible zoonotic origin for rubella. Like many other human viruses, an ancestral animal virus may have jumped species and entered humans where it eventually evolved into modern-day rubella.

Of the three diseases prevented by the MMR vaccine, rubella is by far the mildest illness. Clinically, rubella is a respiratory disease spread primarily by coughing and sneezing. It can also be picked up on the hands from infected surfaces (where it survives for about 2 hours) and then be introduced into the mouth or nose (and possibly the eyes) to initiate an infection. There is an incubation period of 2–3 weeks during which there is a systemic spread of the virus as it is distributed throughout the body via the bloodstream. Importantly, 25–50% of infected individuals are asymptomatic. While they don't develop clinical signs of disease, they can produce and shed viruses making them a significant and hard-to-identify source of viral spread in families and communities. When symptoms do occur they typically present first as a low fever (less than 102 degrees Fahrenheit) followed within a few days by a flat to slightly raised, reddish, itchy rash that begins on the face and spreads downward. The rash can be sporadic to very extensive and may persist from a few hours to several days. Similar to measles, the rash is an immune response to the viral infection and not a site of viral reproduction so the rash itself is not infectious. Accompanying the rash and fever are other symptoms that can include headaches, sore throat, joint aches, and reddened eyes. In some patients, there can be swelling in the lymph nodes of the neck or behind the ears which can be painful. Regardless of which particular constellation of symptoms presents, children generally recover in 3–7 days with no lasting effects. Complications are rare and death is extremely rare so this is considered a fairly harmless childhood illness. Other than a few days of missed school, rubella just isn't very troublesome in children.

An Unsuspected Danger

From the clinical description of acute cases, you might imagine that rubella is an innocuous virus that couldn't possibly cause much misery and controversy, but the reality is just the opposite. Rubella has one property that made its elimination a national priority, a predilection for infecting embryonic and fetal cells and a propensity for causing massive destruction of these infected tissues. Maternal infection during pregnancy allows the virus to cross the placenta, attack the developing cells, and act as a teratogen (a factor that causes malformation of an embryo). This in-utero invasion can result in embryonic death or horrible physical abnormalities and/or neurological impairments ranging from autism to profound intellectual impairment. This constellation of physical and intellectual defects is collectively known as congenital rubella syndrome (CRS) (Gudeloglu et al. 2023). Functionally, CRS results when the virus infects embryonic and fetal tissues (the developing child is an embryo from conception through week 10 of gestation and is a fetus after that) causing both the cessation of cell growth and cell death (George et al. 2019). The result is widespread and permanent damage as critical cells and tissues are destroyed by the virus. Losing embryonic cells early in pregnancy is the most harmful as the loss of certain cells can prevent the production of whole lineages of progeny cells that should have arisen from the progenitor cell. Consequently, the earlier in pregnancy that the infection occurs, the greater the risk of serious consequences for the resulting child. Most estimates suggest that infection during the first trimester has a high incidence of fetal death (stillbirths and spontaneous abortions) and results in 90% of the surviving children having major physical and/or cognitive deficits. Fortunately, the risk drops to 25% for second-trimester infections and even much lower for third-trimester infections. However, children infected at any time in utero may be born infected and continue to shed rubella virus for up to 12 months, making them a source of infection for those around them. Once these devastating effects on pregnancy were established, rubella immediately went from an inconsequential and mostly ignored childhood illness to a major public health concern.

The discovery that rubella is a teratogenic virus is another case of remarkable scientific serendipity. While German measles was recognized as a distinct disease entity in the early 1800s, its role in birth defects wouldn't be discovered until the 1940s following a chance observation Down Under. Australia entered World War II in 1939 and began enlisting and training thousands of young men to be soldiers. In 1940 and 41, a German measles outbreak occurred, often causing numerous cases among the closely quartered young trainees. As the soldiers went on leave they spread the virus widely into the civilian population resulting in a nationwide epidemic. Norman Gregg was a 48-year-old pediatric ophthalmologist practicing in Sydney. A decorated World War I veteran, he was picking up many new patients as younger doctors enlisted and left to join the war effort. One day there were three young mothers in his waiting room with newborns suffering from congenital cataracts. Coincidentally, as they waited and chatted, the mothers shared that they each had German measles early in their pregnancies. We now know that congenital cataracts are a common manifestation of CRS, but otherwise, they are a fairly rare symptom in neonates (less than 1 case in 1000 births). In that era, congenital cataracts were believed to be a genetic problem and the concept of an infectious etiology had not been seriously considered, so Dr. Gregg could easily have ignored this unusual situation of three cases in one day. However, he was intrigued by the potential connection with German measles and began scouring Australia's recent medical records for other congenital cataract cases. He quickly discovered 78 cases with 68 having a known German measles case or exposure. While such retrospective studies are never proof of a cause and effect, he strongly felt that this potential connection needed to be reported to alert others of the danger and stimulate further research; his very detailed observations were published in 1941 (Gregg 1941). Like many reports that challenge accepted scientific dogma, his paper was mostly ignored and dismissed initially but it did have the intended effect. Other mothers with afflicted children began to come forward with stories of German measles during pregnancy, including several with deaf children, another frequent clinical issue arising from in-utero rubella infection. Over the following decades additional sequelae associated with embryonic rubella infections were slowly identified and the accumulating evidence confirmed that rubella was the cause of these various birth defects. Eventually, the serious danger of this virus during pregnancy was widely recognized. It also became clear that women who previously had rubella before pregnancy were immune and not at risk for contracting the disease during pregnancy and passing the virus to the gestating embryo or fetus. Since a prior infection eliminated the possibility of CRS, the practice of "rubella parties" became a social phenomenon in the 1950s and 60s. The strategy was to reduce the chance of a girl reaching childbearing age without having had a rubella infection by intentionally exposing her to rubella. If any child in the neighborhood came down with German measles, some mothers would send their daughters to interact with the sick child in hopes that the girls would become infected and generate immunity, in effect "natural immunization". For a disease as mild as rubella this was an acceptable strategy though the practice was not widespread enough or effective enough to generate herd immunity, leaving a highly vulnerable population at the mercy of the coming epidemic.

A Fateful Epidemic

While rubella is an endemic disease that circulates constantly around the globe, it was once prone to periodic epidemics at fairly regular intervals. In the United States, rubella flared every 6–8 years causing a significant rise in cases. In the early 1960s rubella cases rose in Europe and then spread to the United States causing a major national epidemic from 1964 to 1965. The

Centers for Disease Control (CDC) estimates that 12.5 million people in the country contracted rubella with as many as 50,000 pregnant women becoming infected. The results were devastating. There were over 10,000 miscarriages attributed to rubella, several thousand stillbirths, and 20–25,000 cases of CRS. Some estimates suggest there were roughly 5000 therapeutic abortions performed during that epidemic even though abortion was largely illegal in the United States. Some historians credit the massive national distress and the misery inflicted by rubella during that epidemic as helping change the public attitude towards abortion. Along with women's liberation and the changing sexual mores of the 1960s, the CRS epidemic brought abortion from the back alleys to the national consciousness. Watching mothers dealing with horribly damaged infants invoked a more sympathetic and accepting attitude toward women who wanted a choice about whether or not to carry CRS babies. Long before *Roe v. Wade*, in response to this national tragedy, many states began to legalize abortion to protect the mother's life or when the fetus had profound mental and/or physical defects. Ultimately the rubella vaccine all but eliminated CRS in the United States, drastically reducing this particular need for abortions.

A Culture of Excellence

With all viruses, the trick is finding some way to detect this invisible entity. We can't see viruses, except by electron microscopy and similar cumbersome techniques, so they are typically assayed by their effects on the host. Early virologists often used animal infection as the readout. Samples collected from patients were injected into animals to see if characteristic symptoms developed. Samples could be collected and passed from animal to animal to confirm transmission and generate larger amounts of viral stocks. While feasible, this method required a susceptible animal host, could be very slow as symptoms could take days to weeks to develop, and was somewhat dangerous when handling infected animals. Chicken eggs were a much better alternative host as they were cheap, plentiful, easy to handle, and usually produced signs of infection more quickly than animals. Unfortunately, not all viruses reproduce effectively in eggs so cultured cells provided the third viral assay system. As developed by John Enders, patient samples were incubated with cultured cells. If infection occurred then viral reproduction typically caused the cells to change shape or become damaged looking. These so-called cytopathic effects (CPE) were visible via simple light microscopy and could be easily assessed.

Such CPE often happens within a few days so the presence or absence of a virus could be rapidly determined.

The acceptance of rubella as a dangerous teratogenic virus in the 1950s changed our perception of this virus and converted it from an innocuous childhood illness into a major public health threat worthy of vaccine development. The only problem was that the rubella virus caused little or no CPE in the various tested cell cultures. Without a convenient assay, this virus was difficult to work with and progress was frustratingly slow although multiple groups would eventually isolate rubella from different patients. One of the first to succeed was Thomas Weller, the Nobel recipient along with John Enders and Frederick Robbins for the development of cell culture. Using a urine sample from his 10-year-old son, he cultured the sample on human embryonic cells and was able to detect subtle CPE indicative of viral infection (Weller and Neva 1962). However, a more effective assay that became the standard was developed by Paul Parkman and Mal Artenstein at the Walter Reed Army Institute for Research (Parkman et al. 1962). In another example of scientific serendipity, these two investigators encountered rubella cases while studying adenovirus infections among young army recruits at Fort Dix in New Jersey. Struggling to isolate this virus that exhibited such poor CPE they developed a surrogate test that has become known as an interference assay. When viruses infect cells they trigger the production and release of interferon. You can think of interferon as a burglar alarm that warns the uninfected cells that a virus is in the neighborhood. The uninfected cells respond to interferon by expressing antiviral proteins that protect the cells from subsequent infection, just as a burglar alarm warns the neighbors and allows them time to assemble their defenses against possible robbers. This protection that is induced by interferon is not specific to rubella and is a general state that protects against almost any invading virus. Utilizing this concept, Park and Artenstein rationalized that if rubella was present in their clinical samples, it would induce interferon, create an antiviral state in the surrounding cells, and protect those cells from a challenge infection with a known virus that exhibited readily detectable CPE. For their study, they used African green monkey kidney (AGMK) cells and a highly cytopathic test virus called ECHO 11. Normally, ECHO 11 infection would destroy the AGMK cells, but if rubella was added to the cells first, the production of interferon should interfere with ECHO 11 infection and reduce or eliminate CPE. The interference test worked to perfection and soon became the standard for assaying the rubella virus.

Once it was possible to isolate and assay the rubella virus, the race for a vaccine began in earnest. Parkman and co-workers repeatedly passed their rubella isolate in AGMK cells and after 77 passages they generated an attenuated strain that they called high passage virus 77 (HPV-77). Multiple vaccine manufacturers similarly pursued their proprietary rubella isolates, including the prolific Maurice Hilleman at Merck who had an isolate called the Benoit strain. Typically, each isolate was passaged in animal cell culture to obtain an attenuate version. The exception was Stanley Plotkin at the Wistar Institute who instead of animal cells explored the use of human embryonic cells as a host for culturing human viruses. Human cells isolated from different types of embryonic tissues were collectively known as human diploid cell strains (HDCS). HDCS cells were initially controversial because of the fear that they might harbor unknown human viruses (a legitimate concern) although the same problem existed with animal cells as we have already seen for the polio vaccine (SV40 contamination—Chap. 6) and the avian leukosis virus contamination problems with the mumps vaccine. Ultimately, Plotkin utilized an HDCS cell line known as WI-38 (WI stood for Wistar Institute) and a rubella sample obtained after 3 passages from the 27th rubella aborted fetus sample submitted to their lab, an isolate they designated RA27/3. To attenuate RA27/3, Plotkin continued passaging it in HDCS cells but also added another wrinkle, cold adaptation. Human viruses typically reproduce best at human body temperature which is 37 degrees centigrade. Work with poliovirus showed that growing the virus at 30 °C selected for cold-adapted mutants whose replication was now optimized for this lower temperature. These cold-adapted viruses often replicate poorly at 37 °C, and because they reproduce well at lower temperatures but not at higher temperatures, they are called temperature-sensitive (ts) mutants. The ts mutants generally grow poorly in the human body and are often attenuated enough to be suitable as vaccines. Plotkin utilized this ts approach and converted RA27/3 into a highly attenuated strain that retained excellent immunogenicity. After successful clinical trials in Europe, RA27/3 was licensed in 1970 in the United Kingdom and became the widely used rubella vaccine in Europe.

While Plotkin was pioneering HDCS as a vaccine production platform, resistance to HDCS cells for vaccines was still prevalent in the United States so the pharmaceutical companies elected to utilize animal cell platforms instead. During this early period in rubella vaccine development, Paul Parkman moved from Walter Reed to the Food and Drug Administration (FDA) and continued his work with HPV-77. Being at the FDA gave HPV-77 the imprimatur as the preferred strain most likely to get quick FDA approval, so it became the default strain used by multiple vaccine manufacturers. Most rubella vaccine researchers, including Hilleman at Merck, gave up their own isolates and switched to the freely available HPV-77 strain. To develop their unique vaccine products, each company passaged HPV-77 further by switching from AGMK cells to some other type of animal cell. For example, Merck chose duck embryo fibroblasts and Philips-Roxane used dog kidney cells. By the late 1960s, three companies had licensed and released rubella vaccines in the United States, and large-scale vaccine programs were underway in hopes of eliminating congenital rubella syndrome (CRS) by creating a highly vaccinated population. In 1971, rubella was combined with measles and mumps vaccines into the trivalent MMR vaccine to reduce the number of shots children needed to receive and helped ensure high levels of community protection against all three viral infections.

The rubella vaccines were an immediate success, and in the first decade after their introduction, both the cases of rubella and the number of children born with CRS declined dramatically. However, accumulating data from the vaccinated population in the early 1970s indicated that the HPV-77-based vaccines were not ideal. The dog kidney-derived vaccine had high levels of adverse reactions and all versions showed less protection than a natural infection with rubella, with some vaccinees still being susceptible to infection. In contrast, the results in Europe with the RA27/3 vaccine produced in WI-38 cells showed none of these problems, and eventually, Plotkin's vaccine prevailed. In 1979 the RA27/3 vaccine was licensed in the United States to replace the HPV-77 vaccines, and RA27/3 is now the predominant rubella vaccine used around the world. Similarly, Plotkin's espousal of HDCS cultures as an excellent platform for vaccine production slowly gained acceptance, and HDCS cultures have become widely accepted for this purpose. While none of these three viruses in the MMR vaccine has yet been eliminated, most countries with effective vaccination programs have virtually no endemic cases and typically only see cases imported from areas where these viruses are still more prevalent. Multiple generations of Americans have never experienced measles, mumps, or rubella or even seen cases of these illnesses. The current paucity of cases makes it hard for people to understand that only a few decades ago all three viruses were rampant and caused countless hours of sickness along with many tragic deaths and disabilities. The efforts of pioneers like Enders, Hilleman, and Plotkin are mostly forgotten by the public, but their vaccine legacies remain and continue to protect generation after generation of children.

References

Bennett AJ, Paskey AC, Ebinger A, Pfaff F, Priemer G, Hoper D, Breithaupt A, Heuser E, Ulrich RG, Kuhn JH, Bishop-Lilly KA, Beer M, Goldberg TL (2020) Relatives of rubella virus in diverse mammals. Nature 586(7829):424–428. https://doi.org/10.1038/s41586-020-2812-9

Bonnet MC, Dutta A, Weinberger C, Plotkin SA (2006) Mumps vaccine virus strains and aseptic meningitis. Vaccine 24(49–50):7037–7045. https://doi.org/10.1016/j.vaccine.2006.06.049

Deer B (2011) How the case against the MMR vaccine was fixed. BMJ 342:c5347. https://doi.org/10.1136/bmj.c5347

DeStefano F, Shimabukuro TT (2019) The MMR vaccine and autism. Annu Rev Virol 6(6):585–600. https://doi.org/10.1146/annurev-virology-092818-015515

Drexler JF, Corman VM, Muller MA, Maganga GD, Vallo P, Binger T, Gloza-Rausch F, Cottontail VM, Rasche A, Yordanov S, Seebens A, Knornschild M, Oppong S, Adu Sarkodie Y, Pongombo C, Lukashev AN, Schmidt-Chanasit J, Stocker A, Carneiro AJ, Erbar S, Maisner A, Fronhoffs F, Buettner R, Kalko EK, Kruppa T, Franke CR, Kallies R, Yandoko ER, Herrler G, Reusken C, Hassanin A, Kruger DH, Matthee S, Ulrich RG, Leroy EM, Drosten C (2012) Bats host major mammalian paramyxoviruses. Nat Commun 3:796. https://doi.org/10.1038/ncomms1796

Dux A, Lequime S, Patrono LV, Vrancken B, Boral S, Gogarten JF, Hilbig A, Horst D, Merkel K, Prepoint B, Santibanez S, Schlotterbeck J, Suchard MA, Ulrich M, Widulin N, Mankertz A, Leendertz FH, Harper K, Schnalke T, Lemey P, Calvignac-Spencer S (2020) Measles virus and rinderpest virus divergence dated to the sixth century BCE. Science 368(6497):1367–1370. https://doi.org/10.1126/science.aba9411

Enders JF (1957) The future of virus studies in tissue culture. J Natl Cancer Inst 19(4):735–743; discussion 744–752

Enders JF, Peebles TC, McCarthy K, Milovanovic M, Mitus A, Holloway A (1957) Measles virus: a summary of experiments concerned with isolation, properties, and behavior. Am J Public Health Nations Health 47(3):275–282. https://doi.org/10.2105/ajph.47.3.275

Flores J, Immergluck LC (2023) Measles update in an era of vaccine hesitancy and global pandemic. Pediatr Rev 44(9):529–532. https://doi.org/10.1542/pir.2022-005921

Gastanaduy PA, Goodson JL, Panagiotakopoulos L, Rota PA, Orenstein WA, Patel M (2021) Measles in the 21st century: progress toward achieving and sustaining elimination. J Infect Dis 224(12 Suppl 2):S420–S428. https://doi.org/10.1093/infdis/jiaa793

George S, Viswanathan R, Sapkal GN (2019) Molecular aspects of the teratogenesis of rubella virus. Biol Res 52(1):47. https://doi.org/10.1186/s40659-019-0254-3

Godlee F, Smith J, Marcovitch H (2011) Wakefield's article linking MMR vaccine and autism was fraudulent. BMJ 342:c7452. https://doi.org/10.1136/bmj.c7452

Gokhale DV, Brett TS, He B, King AA, Rohani P (2023) Disentangling the causes of mumps reemergence in the United States. Proc Natl Acad Sci USA 120(3):e2207595120. https://doi.org/10.1073/pnas.2207595120

Gregg NM (1941) Congenital cataract following German measles in the mother. Trans Ophthalmol Soc Aust 3:35–46

Griffin DE (2018) Measles vaccine. Viral Immunol 31(2):86–95. https://doi.org/10.1089/vim.2017.0143

Griffin DE, Oldstone MBA (2009) Measles: history and basic biology. Current topics in microbiology and immunology, vol 329. Springer, Berlin

Gudeloglu E, Akillioglu M, Bedir Demirdag T, Unal NA, Tapisiz AA (2023) Congenital Rubella syndrome: a short report and literature review. Trop Dr 53(1):171–175. https://doi.org/10.1177/00494755221134327

Hilleman MR (1999) Personal historical chronicle of six decades of basic and applied research in virology, immunology, and vaccinology. Immunol Rev 170:7–27. https://doi.org/10.1111/j.1600-065x.1999.tb01325.x

Hilleman MR (2011) The development of live attenuated mumps virus vaccine in historic perspective and its role in the evolution of combined measles–mumps–rubella. In: Plotkin SA (ed) History of vaccine development. Springer, New York, pp 207–218. https://doi.org/10.1007/978-1-4419-1339-5_23

Hinman AR, Orenstein WA, Bloch AB, Bart KJ, Eddins DL, Amler RW, Kirby CD (1983) Impact of measles in the United States. Rev Infect Dis 5(3):439–444. https://doi.org/10.1093/clinids/5.3.439

Horton R (2004) A statement by the editors of The Lancet. Lancet 363(9411):820–821. https://doi.org/10.1016/S0140-6736(04)15699-7

Hubschen JM, Gouandjika-Vasilache I, Dina J (2022) Measles. Lancet 399(10325):678–690. https://doi.org/10.1016/S0140-6736(21)02004-3

Hughes WF, Rubin H, Watanabe DH (1963) Development of a chicken flock apparently free of leukosis virus. Avian Dis 7(2):154. https://doi.org/10.2307/1588044

Institute of Medicine (US) Immunization Safety Review Committee (2001) In: Stratton K, Gable A, Shetty P, McCormick M (eds) Immunization safety review: measles-mumps-rubella vaccine and autism. National Academies Press, Washington (DC). https://doi.org/10.17226/10101

Johnson CD, Goodpasture EW (1934) An investigation of the etiology of mumps. J Exp Med 59(1):1–19. https://doi.org/10.1084/jem.59.1.1

Laksono BM, de Vries RD, Verburgh RJ, Visser EG, de Jong A, Fraaij PLA, Ruijs WLM, Nieuwenhuijse DF, van den Ham HJ, Koopmans MPG, van Zelm MC, Osterhaus A, de Swart RL (2018) Studies into the mechanism of measles-associated immune suppression during a measles outbreak in The Netherlands. Nat Commun 9(1):4944. https://doi.org/10.1038/s41467-018-07515-0

Madsen KM, Hviid A, Vestergaard M, Schendel D, Wohlfahrt J, Thorsen P, Olsen J, Melbye M (2002) A population-based study of measles, mumps, and rubella vaccination and autism. N Engl J Med 347(19):1477–1482. https://doi.org/10.1056/NEJMoa021134

Parkman PD, Buescher EL, Artenstein MS (1962) Recovery of rubella virus from army recruits. Proc Soc Exp Biol Med 111:225–230. https://doi.org/10.3181/00379727-111-27750

Perry RT, Halsey NA (2004) The clinical significance of measles: a review. J Infect Dis 189(Suppl 1):S4–S16. https://doi.org/10.1086/377712

Petrova VN, Sawatsky B, Han AX, Laksono BM, Walz L, Parker E, Pieper K, Anderson CA, de Vries RD, Lanzavecchia A, Kellam P, von Messling V, de Swart RL, Russell CA (2019) Incomplete genetic reconstitution of B cell pools contributes to prolonged immunosuppression after measles. Sci Immunol 4(41). https://doi.org/10.1126/sciimmunol.aay6125

Rivers TM (1927) Filterable viruses a critical review. J Bacteriol 14(4):217–258. https://doi.org/10.1128/jb.14.4.217-258.1927

Rubin S, Eckhaus M, Rennick LJ, Bamford CG, Duprex WP (2015) Molecular biology, pathogenesis and pathology of mumps virus. J Pathol 235(2):242–252. https://doi.org/10.1002/path.4445

Stokes J Jr, Weibel RE, Buynak EB, Hilleman MR (1967) Live attenuated mumps virus vaccine. II Early clinical studies. Pediatrics 39(3):363–371

Wakefield AJ (1999) MMR vaccination and autism. Lancet 354(9182):949–950. https://doi.org/10.1016/S0140-6736(05)75696-8

Wakefield AJ (2010) Callous disregard: autism and vaccines -- the truth behind a tragedy. Skyhorse Pub, New York

Wakefield AJ, Murch SH, Anthony A, Linnell J, Casson DM, Malik M, Berelowitz M, Dhillon AP, Thomson MA, Harvey P, Valentine A, Davies SE, Walker-Smith JA (1998) Ileal-lymphoid-nodular hyperplasia, non-specific colitis, and pervasive developmental disorder in children. Lancet 351(9103):637–641. https://doi.org/10.1016/s0140-6736(97)11096-0

Weller TH, Neva FA (1962) Propagation in tissue culture of cytopathic agents from patients with rubella-like illness. Proc Soc Exp Biol Med 111(1):215

8

Hepatitis B Virus: Blood, Sex, and Drugs

Keywords Australia antigen • Baruch (Barry) Blumberg • Chronic infections • Dane particle • Hepatitis B surface antigen (HBsAg) • Hepatocellular carcinoma • Latent infections • Persistent viruses • Reverse transcription • Subunit vaccine • Virus-like particle (VLP)

Abbreviations

ACIP	Advisory Committee for Immunization Practices
ALT	Alanine aminotransferase
AST	Aspartate aminotransferase
AU	Australia antigen
BCE	Before Common Era
CDC	Centers for Disease Control
GSHV	Ground squirrel hepatitis B virus
HAV	Hepatitis A virus
HBc	Hepatitis B virus core protein
HBsAg	Hepatitis B virus surface antigen
HBV	Hepatitis B virus
HBx	Hepatitis B virus X protein
HCC	Hepatocellular carcinoma
HCV	Hepatitis C virus
HPV	Human papillomavirus
IARC	International Agency for Research on Cancer
IV	Intravenous
NANB	NonA, NonB hepatitis
VLP	Virus-like particle
WHV	Woodchuck hepatitis virus

A Persistent Enemy

The viruses covered in the previous chapters all caused acute infection. In virological parlance, an acute infectious agent enters the host and the infection has only one of two outcomes. Either the infected individual mounts an effective immune response that allows them to recover and completely eliminate the virus from the body or they die. Unfortunately, many viruses are much more insidious and can cause persistent, chronic infections. For persistent viruses, after recovery from the initial infection, instead of being eliminated, these viruses circumvent the immune defenses and remain in the host indefinitely. Since these affected individuals are usually healthy and symptom-free, they are often unaware that they now have a permanent infection. For some viruses, the persistent virus is latent and lurks in the body with little or no replication although it may be triggered to reproduce months to years later. While the virus is latent these carriers have no symptoms and are not infectious. However, when reactivated, these latent viruses can give rise to acute disease again and can be spread to other individuals (see Chap. 10—Chickenpox). In contrast to latent viruses, some persistent viruses are chronic and continue to reproduce in the afflicted individual even though there may be no obvious signs of disease. The continual production of new viruses makes these carriers infectious and a public health risk as they can unknowingly pass the virus on to unsuspecting contacts. In addition, some chronic viral infections slowly cause damage that accumulates and can lead to disease and death decades after the initial infection. Hepatitis B virus is one of the chronic viruses, and lifelong infections are associated with liver damage, including a high incidence of liver cirrhosis and liver cancer. The discovery of this virus and its conquest through vaccine development is a fascinating saga encompassing both triumphs and tragedies as well as introducing a new paradigm for viral vaccines, the subunit vaccine.

Hepatitis B virus (HBV) is an ancient human pathogen, but its history is murky with at least 5 theories proposed to explain its origins and entry into human populations (Littlejohn et al. 2016). The theories range from co-evolution with primates starting 10–35 million years ago to a much more recent entry into humans somewhere between 20 and 50,000 years ago. Unfortunately, the genetics of this virus are complex and don't readily produce easily interpretable phylogenetic trees reaching back beyond 10–15,000 years (Muhlemann et al. 2018). The genetic complexities are compounded by the fact that HBV can infect non-human primates such as chimpanzees, gorillas, orangutans, and gibbons. Cross-species transmission over the millennia likely occurred, leading to recombinational events between human and primate hepatitis viruses which introduces more confounding heterogeneity. The combination of these issues makes deciphering the ancient origins of HBV difficult and not yet resolved in scientific circles. But at least we know that the virus was present and prevalent in human populations as early as the Bronze Age (3000 BCE to 1000 BCE), so this pathogen has a long history of causing human disease and misery. Even today when an effective vaccine is available, the virus remains endemic in some parts of the world with close to three million people having chronic HBV infections and roughly 800,000 victims dying each year.

As a virus, HBV is quite small with a DNA genome of only 3200 base pairs. (The DNA bases are A, C, G, and T, and these four "letters" make up the alphabet of DNA. The A bases pair with T bases and the C bases pair with G bases to make the double-stranded DNA.) Compared to HBV, the smallpox virus (Chap. 1) has a DNA genome of over 186,000 base pairs, nearly 60 times larger. Even the RNA viruses in Chaps. 3, 4, 5, 6, and 7 have genomes two (poliovirus) to five (measles and mumps) times larger than HBV. With its tiny genome, HBV has limited protein-coding capacity and only makes seven proteins. Yet this small size is deceptive as HBV is dangerous and can cause serious acute disease (hepatitis) as well as imperiling chronic infections. HBV is also unique in its genome replication strategy (Fig. 8.1). All other human DNA viruses utilize replication enzymes called DNA polymerases which directly copy the parental virus DNA (the template) into new DNA copies (the progeny), i.e. they replicated directly from DNA to DNA. In contrast, HBV uses a more circuitous process involving an RNA intermediate. Upon infection, HBV uses a host cell enzyme, RNA polymerase (the enzyme we use to make messenger RNAs from our genes), to copy one strand of its DNA into a complementary single-strand RNA version. Then a viral encoded polymerase (one of the seven HBV proteins) uses the RNA strand as a template and copies it back into a new single-stranded DNA. This new DNA strand is itself copied to generate the final double-stranded progeny genome, so the overall process is DNA to RNA to DNA. The RNA-to-DNA step is known as reverse transcription. No other human DNA virus utilizes this reverse transcription approach to replicate its genome, but this is just one of many ways that HBV is a unique human pathogen.

1. DNA to DNA Replication

2. DNA to RNA to DNA Replication

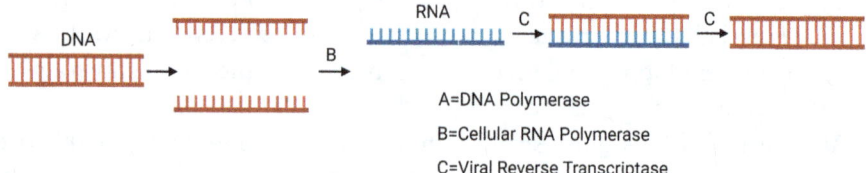

A=DNA Polymerase
B=Cellular RNA Polymerase
C=Viral Reverse Transcriptase

Fig. 8.1 Hepatitis B genome replicationPanel 1 shows the typical replication scheme for double-stranded DNA. The original DNA is separated into single strands and then each single strand is copied by a DNA polymerase (A) to regenerate two double-stranded copies. Panel 2 shows the replication process for the hepatitis B virus. One of the viral DNA strands is copied to a single-stranded RNA by the cellular RNA polymerase (B). This RNA is packaged into the assembling viral particle and within the nascent virion the RNA strand is copied into DNA by the viral reverse transcriptase (RT) enzyme (C). Subsequently, the RNA strand is degraded, and the remaining single-stranded DNA is copied by the RT to generate the double-stranded progeny DNA that is the complete viral genome

A Tale of Two Viruses

The discovery and identification of HBV is a convoluted tale involving the confluence of clinical medicine, basic research, and chance observation. On the clinical side, hepatitis simply means inflammation of the liver, a condition that can be caused by many factors including chemicals, drugs, immune reactions, and infection by bacteria or viruses. Patients typically present with abdominal pain, tiredness, and gastrointestinal symptoms such as vomiting, diarrhea, and poor appetite. Some individuals develop yellowing (jaundice) of the skin and the whites of the eyes when the damaged liver cannot effectively remove bilirubin (a yellowish compound generated during the breakdown of old red blood cells) from the body. This constellation of symptoms was observed as far back as the writing of Hippocrates over 2000 years ago so there is a long history of clinical observations about hepatitis. However, the symptoms don't discriminate between possible causes, and the underlying agents responsible for many hepatitis cases, including HBV, eluded identification until the latter half of the twentieth century.

Even before the specific agents were identified, clues were accumulating that infectious organisms were an important cause of most hepatitis cases. There is a history stretching back to the 1600s of hepatitis outbreaks associated with military campaigns (Feinstone 2019). The close crowding of troops, the lack of hygiene, and the poor sanitary conditions of food and water seemed to favor the development of hepatitis, conditions that we now know can facilitate the person-to-person spread of bacteria and viruses. Likewise, observations as far back as the late 1800s described outbreaks of hepatitis associated with smallpox vaccination campaigns, suggesting that some adventitious agent was present in the vaccine material. Remember, that these early smallpox vaccines were crude and sometimes made from human lesions that could easily be contaminated with other infectious agents present in the blood or fluids of the vaccine source person. Similar vaccine-associated hepatitis outbreaks occurred in the early 1900s, primarily with the yellow fever vaccine. Human serum was typically added to the yellow fever virus preparations as a "stabilizing" material, and with no direct assay for a hepatitis agent, there was no way to ascertain if the serum did or did not contain other infectious organisms. However, since the serum was filtered to remove cells and bacteria, the unknown agent was presumed to be a virus. This yellow fever vaccine problem became critical at the beginning of World War II. To prepare American troops for the South Pacific, mass yellow fever vaccination was undertaken in early 1942, and 50,000 acute cases of hepatitis occurred among the soldiers (Seeff et al. 1987). Jaundice and hepatitis cases were also observed in soldiers receiving serum or blood transfusions, suggesting an infectious agent present in the blood (Grossman et al. 1945). Likewise, hepatitis outbreaks were seen in vaccinated British troops leading to the work of British physician Frederick O. MacCallum. MacCallum already had an interest in hepatitis having published an earlier report of a vaccine-associated hepatitis outbreak (Findlay and Maccallum 1938). Studying the reports of hepatitis cases and outbreaks, as well as using volunteers for experimental infection through various routes (MacCallum 1972b), MacCallum ultimately concluded that there were two forms of viral hepatitis transmission which he called hepatitis A and hepatitis B. Hepatitis A transmission was primarily through fecally contaminated food or water but could also occur through person-to-person contact such as living in close quarters. Since this form could be spread fairly casually in the community, it was designated infectious hepatitis. In contrast, hepatitis B was transmitted via blood or bodily fluids. Since exposure to blood or bodily fluid required much more intimate contact it was designated serum hepatitis to distinguish it from type A. These distinct differences in transmission routes were initially confounding, but we now know that hepatitis A and hepatitis B are actually two different diseases caused by separate viruses. Hepatitis A (Chap. 9) is caused by an RNA virus (HAV) of the picornavirus family (same family as poliovirus) while hepatitis B (HBV) is a DNA virus in the hepadnavirus family. While completely unrelated, both viruses target the liver and cause clinically similar acute hepatitis. Parenthetically, once diagnostic assays were developed for both HAV and HBV, some cases of viral hepatitis didn't seem to be caused by either of these viruses. For many years these unclassified viral hepatitis cases were simply called nonA, nonB (NANB). In 1989 the NANB culprit virus was finally discovered and named hepatitis C virus (HCV) (Choo et al. 1989); this agent turned out to be a novel RNA virus unrelated to either HAV or HBV. To date, no vaccine has been developed against HCV but it will be explored in Chap. 15.

Clarifying the epidemiology of hepatitis A versus hepatitis B was clinically useful and immediately suggested preventative public measures such as maintaining clean food and water supplies and avoiding clinical procedures that allowed the person-to-person exchange of blood and fluids. While simple in theory, implementing these practices remained challenging with no way to detect the viruses or screen samples. Especially in clinical situations, avoidance of blood, fluids, and blood products by doctors, nurses, dentists, and various healthcare personnel, including first responders, was challenging. Modern universal precautions like gloves, masks, goggles, and face shields, weren't routinely used outside of operating rooms so there was ample opportunity for HBV exposure during non-surgical medical procedures. Numerous studies before the implementation of the HBV vaccine found that healthcare workers had a much higher prevalence of HBV infections than the general population (Lewis et al. 2015), putting both them and their patients at risk. It was even common in that era to reuse unsterilized syringe needles before the risk of HBV transmission in minute amounts of blood on the needles was proven (MacCallum 1972a). Compounding these issues, the existence of healthy, chronically infected individuals with active viruses in their blood was unappreciated. These individuals were infectious to others and contributed to the contamination of blood banks with HBV, putting numerous transfusion recipients at risk. What was badly needed was the identification and isolation of these viruses so that diagnostic tests and vaccines could be developed.

While MacCallum's work was an important step in defining and distinguishing the epidemiology of hepatitis A versus hepatitis B infections, the causative agents were still elusive. The agents were filterable, so they appeared to be viruses by the standard definition of that era. However, attempts to culture a virus from patient samples into chicken eggs, cells, or test animals failed (MacCallum and Bauer 1944), and other technologies available in the 1940s were insufficient to isolate the pathogens. Lacking the means to cultivate the virus in cells or model animals, investigators turned to human experimentation (Block et al. 2016). As for the viruses in previous chapters, experiments that would be considered heinous today were conducted and accepted during World War II and the post-war period. For example, soldiers and prisoner volunteers (disproportionately minority individuals due to the high incarceration rate for Blacks in the American prison system), were often utilized for early hepatitis experiments as carefully detailed in Sydney Halpern's book "Dangerous Medicine: The Story Behind Human Experiments with Hepatitis" (Halpern 2021). These ethically challenged experiments helped refine our understanding of the transmission and symptomology of hepatitis A and B but still failed to reveal the viruses themselves.

As morally problematic as it was to use adult soldiers and prisoners for scientific studies, using children to study hepatitis was even more disturbing. One particularly troublesome study was conducted by Saul Krugman at the Willowbrook State School in New York, an institution for children with intellectual disabilities (Krugman et al. 1967). The school was selected because it had a high incidence of endogenous hepatitis cases with most newly admitted residents becoming infected within a year of arrival. The reasoning was that since these children were going to get infected anyway, it was acceptable to utilize them for infectious studies, a rationale completely intolerable today. It was also known that viral hepatitis in children tended to be mild or even asymptomatic, so the risk to the children from intentional infection seemed minimal to the investigators and the children's parents. Unfortunately, this decision to infect handicapped minors for scientific purposes was based on the hubris of ignorance. We now know that HBV infections have an age-dependent outcome (Yuen et al. 2018). HBV specifically infects and replicates in liver cells called hepatocytes. Many viruses destroy the infected cells to release their new virions, but HBV has a different process and does not directly kill infected hepatocytes. Instead, when HBV replicates the new virions bud out of the cells without damaging the integrity of the cell. However, our immune response targets and destroys those infected cells resulting in liver damage. Thus, much of the liver symptomology, i.e. the classic hepatitis presentation, is due to the immune response against HBV-infected hepatocytes. Adults tend to mount a robust immune response to HBV leading to liver injury and acute disease symptoms. While adults get acutely sick, their immune systems usually eliminate the virus completely so that only 2–6% of adult infections become chronic. In contrast, children have less developed immune systems and often fail to respond effectively to HBV infections. Without a vigorous immune response, there is minimal liver damage and consequently little or no acute symptomology. Sadly though, the lack of immune destruction of the infected hepatocytes often leads to chronic, lifelong infections in around 80–90% of young children exposed to HBV, an outcome unanticipated by early investigators. Chronic HBV infections are associated with a high lifetime risk of cirrhosis and/or liver cancer (hepatocellular carcinoma) so this virus is considered a cancer-causing (oncogenic) agent (Rizzo et al. 2022). With little or no follow-up on all the "volunteer" children from decades past, it is unclear how much cancer burden was unknowingly imposed on these kids in the interest of science.

A Mystery Down Under

Viral hepatitis continued to be an ongoing medical issue throughout the 1940s and 1950s with no significant advancements in prevention or treatment. Ongoing experimentation, however, did refine the clinical features and epidemiology of viral hepatitis. Early studies of viral hepatitis could only focus on individuals who developed acute disease as there were no viral markers to identify test subjects who developed asymptomatic infections. Similarly, with no way to screen healthy individuals for evidence of current or previous HBV infection, the existence of chronic infections was initially unrecognized. One of the first clues that chronic infection occurred was studies that found persistent liver function abnormalities in certain patients. The 1940s and 1950s saw the development of several assays that measured liver health. Initially, bilirubin levels were used to assess liver activity, but subsequently, the levels of two liver enzymes, AST (aspartate aminotransferase) and ALT (alanine aminotransferase), were found to be better indicators of liver function. Elevation of either or both AST and ALT reflect ongoing liver damage and testing for these enzyme levels remains a standard part of routine health exams. Utilizing these liver enzyme tests, epidemiological studies in the 1950s began to demonstrate a correlation between previous jaundice/hepatitis illness and liver dysfunction many years post-infection (Stokes et al. 1954; Murray et al. 1954). From these studies, it was estimated that between 2–5 out of every 1000 people chronically carried HBV although the agent itself could not be identified. This finding was highly concerning since it meant that the pooled blood donations used for transfusions had a high likelihood of being contaminated with HBV, explaining the known incidence rate of post-transfusion hepatitis. What was desperately needed was the isolation of the HBV virions so that diagnostic and screening tools could be developed to protect the blood supply and the recipients of blood and blood products. Because acute serum hepatitis cases were relatively few (other than the occasional vaccine-associated outbreaks) and no virus could be cultivated from patient samples by standard approaches, the virological community was stymied. Surprisingly, finally identifying the pathogen would stem from a case of mistaken identity by researchers studying blood proteins.

The post-war era was a time of rapidly advancing science, not only in virology but in all facets of medicine and biology. One burgeoning research area was the identification and characterization of human blood proteins and their variation among different population groups. It was known that individuals receiving multiple blood transfusions often developed complications, some of which might be due to immunological responses against donor blood proteins that were different than those found in the recipient of the transfusion. Identifying and understanding these differences in blood proteins was the passion of a young physician, Baruch (Barry) Blumberg (Blumberg 2002). Starting in medical school at Columbia's College of Physicians and Surgeons, Blumberg was fascinated by human genetic diversity and its relationship to disease susceptibility. After his medical training, he had the opportunity to visit Africa in 1957 to collect blood samples from varied population groups, the start of his large collection of diverse human blood samples. Using immunological methods, he and his colleague, Harvey Alter, began testing human samples from around the world for differences in blood proteins, leading to several seminal papers describing novel proteins in the blood. Their accidental entry into HBV research began in 1964 during the screening of blood from an Indigenous Australian Aboriginal person (Blumberg et al. 1965). This Australian individual carried a unique blood protein not found in 150 samples of blood from healthy Americans, and this new protein was dubbed the Australia antigen (Au) in honor of its discovery in an Australian blood sample. Intrigued by the novel protein, they expanded their studies to a large collection of samples. This large study found the Au protein was not unique to Australia and was present in some persons with numerous blood transfusions (many were hemophiliac cases) as well as about 10% of leukemia patients tested. The leukemia observation was potentially exciting if the Au protein could be connected to the development of this cancer. Knowing that people with Down syndrome had a high incidence of leukemia, they screened blood from Down syndrome patients at a large institution for the intellectually disabled and found that 30% of them were positive for this new protein (Sutnick et al. 1968).

The high frequency of the Au protein in both leukemia patients and Down syndrome individuals initially supported the conjecture that this protein might be associated with susceptibility to leukemia, a hypothesis that turned out to be completely erroneous. Subsequent large-scale testing of stored serum samples found the Au protein in patients with many varied conditions as well as having a high incidence in healthy people from the Pacific Island region (Oceania), none of which indicated any relationship to leukemia. Further studies of people with Down syndrome provided a critical observation that would ultimately be the key to understanding the real nature of the Au protein. Unlike individuals living in large institutions, Down syndrome people living in small institutions or their family homes had a ten-fold lower incidence of the Au protein. If the Au protein was truly a human protein produced by our genes, its prevalence should be constant in the Down syndrome population regardless of the living conditions of the patient. This dramatic difference in prevalence associated with the living environment suggested that having the Au protein was an acquired event rather than being the endogenous expression of a genetically encoded protein. Living in crowded conditions with poor sanitation, such as often found in large institutions for the intellectually disabled, is an ideal situation for infectious pathogen transmission. Perhaps the Au protein was simply a marker for some disease found in populations worldwide, a disease that was prone to spread under institutional conditions. If so, how to connect the Au protein to a specific disease?

The clue to the Au protein's disease association came quickly due to two chance events, but confirmation of its role in serum hepatitis would come more slowly as a preponderance of evidence accumulated. The initial key event was the seroconversion of a technician in the Blumberg lab (Blumberg et al. 1968), an event that we've seen with several previous viruses, reflective of the fairly primitive biosafety conditions in labs of that era. This individual was routinely used as a control since she was Au-negative, but after becoming ill with dark urine (a possible sign of liver dysfunction), she converted to Au-positive. Similarly, one of the Down syndrome patients being monitored in their studies was also initially Au-negative but became ill, had abnormal liver function tests, and then became Au-positive (Sutnick et al. 1968). Both these observations suggested that the Au protein was associated with liver infection, possibly serum hepatitis. Armed with this promising lead, the ideal proof would have been to fulfill Koch's postulates: isolate the agent from a patient with serum hepatitis, propagate the agent in culture to prepare a pure specimen, and then use the specimen to produce the disease in test animal (or human). Unfortunately, none of this was feasible since the agent itself hadn't been isolated, there was no method for culturing it, and there was no appropriate animal model available. Consequently, the "proof" that Au protein reflected a serum hepatitis infection accrued slowly throughout the 1960s by the accretion of clinical and epidemiologic data demonstrating the presence of this protein in the blood of hepatitis B patients (Block et al. 2016). Related studies found the Au protein in biopsied liver tissues, thus making a specific connection to the liver (Huang et al. 1972; Millman et al. 1969). In addition to the direct detection of the Au protein in patients, more classical virological approaches were also being applied to serum hepatitis. Two independent studies used centrifugation methods to concentrate particulate material from the serum of Au-positive patients and then examined the concentrates by electron microscopy (Millman et al. 1969; Bayer et al. 1968). The results showed small, spherical particles resembling viruses, and the particles reacted with antibodies against the Au protein, confirming that Au protein existed on the surface of these particles. The following year, David Dane in London performed similar electron microscopy experiments and visualized not only the small spheres but also a lesser population of larger spheres with surface Au protein that became known as Dane particles (Dane et al. 1970). The Dane particles were proposed as the actual HBV virions with the smaller particles being incomplete, empty virions lacking the viral nucleic acid, an interpretation that proved to be correct. Shortly after Dane's work, investigators were able to show that chimpanzees infected with Au-positive serum became Au-positive, had Au protein in their livers, and exhibited liver abnormalities, a least a partial fulfillment of Koch's postulates. Finally, science was catching up with the disease, and the scientific and medical communities began to accept that the hepatitis B culprit virus had been identified at last. What had started as a mistaken belief that Australian Aboriginal peoples might harbor a unique human blood protein morphed into the groundbreaking discovery of the hepatitis B virus. Having the virus in hand was a great advance for medicine. Now diagnostic and screening tests could be created to identify patients with chronic infections and protect the blood supply from HBV contamination, and the development of drugs and vaccines was a future possibility. For his role in this discovery and its clinical significance, Barry Blumberg received the Nobel Prize for Physiology or Medicine in 1976.

In the decades following its initial discovery, we've learned a great deal about the properties of HBV (Tsukuda and Watashi 2020). The large particles first observed by Dane are the complete, infectious virions. This complete virion consists of the viral DNA genome packaged within a protein coat comprised of the HBV core protein (HBc), forming the viral nucleocapsid. Within the infected cell, after the nucleocapsid assembles it traverses through cellular membranes as it exits the cell and acquires these membranes as an envelope that surrounds the nucleocapsid. The envelope is studded with many molecules of the virus-encoded Au protein which has been renamed the HBV surface antigen (HBsAg). The HBsAg exists in three lengths referred to as the large, middle, and small forms, and all three forms are present on the virion surface. When introduced into the bloodstream, as few as 10 complete virions can establish an infection in chimpanzees (Komiya et al. 2008), so the virus is considered highly infectious.

In addition to the complete virion, infected cells excrete enormous quantities of empty virions into the bloodstream. The ratio of empty to complete virions can be as high as 100,000 to 1 (Hu and Liu 2017), and these are the small particles observed in the early electron microscope studies. The empty virions consist of spherical HBsAg-containing membranes that lack the nucleocapsid (no HBc or DNA). Without the viral DNA, the small particles are inert and cannot replicate or cause disease if transmitted to new patients. Given their huge quantity and lack of infectious function there is an obvious question. Why would a virus waste resources in the production of these seemingly worthless small particles? At least one answer appears to be that the small particles act as a defense mechanism for the virus against host antibodies. Normally, once a virus is present in the bloodstream its surface proteins (its antigens) will be recognized by circulating B cells. Once a B cell encounters its cognate antigen that cell will be stimulated to multiply and produce antibodies. Many of the resulting antibodies will be neutralizing antibodies that bind to their target protein on the virion surface and coat the virion. This antibody coat blocks the virion from binding to its entry receptors on cells so the coated virion is neutralized and unable to enter a new cell. As neutralizing antibody levels increase in response to virion antigens, eventually all the free virions are neutralized. This process stops the infection of new cells and is one of the primary mechanisms by which viral infections are prevented or resolved. HBV has evolved to evade the antibody response by overwhelming it with the surface antigen. There is so much HBsAg present in the empty particles that there isn't enough antibody produced to bind and neutralize all of this protein. Consequently, as the small particles soak up most of the antibodies, some of the complete virions escape neutralization and remain infectious in the blood. Thus, chronic HBV carriers can remain infectious to others for years even though they have high anti-HBsAg antibody levels. This evasive scheme allows infectious viruses to persist indefinitely in these carriers, giving the virus years to decades of opportunity for spread to new victims. With no symptoms or any cognizance of their carrier state, these chronically infected individuals have helped the virus perpetually maintain itself in the human population for millennia. It was only with the advent of the HBV vaccine that we began to interdict this nefarious relationship with humans.

Chronic Carriers, Cancer, and a Vexing Causation

Both the isolation of the HBV virion and the confirmation that the Au protein was the virion surface protein (HBsAg) were a significant boon to medicine and public health. Assays to detect HBsAg were rapidly incorporated into medical practice where they were useful for distinguishing acute hepatitis B from hepatitis A cases. Even more importantly, blood, serum, and other blood-derived products could be tested for the presence of HBV, allowing contaminated batches to be discarded. Likewise, donors could now be screened for chronic HBV infections that would disqualify them. Previously, it was known that blood banks were contaminated, but there was no means to identify who among the seemingly healthy donors was a chronic HBV carrier. Interestingly, one outcome of HBV testing was a change in the policy for blood and serum donations. Originally, blood banks depended mostly on paid donors. Once HBV testing became widespread it was discovered that paid donors had a much higher prevalence of HBV positivity than unpaid volunteer donors. This was likely because many individuals needing cash for their donations tended to have lifestyles with a higher risk for HBV transmission (e.g. drug abusers and former inmates of prisons and institutions) than altruistic donors. This led to the greater screening of donors for their background history and lifestyle information as well as the transition to mostly volunteer donors. Using volunteer donors is still the mainstay today for the Red Cross, hospitals, and other blood collection organizations.

As important as HBV screening was for protecting the blood supply, the ability to identify chronic carriers finally allowed an important medical question to be addressed. Many if not most HBsAg-positive carriers seemed perfectly healthy but did chronic HBV infection impose any long-term health risks on these individuals? The diagnostic and analytical tools for HBV developed in the 1970s were finally sufficient to accurately identify and follow chronic HBV carriers, allowing both retrospective and prospective clinical studies. In retrospective studies, a cohort of patients with a specific condition is identified from medical records and then a control cohort without the condition is matched as closely as possible for other variables such as gender, age, and general health. For the HBV studies, since HBV targets the liver, patients with various types of liver disease formed the disease cohort. Once the disease and control cohorts were identified, they were then compared for the presence or absence of the experimental variable which in this case was HBV infection. Multiple studies from various countries found that patients with cirrhosis and a type of liver cancer called hepatocellular carcinoma (HCC) were many times more likely to be HBV positive than control groups (Turbitt et al. 1977; Nayak et al. 1977; Tan et al. 1977). Unfortunately, retrospective studies don't prove that the experimental variable, HBV, causes the disease condition, they only indicate a possible correlation. Many unknown and confounding variables could influence the results and show a correlation without causation. For example, it wouldn't necessarily be known when the patients became HBV-positive. Perhaps they became infected after their disease presented which would eliminate HBV from having a causative role. Nonetheless, the observation that HBV infection correlated with cirrhosis and HCC in diverse population groups from multiple countries was a compelling observation requiring further study.

The more definitive type of clinical study is the prospective study. For the prospective studies, patients with and without HBV chronic infections were enrolled. These two cohorts were then followed for months to years to see if there were differences in the incidence of cirrhosis and/or HCC between these two groups. Since the test variable, HBV chronic infection, was present before the liver disease presentation, this provides a stronger indicator of causality. It is also easier to match the test and control cohorts to eliminate other variables since the investigators can selectively enroll appropriate participants with the right characteristics rather than just relying on whatever medical records are available. The downside of prospective studies is that they can take years to complete, especially when the ultimate disease endpoint may occur years after the initiating event. Consequently, much of the decade of the 1970s was spent assembling prospective study cohorts and following these populations to see if statistically significant differences arose. Ultimately though, as with the retrospective studies, early prospective studies found that chronic HBV infections were a large risk factor for the subsequent development of liver diseases, primarily cirrhosis and HCC (Obata et al. 1980; Lehmann and Wegener 1979; Sumithran and MacSween 1979). Perhaps the most convincing report was the landmark study of over 22,000 Taiwanese men conducted by Robert Palmer Beasley and colleagues (Beasley et al. 1981). Published in 1981, their study enrolled over 3000 HBV-positive and 19,000 HBV-negative individuals mostly between the ages of 40 and 59. Individuals in the study were followed for multiple years and assessed for incidence of HCC. The most striking finding was that the relative risk of developing HCC was 233 times greater in the HBV-positive cohort. Numerous subsequent prospective studies over the next 2 decades confirmed Beasley's results and revealed the horrendous damage inflicted by chronic HBV infections (Schweitzer et al. 2015; Tang et al. 2018). Based on the accumulating evidence, in 1994 the hepatitis B virus was designated a Group 1 human carcinogen by the International Agency for

Research on Cancer (IARC—a branch of the World Health Organization). Current estimates suggest that roughly 40% of untreated patients with chronic HBV infections will progress to cirrhosis, and between 10% and 25% will develop HCC (McGlynn et al. 2015). With the Centers for Disease Control (CDC) estimating that nearly three hundred million people worldwide are living with chronic HBV infections and another 1.5 million new cases occur each year, the contribution of this virus to serious liver disease is enormous.

Once the connection between HBV and HCC was established by epidemiological studies, attention focused on understanding the molecular mechanisms by which this virus subverts cell function and promotes the formation of cancer in the liver. After, 50 years of study we know that HBV's role in HCC is complex, multifactorial, and likely not yet completely understood (D'Souza et al. 2020). We do know that there are both nonspecific and specific effects caused by chronic HBV infection. The primary nonspecific effect that contributes to cancer initiation is the induction of continual inflammation in the liver. Persistent infection of hepatocytes by HBV triggers a constant immune attack that causes chronic inflammation along with liver cell death and regeneration. Chronic inflammation and liver proliferation, regardless of the cause, promote mutational changes in the liver cells that significantly increase the risk of an oncogenic event occurring. Beyond this nonspecific role as an agent of inflammation, several virus-specific factors also contribute to cancer development. Integration of the HBV DNA into cellular chromosomal DNA occurs at a low frequency in infected cells, but when it does occur it can cause chromosomal instability and disrupt the normal expression of cellular genes. These events can dysregulate the cell's biology and lead to a normal cell transforming into a cancerous cell (Tu et al. 2017). Likewise, the expression of virally encoded proteins within infected cells, particularly the HBsAg and the core protein (HBc), can induce cellular stress, disrupt cellular gene expression pathways, and cause DNA damage, all events that can facilitate cellular transformation. As any or all of these events may be ongoing in thousands of infected cells, the chances of an oncogenic change happening during a patient's lifetime are quite high.

In conjunction with the cancer risk factors described above, HBV also expresses a viral protein with direct oncogenic activity (Sivasudhan et al. 2022). Named the X protein (HBx) because it lacked homology with any known protein, this small protein of only 154 amino acids has a huge biological impact on the host cell. Evidence for the oncogenic potential of the X protein comes from both human and animal studies. While HBV is strictly a human pathogen and won't infect common laboratory animals except for some primates, other animal species have been found that have viruses related to HBV. For example, ground squirrels and woodchucks each have an HBV type virus, ground squirrel hepatitis B virus (GSHV) and woodchuck hepatitis virus (WHV), respectively. While infection with these viruses is specific to their host animal, GSHV and WHV are genetically related to HBV and express very similar proteins, including an HBx equivalent protein. When squirrels or woodchucks are infected with their respective virus, they can develop chronic liver infections that lead to HCC just as in humans. However, when infected with a mutant form of the virus that cannot express the X protein then HCC is greatly reduced, demonstrating the importance of X for cancer development in these animals and by analogy in humans.

HBx is a multifunctional protein that interacts with numerous cellular systems to create an intracellular environment favorable for viral maintenance and reproduction. Known and suspected effects of HBx include activating cellular proliferation, preventing cellular senescence, and enhancing chromosomal instability, all events that can lead to transformed cells. Transformation into a cancer cell not only requires enhanced growth properties but also entails overcoming specific anti-cancer proteins and mechanisms inherent in our cells, and HBx attacks several of these critical defenses. HBx helps infected cells evade the immune defenses, defeat growth suppression systems that normally stop aberrant cells from dividing, and resist cell death pathways that try to cause abnormal cells to self-destruct. This panoply of HBx activities, along with its other functions that are less well-characterized, makes this protein an important contributor to HBV's oncogenic potential. The combination of HBx functions with the other multiple HBV-mediated effects makes chronically HBV-infected cells ripe for transformation. Still, tumor formation is a complex event that requires many steps before a normal cell becomes a cancer cell that can multiply and become an actual tumor. While chronic HBV infection raises the odds that these transforming cellular changes can occur, it does not guarantee that HCC will manifest in every chronic patient. Some individuals are unlucky and the tumor ultimately develops during their lifetime while in other people this never happens. We know that there are risk factors that enhance the oncogenicity of chronic HBV, for example, heavy use of alcohol or tobacco, but even with these extra risks, the development of HCC in any individual is still not certain. Since we can't yet predict whom the cancer will strike, efforts focused on treating chronic infections through antiviral drugs or preventing them through vaccination are the only approaches to preventing this pervasive viral cancer and other liver damage.

A New Vaccine Paradigm

Developing a vaccine to prevent acute HBV infections was a high priority even before there was an awareness of the long-term liver sequelae of chronic HBV infections. The risks of acute infection to military personnel, healthcare workers, and the general population were sufficiently great to stimulate vaccination attempts as early as the late 1960s when very little was known about the actual properties of the hepatitis B virus itself. There was, however, a major scientific roadblock to vaccine development. Beginning with Louis Pasteur's seminal work on rabies, for the next 100 years there were only two categories of viral vaccines: (1) killed vaccines that used chemical treatments to inactivate viral particles while retaining their shape and immunogenicity and (2) attenuated viral strains that had lost their pathogenicity due to mutational changes but still invoked a protective immune response. Unfortunately, neither of these methods was suitable for the hepatitis B virus. Since HBV could not be grown in cell cultures or any convenient animal model, there was no way to generate attenuated mutants or prepare large quantities of virions for chemical inactivation. A new approach was needed, and by the late 1970s molecular biological techniques had advanced sufficiently to conceive and implement a third vaccine strategy, the so-called "subunit" vaccine. This concept would be successfully used to produce an HBV vaccine as well as subsequent vaccines for human papillomaviruses (Chap. 12) and the COVID-19 virus (Chap. 13).

Unable to implement the standard vaccine development approaches for HBV, clever researchers began to recognize and exploit the unique features of this virus. Once assays were developed to detect HBV, it was quickly recognized that patients with chronic HBV infections had significant amounts of infectious virus in their blood, as well as large quantities of the HBsAg-containing empty virions. This vast overproduction of hepatitis B surface antigen that the virus used to avoid antibody neutralization would eventually provide the clues for vaccine development and immunological protection against HBV infection. Initially, several groups reasoned that human blood from these chronic carriers might serve as a sufficient source of viral particles for vaccines if the viruses could be inactivated and still retain the ability to induce a protective immune response. One of the first to implement this idea was Saul Krugman, still conducting studies on the patients living at the Willowbrook school (Krugman et al. 1971). In a small study that would be considered completely heinous today, Krugman injected healthy children with serum from an HBV patient as the source of HBV. The control group received untreated serum and all 25 children in this group showed evidence of HBV infection. The test group of 10 children received serum that had been boiled to inactivate the virus, making it essentially a killed virus vaccine. Some children received one dose of the inactivated serum and some received two doses. Several months later the test children were challenged with unheated infectious serum to assess their resistance to HBV. All the children receiving two doses were fully protected, and the children receiving one dose were either fully or partially protected. These morally troubling studies demonstrated that a killed vaccine against HBV was feasible and encouraged other groups to pursue this approach.

Like Dr. Krugman, other research groups, including those led by Barry Blumberg (Millman et al. 1971) and Robert Purcell (Purcell and Gerin 1975), started with human serum as the source of HBV virions. Rather than use whole serum that contained numerous human proteins in addition to the HBsAg, these groups took a more biochemical approach. They used extensive purification methods to isolate the viral material from the serum and eliminate the extraneous serum proteins. Given that the amount of empty HBsAg vesicles greatly exceeded the number of complete virions, these preparations were mostly HBsAg particles rather than whole viruses. Now the key question was whether or not HBsAg, a subunit of the virus, was sufficient for inducing protective immunity. Fortuitously, human subjects were no longer required for the initial trials as chimpanzees were recently shown to be susceptible to human HBV (Maynard et al. 1972). Following a successful vaccination trial in chimpanzees (Purcell and Gerin 1975), a subsequent human study confirmed that vaccination with an HBsAg preparation could protect from HBV infection (Szmuness et al. 1980). While this subunit approach was a unique and novel concept for its time, we fully understand the science today. HBsAg is the only protein on the HBV virion surface, and it is the protein that interacts with the host cell receptor to initiate binding and viral entry. Antibodies against HBsAg block its interaction with receptors and neutralize the virus. Consequently, there is no need to use the whole virion as the immunogen in the vaccine and HBsAg protein alone effectively generates neutralizing antibodies that protect the vaccinated individual.

Blumberg patented his protocol for preparing and inactivating HBsAg and licensed it to Merck where once again Maurice Hilleman helped develop it into a commercial vaccine. With the success of various vaccine trials in the late 1970s, Merck's HBV vaccine was approved and licensed for the public in 1981 as Heptavax-B (Krugman 1982). Then, after over a decade of work, there was a safe and effective vaccine to protect high-risk individuals. The research community fervently believed that this vaccine would block acute

HBV infection which would prevent chronic infections and the subsequent risk of cirrhosis and HCC. Sadly, what should have been a celebrated scientific achievement only had a short period of glory and acceptance due to two factors. First, there was an inherent flaw in this vaccine approach. To make the vaccine there had to be healthy serum donors with chronic HBV infections. If the vaccine became widely used then eventually the population of potential donors would shrink and the source material for vaccine production would diminish or be eliminated. Having a nonrenewable source as your starting material is distinctly not a long-term strategy for production. Second, and even more importantly for the demise of this vaccine, the early 1980s saw the appearance of a mysterious and ravaging new disease, AIDS (acquired immunodeficiency syndrome) (Sellers 1982). Scientists rapidly homed in on an unknown pathogen as the likely etiological agent of AIDS, and epidemiological studies associated the transmission of this agent with blood and blood products, as well as sexual activity. Suddenly, making the HBV vaccine from the serum of donors was highly problematic. In the United States, people with chronic HBV infections typically fell into four groups, healthcare providers exposed to blood and bodily fluids, individuals with a history of intravenous (IV) drug abuse, people with risky sexual behaviors, and persons who received multiple blood transfusions and/or used blood-derived medical products (e.g. hemophiliacs who use human-derived clotting factors). These four groups were exactly the same groups at high risk for AIDS. Lacking any screening method to discriminate between HBV carriers who did or didn't have the AIDS agent, the source material for vaccine production became suspect and the public rightly rejected the vaccine.

While the death knell was sounding for the serum-derived HBV vaccine, molecular biology was about to introduce a completely new way to produce pathogen proteins. Up until the 1970s, the only way to produce usable amounts of a particular protein was to purify it from some source that naturally contained it. This source could be a plant, animal, or human as long as the source produced an adequate supply of the target protein. Scientists would begin with a large quantity of the starting material, for example, insulin was produced from the pancreas of cows or pigs. The harvested organs would be ground up and the insulin extracted and purified using biochemical separation methods. As long as there was a sufficient source, a method to isolate the target protein could usually be developed. There were, however, serious limitations to this approach. The source material could be rare or expensive and the purification process could be difficult and/or costly. Additionally, proteins from non-human sources could elicit complications when used in humans due to immunological recognition of these non-human proteins as different than our endogenous human versions. The development of cloning technologies in the 1970s was about to eliminate all of those complications and greatly expand the realm of pharmaceutical proteins for clinical usage.

Cloning is fairly simple in concept although it took years to develop all the tools and protocols needed to implement cloning in practice. The idea is to take a gene from one source and put it into another organism, typically a bacterium such as the common *E. coli*. If the target gene encodes a protein, for example, the human insulin gene, then the human insulin protein can be produced in *E. coli*. The resulting protein is called a recombinant protein since it is the product of two (or more) combined species. Since cultures of *E. coli* can be grown in large quantities, this provides a cheap, abundant, constant, and convenient source material for the purification of the recombinant protein. Of course, the process of isolating a gene and permanently inserting it into a bacterium under conditions where the foreign gene is expressed into protein is not trivial, but by the late 1970s, the methodology was well established. Even as the serum-derived HBV vaccine was in development, William Rutter and his team at the University of California-San Francisco were exploring the possible expression of HBsAg as a recombinant protein (Edman et al. 1981). A recombinant HBsAg might be suitable for a vaccine and would eliminate both the problems inherent with the serum-derived HBsAg material. Ultimately they determined that *E. coli* was not suitable for HBsAg production but that yeast cells were (bacteria are prokaryotes and some eukaryotic proteins aren't produced well in prokaryotes; yeast and humans are both eukaryotes) (Valenzuela et al. 1982). Working with Maurice Hilleman at Merck, the yeast-expressed HBsAg was developed into a vaccine product that performed well in animal studies (McAleer et al. 1984) and early-stage clinical trials that measured the safety and immunogenicity of the recombinant protein (Zajac et al. 1986). Importantly, the HBsAg expressed in yeast didn't just remain as a monomeric, independent protein molecule. Instead, the HBsAg molecules aggregated and self-assembled into spherical particles resembling the actual virion that were termed virus-like particles (VLPs) (Valenzuela et al. 1982). With no viral genome present, these recombinant VLPs were completely non-infectious and posed no risk to the recipient of hepatitis disease. Furthermore, being composed of a single protein antigen, there generally were fewer side effects of VLP vaccination compared to the more complex antigenic mixtures of attenuated or killed vaccines. Lastly, presenting HBsAg to the immune system in a VLP configuration resembling the authentic virion induced a more robust immune response than if the protein had remained monomeric. Based on these results, the FDA licensed the vaccine in 1986 as Recombivax HB to replace the suspect serum-derived vaccine. This new product was the first recombinant vaccine ever licensed in the United States, and it confirmed that recombinant technology was a viable strategy for future vaccine development. Eventually, other companies would also make recombinant HBsAg vaccines using a similar approach, some designed for adults and some for pediatric use. Over the four decades after the original Recombivax HB was marketed, more than a dozen recombinant protein vaccines against viruses and bacteria were licensed, including one of the COVID-19 vaccines (Chap. 13) that were critical for reducing the global pandemic and protecting vulnerable populations.

Beyond Recombivax HB

With safe and effective HBV vaccines on the market, researchers and clinicians around the world were enthusiastic about controlling HBV infections and were optimistic that HCC could be drastically reduced. Now the key question was what the most effective vaccination strategy would be to reduce the incidence of hepatitis B. In the United States, the initial thrust targeted high-risk groups rather than trying to vaccinate the general population. While the high-risk groups included healthcare workers who were in jeopardy of occupational exposure, the largest number of hepatitis B cases occurred in IV drug users, homosexual men, and individuals with multiple sex partners and unsafe sexual practices (Alter et al. 1990). Vaccine acceptance made inroads with the healthcare community, but the other high-risk populations were challenging. Identifying these individuals and getting the vaccine to them was not easy, and even when they were identified, motivating them to complete the vaccination series (typically 3 shots initially) was difficult. Social stigma, absence of insurance, and lack of awareness of the HBV infection risk conspired to keep HBV vaccination rates low. Furthermore, these populations were all adults and chronic infections develop in less than 10% of adults, so targeting these groups didn't significantly reduce the permanent infectious burden of chronic carriers. After several years with limited progress (Coleman et al. 1998), the CDC and the Advisory Committee for Immunization Practices (ACIP) changed strategies and recommended that every child be vaccinated against HBV to give them immunity before they were old enough to engage in risky behaviors. This was quickly followed by the American Academy of Pediatrics endorsing the ACIP recommendations in early 1992. After reviewing the data, all these groups concluded that the HBV vaccine was safe and effective enough for widespread use. Vaccinating children should prevent whole generations from developing chronic infections, would generate more widespread herd immunity, and ultimately should reduce HBV infections nationwide.

The consensus recommendation for universal vaccination was an important step, but not one that was immediately translated into action. Unlike many of the other communicable viral diseases, hepatitis B was not an obvious and imminent threat to American children and was perceived as an adult disease. In the social climate of the 1980s, vaccine hesitancy and worries about vaccine overload were becoming more prevalent. There were even accusations that the HBV vaccine might cause autoimmune problems in some recipients (Marshall 1998), an allegation later found to have no merit. With these issues initially unresolved, many parents and pediatricians were resistant to introducing another vaccine to prevent an infection that didn't seem critical for children. Weighed against the immediacy of protecting their children, adding a vaccine to address a societal problem had limited appeal to the public. On top of this general concern, the transmission of HBV by sexual activity and drug use became an issue. Some parents feared that vaccinating their children against HBV would encourage promiscuous behavior or drug abuse as the children grew up, an issue that similarly occurred with the later human papillomavirus (HPV) vaccine (Chap. 12). The combination of these problems made HBV vaccine adoption and coverage a much slower process than vaccines against more urgent disease threats like smallpox and polio. Nonetheless, the importance of this vaccine for protecting current and future generations slowly gained momentum through education and legislative support. By the year 2000, only 4 states had no HBV vaccination requirement in place with the other 46 states having requirements for public school enrollment and/or daycare attendance. By 2023, only four states lacked HBV vaccination requirements for school enrollment, and three of those states did require this vaccine for children attending childcare centers. Only the state of Alabama has no HBV vaccination requirements for children. The effectiveness of this vaccination campaign is evident from the decline in hepatitis B cases over the last four decades (Bixler et al. 2023). There has been a 99% decline in cases from around 200,000 reported in 1980 to roughly 2000 cases in 2019. With the high HBV vaccine coverage in the United States, this type of hepatitis has become a rare disease.

In contrast to the United States where HBV infections were primarily seen in adult populations, some regions of the world, for example, Taiwan, had high endemic rates among children (Liu and Chen 2020). In addition to adult-adult transmission via blood and intimate activities, HBV is also readily transmitted perinatally from mother to infant. In Taiwan, before the vaccine was developed about 10–20% of adults were chronically infected. As chronically infected women had children, they exposed their offspring to the virus and their children had a high risk of developing a chronic infection. This established a perpetual cycle of early-life infection leading to chronically infected adults who passed the virus onto the next generation. Not surprisingly, countries like Taiwan with a high endemic rate of chronic HBV infections had extremely high rates of liver disease and liver cancer (Lin et al. 1986). To confront this chronic problem, Taiwan initiated mass HBV vaccination for newborns in 1984 with remarkable results (Liu and Chen 2020). The vaccine coverage rate quickly reached greater than 95% and has approached 98% in recent years. Various population studies in Taiwan confirmed that within 10 years of beginning the vaccination campaign, the chronic carrier rate in children dropped from over 10% to less than 1% and this low incidence has been maintained ever since. Importantly, the success of vaccination in Taiwan allowed an important scientific prediction to be tested. All previous evidence strongly suggested that chronic HBV infections played a role in the development of cirrhosis and hepatocellular carcinoma (HCC). If HBV was truly an oncogenic virus, then a reduction in chronic infections should produce a corresponding reduction in liver cancer, although this effect would take decades to confirm as liver disease typically manifests in middle-aged or older. While the epidemiological studies are still ongoing, mortality due to liver disease, including HCC, has shown a steady decline in Taiwan since the HBV vaccine's introduction (Chiang et al. 2022). Similar studies in other regions with formerly high endemic rates of chronic HBV infections are beginning to show the same promising trend—whenever HBV vaccination coverage increases, HCC cases begin to decrease. These epidemiological observations strongly support the conclusion that HBV is oncogenic and that preventing infection removes this cancer risk. Basic science exploration coupled with investigator curiosity led from the Australia antigen to the hepatitis B virus to the world's first anti-cancer vaccine. Although the global burden of HCC still remains high due to other factors (hepatitis C virus infection, excessive alcohol consumption, environmental exposures), we now have confidence that one major cause of liver cancer could be eliminated through widespread HBV vaccination.

References

Alter MJ, Hadler SC, Margolis HS, Alexander WJ, Hu PY, Judson FN, Mares A, Miller JK, Moyer LA (1990) The changing epidemiology of hepatitis B in the United States: need for alternative vaccination strategies. JAMA 263(9):1218–1222. https://doi.org/10.1001/jama.1990.03440090052025

Bayer ME, Blumberg BS, Werner B (1968) Particles associated with Australia antigen in the sera of patients with leukaemia. Down's Syndrome and hepatitis. Nature 218(5146):1057–1059. https://doi.org/10.1038/2181057a0

Beasley RP, Hwang LY, Lin CC, Chien CS (1981) Hepatocellular carcinoma and hepatitis B virus. A prospective study of 22 707 men in Taiwan. Lancet 2(8256):1129–1133. https://doi.org/10.1016/s0140-6736(81)90585-7

Bixler D, Roberts H, Panagiotakopoulos L, Nelson NP, Spradling PR, Teshale EH (2023) Progress and unfinished business: hepatitis B in the United States, 1980-2019. Public Health Rep. https://doi.org/10.1177/00333549231175548

Block TM, Alter HJ, London WT, Bray M (2016) A historical perspective on the discovery and elucidation of the hepatitis B virus. Antivir Res 131:109–123. https://doi.org/10.1016/j.antiviral.2016.04.012

Blumberg BS (2002) The discovery of the hepatitis B virus and the invention of the vaccine: a scientific memoir. J Gastroenterol Hepatol 17(Suppl):S502–S503. https://doi.org/10.1046/j.1440-1746.17.s4.19.x

Blumberg BS, Alter HJ, Visnich S (1965) A "new" antigen in Leukemia sera. JAMA 191:541–546. https://doi.org/10.1001/jama.1965.03080070025007

Blumberg BS, Sutnick AI, London WT (1968) Hepatitis and leukemia: their relation to Australia antigen. Bull N Y Acad Med 44(12):1566–1586

Chiang CJ, Jhuang JR, Yang YW, Zhuang BZ, You SL, Lee WC, Chen CJ (2022) Association of nationwide hepatitis B vaccination and antiviral therapy programs with end-stage liver disease burden in Taiwan. JAMA Netw Open 5(7):e2222367. https://doi.org/10.1001/jamanetworkopen.2022.22367

Choo QL, Kuo G, Weiner AJ, Overby LR, Bradley DW, Houghton M (1989) Isolation of a cDNA clone derived from a blood-borne non-A, non-B viral hepatitis genome. Science 244(4902):359–362. https://doi.org/10.1126/science.2523562

Coleman PJ, McQuillan GM, Moyer LA, Lambert SB, Margolis HS (1998) Incidence of hepatitis B virus infection in the United States, 1976-1994: estimates from the National Health and Nutrition Examination Surveys. J Infect Dis 178(4):954–959. https://doi.org/10.1086/515696

Dane DS, Cameron CH, Briggs M (1970) Virus-like particles in serum of patients with Australia-antigen-associated hepatitis. Lancet 1(7649):695–698. https://doi.org/10.1016/s0140-6736(70)90926-8

D'Souza S, Lau KC, Coffin CS, Patel TR (2020) Molecular mechanisms of viral hepatitis induced hepatocellular carcinoma. World J Gastroenterol 26(38):5759–5783. https://doi.org/10.3748/wjg.v26.i38.5759

Edman JC, Hallewell RA, Valenzuela P, Goodman HM, Rutter WJ (1981) Synthesis of hepatitis B surface and core antigens in E. coli. Nature 291(5815):503–506. https://doi.org/10.1038/291503a0

Feinstone SM (2019) History of the discovery of hepatitis A virus. Cold Spring Harb Perspect Med 9(5). https://doi.org/10.1101/cshperspect.a031740

Findlay GM, MacCallum FO (1938) Hepatitis and jaundice associated with immunization against certain virus diseases: (section of comparative medicine). Proc R Soc Med 31(7):799–806

Grossman EB, Stewart SG, Stokes J Jr (1945) Post-transfusion hepatitis in battle casualties and a study of its prophylaxis by means of human immune serum globulin. JAMA J Am Med Assoc 129:991–994. https://doi.org/10.1001/jama.1945.02860490003002

Halpern SA (2021) Dangerous medicine: the story behind human experiments with hepatitis. Yale University Press, New Haven

Hu J, Liu K (2017) Complete and incomplete hepatitis B virus particles: formation, function, and application. Viruses 9(3). https://doi.org/10.3390/v9030056

Huang SN, Millman I, O'Connell A, Aronoff A, Gault H, Blumberg BS (1972) Virus-like particles in Australia antigen-associated hepatitis. An immunoelectron microscopic study of human liver. Am J Pathol 67(3):453–470

Komiya Y, Katayama K, Yugi H, Mizui M, Matsukura H, Tomoguri T, Miyakawa Y, Tabuchi A, Tanaka J, Yoshizawa H (2008) Minimum infectious dose of hepatitis B virus in chimpanzees and difference in the dynamics of viremia between genotype A and genotype C. Transfusion 48(2):286–294. https://doi.org/10.1111/j.1537-2995.2007.01522.x

Krugman S (1982) The newly licensed hepatitis B vaccine. Characteristics and indications for use. JAMA 247(14):2012–2015

Krugman S, Giles JP, Hammond J (1967) Infectious hepatitis. Evidence for two distinctive clinical, epidemiological, and immunological types of infection. JAMA 200(5):365–373. https://doi.org/10.1001/jama.200.5.365

Krugman S, Giles JP, Hammond J (1971) Viral hepatitis, type B (MS-2 strain). Studies on active immunization. JAMA 217(1):41–45

Lehmann FG, Wegener T (1979) Etiology of human liver cancer: controlled prospective study in liver cirrhosis. J Toxicol Environ Health 5(2–3):281–299. https://doi.org/10.1080/15287397909529750

Lewis JD, Enfield KB, Sifri CD (2015) Hepatitis B in healthcare workers: transmission events and guidance for management. World J Hepatol 7(3):488–497. https://doi.org/10.4254/wjh.v7.i3.488

Lin TM, Tsu WT, Chen CJ (1986) Mortality of hepatoma and cirrhosis of liver in Taiwan. Br J Cancer 54(6):969–976. https://doi.org/10.1038/bjc.1986.269

Littlejohn M, Locarnini S, Yuen L (2016) Origins and evolution of hepatitis B virus and hepatitis D virus. Cold Spring Harb Perspect Med 6(1):a021360. https://doi.org/10.1101/cshperspect.a021360

Liu CJ, Chen PJ (2020) Elimination of hepatitis B in highly endemic settings: lessons learned in Taiwan and challenges ahead. Viruses 12(8). https://doi.org/10.3390/v12080815

MacCallum FO (1972a) 1971 International symposium on viral hepatitis. Historical perspectives. Can Med Assoc J 106(Spec Issue).:Suppl::423–426

MacCallum FO (1972b) Hepatitis. Am J Dis Child 123(4):332–335. https://doi.org/10.1001/archpedi.1972.02110100064026

MacCallum FO, Bauer DJ (1944) Homologous serum jaundice - transmission experiments with human volunteers. Lancet 1:622–627

Marshall E (1998) A shadow falls on hepatitis B vaccination effort. Science 281(5377):630–631. https://doi.org/10.1126/science.281.5377.630

Maynard JE, Berquist KR, Krushak DH, Purcell RH (1972) Experimental infection of chimpanzees with the virus of hepatitis B. Nature 237(5357):514–515. https://doi.org/10.1038/237514a0

McAleer WJ, Buynak EB, Maigetter RZ, Wampler DE, Miller WJ, Hilleman MR (1984) Human hepatitis B vaccine from recombinant yeast. Nature 307(5947):178–180. https://doi.org/10.1038/307178a0

McGlynn KA, Petrick JL, London WT (2015) Global epidemiology of hepatocellular carcinoma: an emphasis on demographic and regional variability. Clin Liver Dis 19(2):223–238. https://doi.org/10.1016/j.cld.2015.01.001

Millman I, Zavatone V, Gerstley BJ, Blumberg BS (1969) Australia antigen detected in the nuclei of liver cells of patients with viral hepatitis by the fluorescent antibody technic. Nature 222(5189):181–184. https://doi.org/10.1038/222181b0

Millman I, Hutanen H, Merino F, Bayer ME, Blumberg BS (1971) Australia antigen: physical and chemical properties. Res Commun Chem Pathol Pharmacol 2(4):667–686

Muhlemann B, Jones TC, Damgaard PB, Allentoft ME, Shevnina I, Logvin A, Usmanova E, Panyushkina IP, Boldgiv B, Bazartseren T, Tashbaeva K, Merz V, Lau N, Smrcka V, Voyakin D, Kitov E, Epimakhov A, Pokutta D, Vicze M, Price TD, Moiseyev V, Hansen AJ, Orlando L, Rasmussen S, Sikora M, Vinner L, Osterhaus A, Smith DJ, Glebe D, Fouchier RAM, Drosten C, Sjogren KG, Kristiansen K, Willerslev E (2018) Ancient hepatitis B viruses from the Bronze Age to the Medieval period. Nature 557(7705):418–423. https://doi.org/10.1038/s41586-018-0097-z

Murray R, Diefenbach WC, Ratner F, Leone NC, Oliphant JW (1954) Carriers of hepatitis virus in the blood and viral hepatitis in whole blood recipients. II. Confirmation of carrier state by transmission experiments in volunteers. J Am Med Assoc 154(13):1072–1074. https://doi.org/10.1001/jama.1954.02940470024005

Nayak NC, Dhar A, Sachdeva R, Mittal A, Seth HN, Sudarsanam D, Reddy B, Wagholikar UL, Reddy CR (1977) Association of human hepatocellular carcinoma and cirrhosis with hepatitis B virus surface and core antigens in the liver. Int J Cancer 20(5):643–654. https://doi.org/10.1002/ijc.2910200502

Obata H, Hayashi N, Motoike Y, Hisamitsu T, Okuda H, Kobayashi S, Nishioka K (1980) A prospective study on the development of hepatocellular carcinoma from liver cirrhosis with persistent hepatitis B virus infection. Int J Cancer 25(6):741–747. https://doi.org/10.1002/ijc.2910250609

Purcell RH, Gerin JL (1975) Hepatitis B subunit vaccine: a preliminary report of safety and efficacy tests in chimpanzees. Am J Med Sci 270(2):395–399

Rizzo GEM, Cabibbo G, Craxi A (2022) Hepatitis B virus-associated hepatocellular carcinoma. Viruses 14(5). https://doi.org/10.3390/v14050986

Schweitzer A, Horn J, Mikolajczyk RT, Krause G, Ott JJ (2015) Estimations of worldwide prevalence of chronic hepatitis B virus infection: a systematic review of data published between 1965 and 2013. Lancet 386(10003):1546–1555. https://doi.org/10.1016/S0140-6736(15)61412-X

Seeff LB, Beebe GW, Hoofnagle JH, Norman JE, Buskell-Bales Z, Waggoner JG, Kaplowitz N, Koff RS, Petrini JL Jr, Schiff ER et al (1987) A serologic follow-up of the 1942 epidemic of post-vaccination hepatitis in the United States Army. N Engl J Med 316(16):965–970. https://doi.org/10.1056/NEJM198704163161601

Sellers T (1982) CDC warns of possible pathogen as AIDS cause. Emerg Dep News 4(12):11

Sivasudhan E, Blake N, Lu Z, Meng J, Rong R (2022) Hepatitis B viral protein HBx and the molecular mechanisms modulating the hallmarks of hepatocellular carcinoma: a comprehensive review. Cells 11(4). https://doi.org/10.3390/cells11040741

Stokes J Jr, Berk JE, Malamut LL, Drake ME, Barondess JA, Bashe WJ, Wolman IJ, Farquhar JD, Bevan B, Drummond RJ, Maycock WD, Capps RB (1954) The carrier state in viral hepatitis. JAMA J Am Med Assoc 154(13):1059–1065. https://doi.org/10.1001/jama.1954.02940470011003

Sumithran E, MacSween RN (1979) An appraisal of the relationship between primary hepatocellular carcinoma and hepatitis B virus. Histopathology 3(6):447–458. https://doi.org/10.1111/j.1365-2559.1979.tb03027.x

Sutnick AI, London WT, Gerstley BJ, Cronlund MM, Blumberg BS (1968) Anicteric hepatitis associated with Australia antigen. Occurrence in patients with Down's syndrome. JAMA 205(10):670–674

Szmuness W, Stevens CE, Harley EJ, Zang EA, Oleszko WR, William DC, Sadovsky R, Morrison JM, Kellner A (1980) Hepatitis B vaccine: demonstration of efficacy in a controlled clinical trial in a high-risk population in the United States. N Engl J Med 303(15):833–841. https://doi.org/10.1056/NEJM198010093031501

Tan AY, Law CH, Lee YS (1977) Hepatitis B antigen in the liver cells in cirrhosis and hepatocellular carcinoma. Pathology 9(1):57–64. https://doi.org/10.3109/00313027709085239

Tang LSY, Covert E, Wilson E, Kottilil S (2018) Chronic hepatitis B infection: a review. JAMA 319(17):1802–1813. https://doi.org/10.1001/jama.2018.3795

Tsukuda S, Watashi K (2020) Hepatitis B virus biology and life cycle. Antivir Res 182:104925. https://doi.org/10.1016/j.antiviral.2020.104925

Tu T, Budzinska MA, Shackel NA, Urban S (2017) HBV DNA integration: molecular mechanisms and clinical implications. Viruses 9(4). https://doi.org/10.3390/v9040075

Turbitt ML, Patrick RS, Goudie RB, Buchanan WM (1977) Incidence in South-west Scotland of hepatitis B surface antigen in the liver of patients with hepatocellular carcinoma. J Clin Pathol 30(12):1124–1128. https://doi.org/10.1136/jcp.30.12.1124

Valenzuela P, Medina A, Rutter WJ, Ammerer G, Hall BD (1982) Synthesis and assembly of hepatitis B virus surface antigen particles in yeast. Nature 298(5872):347–350. https://doi.org/10.1038/298347a0

Yuen MF, Chen DS, Dusheiko GM, Janssen HLA, Lau DTY, Locarnini SA, Peters MG, Lai CL (2018) Hepatitis B virus infection. Nat Rev Dis Primers 4:18035. https://doi.org/10.1038/nrdp.2018.35

Zajac BA, West DJ, McAleer WJ, Scolnick EM (1986) Overview of clinical studies with hepatitis B vaccine made by recombinant DNA. J Infect 13 Suppl A:39–45. https://doi.org/10.1016/s0163-4453(86)92668-x

9

Hepatitis A Virus: Feces, Food, and Fomites

Keywords IgM antibodies • Immunoelectron microscopy • VP proteins • Yellow berets

Abbreviations

CDC	Centers for Disease Control
CR326	Costa Rican HAV strain
EM	Electron microscopy
FDA	Food and Drug Administration
FRhK-4	Fetal rhesus monkey kidney cells
HBsAg	Hepatitis B surface antigen
HBV	Hepatitis B virus
HCV	Hepatitis C virus
HepA-I	Inactivated hepatitis A vaccine
HepA-L	Live, attenuated hepatitis A vaccine
IEM	Immunoelectron microscopy
Ig	Immunoglobulin
LID	Laboratory of Infectious Diseases
NIH	National Institutes of Health
PHS	Public Health Services
VP	Viral protein
WHO	World Health Organization

© The Author(s), under exclusive license to Springer Nature Switzerland AG 2025
V. G. Wilson, *The Conquest of Viruses*, https://doi.org/10.1007/978-3-031-87562-5_9

The Other Hepatitis Virus

In Chap. 8, the critical studies by the army and at the Willowbrook school revealed the existence of two types of suspected viral hepatitis that were designated type A and type B. Even without isolating the agents themselves, much information was obtained about the clinical features of both diseases. It was quickly determined that type B was typically associated with exposure to blood or bodily fluids while type A was transmitted more casually through household contacts and by ingestion of contaminated food or water. This pattern of transmission observed for the type A disease was consistent with a virus shed in the feces, a conjecture that would eventually prove correct. However, even with these distinct routes of transmission, the clinical presentations of type A or type B infections are very similar. Both types of infection produced an assortment of manifestations from frank jaundice to less specific indicators such as fever, anorexia, vomiting, diarrhea, fatigue, abdominal pain, and elevation of certain liver enzyme levels. The range of frequency and severity of these symptoms sufficiently overlaps between type A and type B such that the types cannot be distinguished based on patient presentation alone. Nonetheless, experiments at Willowbrook and in other "volunteer" settings did define some differences between these two diseases. For example, the two diseases differ in their incubation periods with type A having a shorter duration from infection to clinical disease. Type A incubation is between two to four weeks while type B has an incubation period of two to five months. Additionally, although both type A and type B infections tended to be mild or asymptomatic in young children, type A never became chronic while type B in children often resulted in permanent, lifelong infections. These different characteristics were consistent and repeatable when infections were passed from volunteer to volunteer. Based on these early studies, the medical and research communities concluded that there were two distinct types of disease presumably caused by different viruses. But even with this clear clinical distinction established in the 1950s, the inability to readily propagate either virus outside of humans handicapped the research into these viruses throughout most of the 1960s. While the discovery of the Australia antigen was the breakthrough that keyed the identification of the hepatitis B virus in 1970, the presumptive hepatitis A virus remained elusive.

Yellow Berets to the Rescue

Ultimately, the discovery of the hepatitis A virus would owe a debt not only to science but also to the social forces of that era. The 1960s were a tumultuous decade in American history where the early members of the "baby boomer" generation (individuals born between 1946 and 1964) were coming of age and casting off the norms and conformity of previous generations. There was a growing social consciousness among young people with a rising demand both for personal freedom of expression and for equality across traditional racial and sexual boundaries. We often think of the 1960s as the decade that birthed the "rights" movements – civil rights, women's rights, and gay rights, all notions that frequently alarmed and alienated older generations. On top of this generational strife, the Cold War with the USSR was a pervasive threat and the U.S. involvement in Vietnam was turning into an actual war. Many boomers identified with the 1960s catchphrase "make love, not war", and were dismayed by the continuous threats of military hostilities around the world. The Vietnam War in particular was a divisive issue, especially because of the mandatory military draft. Enacted in 1940 as Europe entered World War II, the draft initially required all eligible men between the ages of 21 and 34 to register for possible induction into the Armed Services. Over the next 15 years, the draft would keep the Army adequately supplied, but after World War II and the Korean War, there was a period of minimal draft needs. However, the escalation of American military activities in Vietnam beginning in 1964 under President Johnson required a resurgence in the draft to fulfill the demand for troops on the ground. The threat of being drafted into an unpopular war led to years of protests and demonstrations, particularly on college campuses where young men wanted an academic education, not martial training. This resistance only intensified in 1969 when a draft lottery was implemented for all 18-year-olds (although college deferments would remain for another 2 years). The antipathy towards the war and the draft created an environment where many young men looked for alternatives that would help them avoid going to Vietnam.

In addition to this standard draft, there was a parallel "doctor draft" (Khot et al. 2011). The doctor draft was a federal policy enacted during the Korean War that required all physicians and dentists under the age of 51 to register and potentially be drafted in a separate process to fulfill U.S. military needs. Draftees were obligated for two years of service in the Navy, Air Force, Army, or the Public Health Services (PHS). In the decade after the Korean War, the military needs were modest and this doctor draft had little impact on the young physician population. However, just as the expansion of U.S. military operations in Vietnam in the mid-1960s caused a rapid and dramatic surge in the need for troops, there was a corresponding need for more military physicians in all branches of the armed services. As with the general population, many young doctors coming out of their training were not eager to put their lives and careers on hold to serve in Vietnam, and the PHS became an attractive alternative.

The PHS oversees several agencies including the National Institutes of Health (NIH), the Food and Drug Administration (FDA), and the Centers for Disease Control (CDC). Since the 1950s, the NIH has had an intramural Research Associate program that trains young physicians in investigative research, and acceptance into the NIH Research Associate program was an alternative means of fulfilling the military obligation of drafted physicians. Instead of being sent wherever the armed services had a need, the Research Associate could stay in Bethesda, Maryland doing laboratory research. This experience provided them with unique training while working with some of the top clinicians in the country. Initially, this was a small program with only a few Associates in each institute at NIH, but the Vietnam War triggered a significant expansion of this program. Many young physicians during the 1960s were caught in the doctor draft and were looking for ways to serve the country other than in the military. Correspondingly, the NIH was in a period of rising funding and the various institutes were keen to attract the brightest young scholars to join their research activities. Expanding the Research Associate program seemed like an ideal way to bring in an influx of skilled talent and provide a way to serve the country without being in the military. Still, some considered the Research Associate option as a cowardly or unpatriotic alternative to military service, and these men collectively became known as the yellow berets. (Note that there were almost no women Associates because there were few women physicians in that era and they weren't subject to the doctor draft, so the Research Associate spots were given mostly to men.) No one knows the origins of this pejorative nickname, but it arose during the Vietnam era as a label for draft dodgers to contrast them with the renowned and heroic Green Berets of the Army Special Forces. The nickname was popularized in a 1966 song penned by Bob Seger called the "Ballad of the Yellow Berets" (a nod to the earlier "Ballad of the Green Berets"), though the term was likely already in public parlance. What is known is that the doctors in the Research Associate program of that era accepted the term and turned it into a badge of honor. They were serving their country while doing important work, and they were proud to be at NIH and be advancing science.

Records from the early years of the Associate's program are believed to be incomplete, but one estimate suggested that there were just 68 NIH Research Associates total in 1960, while by the end of 1973 (near the war's end) there were 229 Associates, an over three-fold expansion during the Vietnam conflict (Klein 1998). Although these absolute numbers are insignificant compared to the numbers drafted into the military, most of the Associates came from prestigious medical schools like Columbia and Harvard and were truly among the best in their generation. This influx of brilliant, well-trained, and highly motivated young physicians helped supply manpower for the flourishing NIH research engine. Not only did this lead to important scientific advances, but it also provided these young investigators with a nurturing environment that developed their research skills and helped launch their scientific careers. A retrospective analysis of the Research Associates from the program's implementation in 1953 through 1973 revealed its remarkable success (Khot et al. 2011). Among this 20-year cadre of NIH Research Associates, 9 went on to win Nobel Prizes, 10 received the National Medal of Science, 64 were elected members of the prestigious National Academy of Sciences, and 125 became members of the equally prestigious Institute of Medicine. Some stayed on at NIH where four eventually served as the directors of NIH, ten others became directors of various NIH institutes, and many others went on to become department heads and deans of medical schools around the country. This impressive resume unambiguously demonstrated that the NIH Research Associate program was a major force during the Vietnam War era and impacted U.S. biomedical research for decades afterward.

Into this high-powered program came Stephen Feinstone (2019). Dr. Feinstone was an undergraduate at Johns Hopkins University and did his medical training at the University of Tennessee. After two years of clinical training, he became a Research Associate in 1971, landing in the Laboratory of Infectious Diseases (LID). While not from one of the more elite medical schools, Dr. Feinstone had a strong laboratory research background with two publications from his work in medical school, likely factors in his acceptance into this highly competitive program. As an aspiring internist, he was assigned to the laboratory of Robert Purcell to work on viral hepatitis. He initially worked on the newly identified hepatitis B virus but was eventually assigned to find the elusive viral agent of hepatitis A. Infection studies with human volunteers (Krugman et al. 1967) and in marmosets (Deinhardt et al. 1967) had unequivocally established that the hepatitis A agent was present in the stools of patients but attempts to propagate the virus from stool samples into cell culture had never been successful. Culturing was usually necessary to produce large quantities of highly concentrated virions for identification and study. Without this option, investigators were stymied and Feinstone had an unenviable task ahead of him.

Hepatitis A virus was far from the only recalcitrant virus, and investigators worldwide were struggling to find new methods to identify unculturable viruses. One existing approach was electron microscopy (EM) which magnifies samples sufficiently that individual virions can be seen. However, this is very much the proverbial search for a needle in a haystack due to the incredibly minuscule size of virions. For this approach to succeed that needs to be a very high concentration of the virions in the solution to have any chance of detecting them. For example, finding one hepatitis A virus particle in a very small drop of water would be roughly like trying to find a single grain of rice in an Olympic swimming pool, a daunting and nearly impossible task! Samples from hepatitis A patients, and from patients with many other types of viral diseases, simply didn't have a high enough concentration of viral particles to make traditional EM successful. Sometimes samples could be concentrated by centrifugation or chromatography, but these we cumbersome approaches and not universally applicable. To overcome this barrier, June Almeida of Great Britain developed a new approach for visualizing viruses in clinical samples called immunoelectron microscopy (IEM) and published this procedure in 1969, barely two years before Feinstone arrived at NIH (Almeida and Waterson 1969). Her simple yet elegant approach was to use antibodies to aggregate the virions present in a dilute solution. When added in the right proportions, antibodies will bind their target and form lattice structures that anchor the targets together (Fig. 9.1). If you think of the virions as individual grapes, then a bunch of grapes is like the antibody-virion lattice with the stems being the antibodies. This antibody clumping effect collects and concentrates the virions into large bunches that are much more easily found by EM.

Sometimes being at the right place and time is critical to scientific success, and being at the NIH at that moment in time positioned Dr. Feinstone to be the discoverer of the hepatitis A virus. As he embarked on the hepatitis A project, one of the other investigators in the LID had just returned from a sabbatical with Dr. Almeida and brought the IEM technique back to NIH. Realizing its potential for their hepatitis A work, Feinstone and colleagues adopted this technique for their studies. They were also fortunate to have a collection of stool samples derived from Krugman's Willowbrook studies. One of Krugman's hepatitis type A samples collected from a Willowbrook student had been used as the source inoculum for a transmission study conducted on volunteers at the Joliet Prison in Illinois (Boggs et al. 1970). Stool

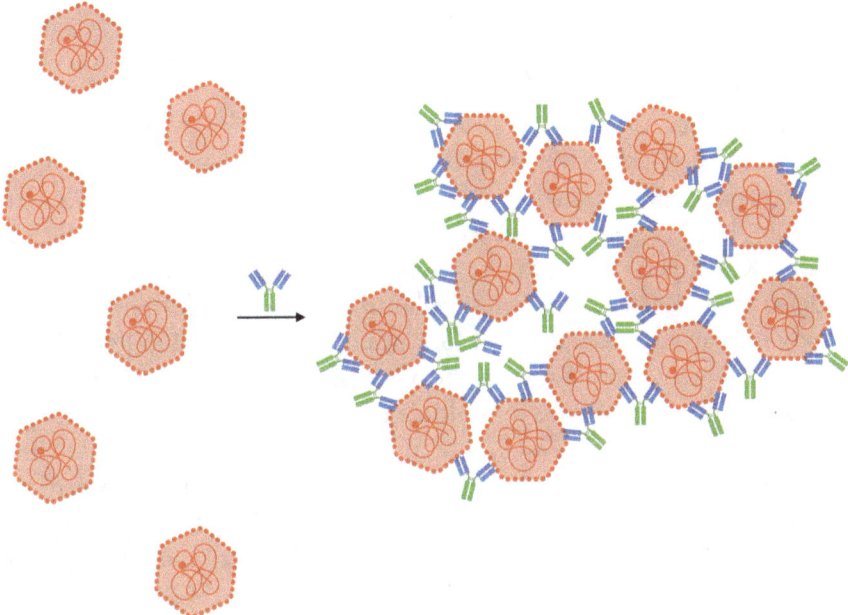

Fig. 9.1 Virion-antibody lattice. Shown is an illustration of a lattice structure formed from hepatitis A virus virions (red) and antibodies (green and blue). Since each antibody has two antigen-binding arms (the blue segments), each arm can bind to a separate virion. Adding antibodies to a virion solution can concentrate the virions by creating a linked network of virions bound together by multiple antibodies. The green region of the antibodies is the constant region that is not involved in antigen binding

samples had been collected from each of the infected inmates and Feinstone's lab had a set of these samples that were available for testing by IEM. In a eureka moment, one sample, designated F33, showed grape-like clusters when treated with the anti-HAV antibodies. The visualized virions were small, spherical, and lacked an envelope as expected for an enteric virus (membrane envelopes cannot survive the digestive system so gastrointestinal viruses typically lack this structure). After a long series of careful tests with numerous controls, they were convinced that the HAV had finally been identified and they quickly published their findings in a December 1973 issue of the journal *Science* (Feinstone et al. 1973). Now with both HBV and HAV in hand, further discoveries could progress more rapidly although an HAV vaccine was still twenty years in the future.

Polio's Cousin

Research in the ten years following Feinstone's discovery of HAV saw the accumulation of extensive information about the virus and important advances in diagnosis. HAV was eventually classified as a member of the Picornavirus family, the family that also contains the infamous poliovirus (Chap. 6), but in a genus named *Hepatovirus* that is distinct from poliovirus. As for all members of the Picornavirus family, HAV has a positive-sense, single-stranded RNA genome packaged into a highly durable, non-enveloped capsid. The capsid is comprised of four proteins termed VP1–VP4 (VP simply stands for "viral protein" and the VP designation is commonly used to denote virion capsid proteins in multiple RNA viruses). VP1, VP2, and VP3 are much larger than VP4, and these first three populate most of the virion surface. Being surface proteins, they are accessible to our immune system and are recognized as foreign antigens. Antibodies against these proteins are important both for recovery from infection and for vaccine-induced protection and will be discussed further below.

The origins of HAV remain obscure, both in its evolutionary history and its emergence as a human virus (Sander et al. 2018). It appears to be an ancient pathogen only able to infect humans and certain primates, but when it entered human populations has not been determined. Interestingly, a recent study of 209 small mammalian species, including bats, found 13 novel *Hepatovirus* species related to HAV (Drexler et al. 2015). The relatively widespread distribution of *Hepatoviruses* in rodents and other small mammals suggests a zoonotic origin for the human HAV. At some point in the distant past, a precursor virus likely jumped from its natural host into primates and humans, although the source animal(s) and the timing of the jump are unknown. Another study looking at the capsid structure of HAV noted that there was relatedness to a picorna-like virus that infects insects (Wang et al. 2015). This observation raises the possibility that picornaviruses in general might have arisen initially as an insect virus and then eventually spread into mammals and diversified into the various genera and species found today.

Like poliovirus, HAV is an enteric virus that persists through an oral-fecal transmission cycle (Van Damme et al. 2023). As a virus that has to pass through the acidic conditions of the stomach and survive the numerous digestive enzymes that reside there, HAV has evolved a highly durable capsid that persists remarkably long in the environment (Sattar et al. 2000). This capsid is resistant to drying, freezing, and many chemicals; retains its infectivity for days to weeks in dried feces; and can survive for possibly a year in water.

Infection is initiated when someone ingests viral particles, typically from contaminated food or water. Additionally, since the virus can survive for long periods on hard surfaces, unsuspecting people can pick up viruses on their hands and then transfer viruses to their mouths, either directly or via food that they are handling. Once ingested, the precise progression of the virus within the body is only somewhat understood due to the lack of adequate animal model systems that faithfully mimic human infection. In general, it is believed that ingested virions pass into the intestines where some replication occurs in infected epithelial cells. Newly made viruses eventually travel to their primary target organ, the liver, either via the bloodstream and/or the biliary system (the pancreas, gallbladder, and bile ducts). As for HBV, HAV infects the hepatocyte cells of the liver and replicates in them without killing these cells (noncytopathic replication). Liver cell damage and hepatitis symptoms are believed to be due to the immunological attack (T cells) on these infected hepatocytes. In conjunction with this immune cell attack on the infected hepatocytes, neutralizing antibodies (humoral immunity) develop which stop viral spread and prevent the infection of new cells. Eventually, this combination of humoral and cellular immunity eliminates the virus and returns the patient to health.

Given its physical properties and transmission route, HAV's epidemiology is similar to that of its relative, poliovirus (Migueres et al. 2021). In less developed nations where food and water sanitation are poor, HAV infections tend to be prevalent and spread easily among family groups. Under these conditions, most individuals acquire the infection early in life. For both polio and hepatitis A viruses, symptoms in young children are generally minimal or completely inapparent. Importantly though, HAV lacks the neurovirulence that makes poliovirus so dangerous. Additionally, HAV is very effectively eliminated by our immune system and never establishes chronic infections. Without the possibility of chronic infections that make HBV so problematic, HAV cases in young people resolve completely and appear to provide lifetime immunity against reinfection. In these less developed countries, while there may be a high incidence of infection, there is very little incidence of hepatitis from this virus or any acute disease. In contrast, citizens of more developed nations have less likelihood of early life exposure and so often enter adulthood completely naïve to this virus. Without childhood-acquired natural immunity, older children and adults who become infected develop more aggressive immune responses that can cause extensive liver damage which manifests as hepatitis symptoms. Fortunately, while adults are more susceptible to acute disease from HAV infections, fatalities and serious complications are rare and patients almost always recover and develop protective immunity. It is important to note that both infected children and adults excrete HAV in their feces, often for weeks. Especially in the absence of symptoms, infectious individuals may be inadvertently contaminating their surroundings, including their food and water supplies. When contaminated water is used for irrigation, this can further spread the virus onto crops where it lurks. Notably, vegetables that can be eaten raw are a common source of HAV outbreaks. This silent spread of these very persistent virions makes environmental contamination a serious and difficult-to-control problem. While the death rate from HAV is thankfully much less than for many serious viral diseases, the impact of HAV is not insignificant. The WHO estimates that there are over 150 million cases each year globally (mostly unreported) with roughly 40,000 deaths and a substantial burden of morbidity and associated healthcare costs (WHO position paper on hepatitis A vaccines – October 2022 2022). Sadly, this could all be prevented with the HAV vaccine which has been available since 1994.

The Path to the HAV Vaccine

The 1970s were a remarkable decade for biological science. Both DNA sequencing and DNA cloning, Nobel Prize-winning discoveries that revolutionized studies of molecular and cellular biology, were invented in the 70s. Similar impactful innovations were being made in virology and immunology, and Dr. Feinstone's visualization of the HAV virions in patient samples was one such important advance. While his observation didn't immediately overcome the major impediment to vaccine development, the inability to culture the virus, his work spurred renewed interest in understanding this virus. Multiple labs throughout the 1970s made progress in many significant areas of HAV research. The original demonstration that HAV could cause infection in marmosets (Deinhardt et al. 1967) led to extensive development of this species as a model system for HAV studies (Deinhardt et al. 1972; Ebert et al. 1978). Other primate species were investigated as possible model hosts for HAV, and chimpanzees and certain other monkeys were also found to be infectable with this virus (Dienstag et al. 1975; Deinhardt et al. 1972). Marmosets were especially useful as the virus accumulated to high levels in their livers where it could be harvested and partially purified (Provost et al. 1975). The ability to infect and produce the virus in marmosets would provide the first reliable source of viral strains for subsequent studies. Marmosets could be inoculated with a sample from a single patient and that single-source virus then propagated from animal to animal to produce a pure strain. Among the first strains isolated this way was one derived from the blood of a Costa Rican patient (Mascoli et al. 1973). This patient's virus was successfully passed through several successive marmosets and was ultimately designated CR326, a strain that would later be used for vaccine development. Having a consistent source of the virus allowed labs to characterize the physical properties of the virus, helping to classify it eventually into the picornavirus family. Other labs used this model to develop a variety of immunological tests to detect either the virus itself or antibodies made against the virus. Such serological tests were critical for following infections in animals and for assessing HAV infections in human patients.

One particularly salient discovery was the observation that two types of antibodies were made during an HAV infection. Antibodies are protein molecules whose formal name is immunoglobulins (Ig). While we've made many references to antibodies in the chapters of this book, that term has been used generically. There are actually five distinct types or classes of immunoglobulins known as IgA, IgD, IgE, IgG, and IgM. The class that is most relevant to vaccines is IgG, and neutralizing antibodies in the bloodstream that are generated by infection or vaccination are usually of this type. However, responding to a pathogen is more complicated than just producing IgG. It is a two-step process where early in infections the first antibodies to be made are of the IgM type. These IgM antibodies are short-lived, are only present during the initial infections, and do not persist for long after recovery. At some point during the acute phase of an infection, there is a class-switching event and the IgM antibodies stop being produced and are replaced by the IgG antibodies that persist through memory B cells. The discovery of both IgM and IgG antibodies during HAV infection provided a tool to discriminate between current and previous infections (Bradley et al. 1977). Testing for IgG antibodies against HAV only reveals whether or not the person has had an exposure to HAV (or a vaccination), but these antibodies could be due to either a recent infection or one long in the past. In contrast, if a patient has IgM antibodies to HAV then they are currently or very recently infected as IgM antibodies rarely persist for more than six months after an HAV infection. This difference makes IgM antibody detection a specific diagnostic test for acute HAV that is still used today (Gholizadeh et al. 2023).

In addition to the basic science advances, the ability to produce viruses in marmosets led to the first prototype HAV vaccine, once again an effort spearheaded by Maurice Hilleman at Merck (Provost and Hilleman 1978). Using the Costa Rican CR326 strain, Provost and Hilleman infected marmosets and produced viral stocks by extracting virions from the livers of infected animals. The crude viral stocks were inactivated by a prolonged (four-day) treatment with formalin then used to vaccinate naïve marmosets. The vaccination generated antibodies against HAV and protected the animals against infection when challenged with live HAV. Even though this crude vaccine material was unsuitable for use in humans and was only tested in marmosets, their results demonstrated that vaccination with an inactivated HAV should protect against an HAV infection, a finding that bolstered efforts to turn this pilot study into a human vaccine.

While the marmoset was sufficient for small studies, without a cell culture system for producing large quantities of viral preparations, vaccine production would never be feasible. However, the "holy grail" of HAV research, cell culture propagation, eluded investigators for decades. There are likely many factors that contributed to early failures with HAV reproduction in cell culture: poor culture techniques, limited assays for following HAV infection (remember HAV is noncytopathic so it doesn't kill or visibly alter cells), contamination of cells with other endogenous viruses, use of cell types that were refractory to HAV infection, and inherent differences in HAV isolates from different patients or animals. Whatever the combination of causes, it wasn't possible to culture HAV until it was. It may have been like the four-minute mile. Once thought to be impossible, after Roger Bannister broke the four-minute barrier in 1954, sub-four minutes became an accepted standard that has been achieved by many elite runners. Again using the Costa Rican CR326 HAV strain, Hilleman and his colleague, Philip Provost, broke the culture barrier in 1979 (Provost and Hilleman 1979). First, they passaged the CR326 strain through 31 additional marmosets and then used the final viral stock to infect a marmoset liver cell culture. They were able to successively passage the virus through multiple rounds of cultures, a transmission process that could only work if the virus was truly replicating in these cultures. Their success was likely due to two elements, the extensive passaging in marmosets so that the virus was well adapted to this host, and using liver cells, some of which were hepatocytes, HAV's natural target cell type. In the same paper, they also demonstrated that HAV could be propagated in fetal rhesus monkey kidney cells (FRhK-4) that seemed to be highly susceptible to HAV infection. As with Hilleman's vaccine prototype, these two culture systems were proof of principle but not the ultimate solution to the culturing problem. Marmoset liver and rhesus kidney primary cultures were short-lived and could only be carried for a limited number of passages before they senesced and died, so the long-term maintenance need for large quantities of virus for vaccine production was not possible with these cell systems.

Breaking the culture barrier for HAV rapidly led to a plethora of advances in propagating HAV by the Hilleman team as well as several other research groups. Within a year, Provost and Hilleman had adapted the CR326 strain for growth in human cell lines WI-38 and MRC-5; both are embryonic lung fibroblast cultures commonly used for vaccine production (Provost and Hilleman 1979). More importantly, a European group propagated HAV directly from the feces of a human patient, proving that animal adaptation was not required for establishing HAV in culture (Frosner et al. 1979). There were two critical findings from these early studies that greatly facilitated the subsequent propagation of HAV. First, HAV's reproduction in cell culture was extremely slow and initially often took weeks to months to accumulate high levels of progeny virions. This time shortened as the virus became cell culture adapted, but it still took weeks, unlike other common viruses that replicated within a few days. The second observation was that HAV was not effectively released from the infected cells and instead remained cell-associated. Many viruses commonly lyse their host cells to release progeny virions out into the media. Typical transmission studies were done by taking this virus-containing liquid media from one plate of cells and using it to infect a new plate of cells. However, with HAV this culture media contains very little virus which likely hindered attempts to propagate HAV by standard protocols. A technical fix used by Hilleman and subsequent HAV researchers was to freeze-thaw the infected cultures first. Subjecting the infected cells to several rounds of freezing and thawing burst the cells and released the virions into the media which could now be used to infect the next round of cells. With a better understanding of HAV's growth kinetics and cell-associated nature, labs from around the world were quickly able to establish HAV in culture (Flehmig et al. 1981; Gauss-Muller et al. 1981; Kojima et al. 1981; Daemer et al. 1981). Culturing HAV directly from patient samples and propagating it in a variety of human and primate cells finally became routine in the early 1980s.

The ability to propagate HAV in vaccine-appropriate cell lines was a critical hurdle, and overcoming this technical challenge set the foundation for finally developing HAV vaccines. As seen with previously presented vaccines, the next key decision for a potential HAV vaccine was what type of vaccine would be most suitable, inactivated or attenuated. Hilleman's pilot study in marmosets with an inactivated vaccine suggested that this approach was feasible (Provost and Hilleman 1978) but didn't address the potential of an attenuated vaccine. With no a priori basis for choosing one or the other, Hilleman's group created and investigated both types of vaccine, with other researchers also joining the vaccine race. By 1982, the Provost-Hilleman team had produced and tested an attenuated HAV strain that was created by passaging marmoset-adapted virus (CR326) first in fetal rhesus monkey kidney cells and then in human cells at 32 °C (Provost et al. 1982). This attenuated virus-induced anti-HAV antibodies in marmosets and protected them from disease when challenged with wild-type HAV, clearly demonstrating that an attenuated vaccine was feasible. Feinstone and Purcell at the NIH were conducting similar experiments with similar results using their Australian isolate, HM-175, as the starting virus for producing an attenuated strain (Feinstone et al. 1983). One problem that emerged from these studies was calibrating the degree of attenuation. Unlike some other viruses that could be passaged until they reached an attenuated state that was stable and suitable for a vaccine, the attenuation of HAV required more finesse. Too little attenuation and the virus could still cause pathology in the liver even though it failed to produce frank hepatitis, a result that was unacceptable for a vaccine used in humans. With too much attenuation the virus could still infect the test animals but failed to elicit a protective immune response and was not useful as a vaccine. These early studies suggested that an attenuated vaccine could potentially be developed for humans although it would take careful development and testing to find the right candidate strain. Thus, like its relative, poliovirus, HAV seemed amenable to developing both an inactivated and an attenuated vaccine. With no clear favorite, researchers continued exploring vaccine types throughout that decade, testing them in a variety of animal model systems (Karron et al. 1988; Flehmig et al. 1987; Binn et al. 1986; Provost et al. 1986b).

With encouraging results in animals and no indications of serious adverse effects, the next step was to finally move into phase I human testing. By 1990, several groups had reported studies of either attenuated or inactivated vaccines in small numbers of human volunteers, with many more trials in progress. (Provost et al. 1986a; Flehmig et al. 1989; Mao et al. 1989). Phase I trials are designed to test new drugs or vaccines on a small number of healthy people to see if there are any adverse effects. In addition, for these early HAV vaccines, the development of anti-HAV antibodies was measured to determine if the vaccine could elicit an immune response in the volunteers. Both the inactivated and attenuated candidate vaccines generated neutralizing antibodies against HAV, although the attenuated vaccines seemed to produce lower levers of HAV antibodies, suggesting that they might be less effective as preventative vaccines (Wiedermann et al. 1990; Midthun et al. 1991; Sjogren et al. 1991; Lewis et al. 1991; Ellerbeck et al. 1992). Nonetheless, the combined results from the phase I studies over several years were encouraging with no significant adverse reactions and induction of anti-HAV antibodies in almost all the recipients. However, what was not tested was the ability of the vaccine to protect the recipients from disease. Clinical efficacy is usually assessed in phase II and III trials once the safety of the product has been established. With no contraindications from phase I reports, by the early 1990s, it was time to move into human protection studies. Given the lesser response to attenuated vaccines, along with the ever-present possibility that an attenuated vaccine can revert to virulence, Merck (formally Merck Sharp & Dohme) and SmithKline Beecham elected to pursue the inactivated vaccine. Other countries, primarily China, continued the development of an attenuated vaccine (Mao et al. 1989), and eventually, this live vaccine became predominantly used in China and a few other countries (Cui et al. 2014).

There are two approaches to determining vaccine efficacy, challenge studies and community-based studies. In challenge studies, vaccinated volunteers are intentionally infected with the virulent disease virus to see whether or not the disease develops. In community-based studies, vaccinated and control individuals go about their normal daily lives, and the number of disease cases in both populations is monitored over time to see if the vaccinated population is protected from community-acquired infections. While challenge experiments had been done previously for some viral vaccines, by the 1990s such studies were becoming uncommon for ethical reasons, especially if they involved children. Consequently, both the Merck and SmithKline vaccines were tested predominantly in placebo-controlled community studies involving either children or adults. There was one practical issue with this approach and that was the lack of HAV cases in most developed countries. For a community study to be significant, there have to be enough naturally occurring cases so that a statistical difference can be shown between case numbers in the control versus the vaccinated cohorts. Merck targeted a small community in upstate New York with a large Hasidic Jewish population (Werzberger et al. 1992). The community had a history of recurring HAV outbreaks during the summer months and offered a natural source of community-acquired infections. In 1991, 1037 children (2–16 years of age) whose parents consented were enrolled in the study. They were randomly divided into two groups with one group receiving the HAV vaccine and one the placebo injection. This was a double-blind study where neither the children nor the personnel administering the shots knew which was the vaccine group and which was the control group. Ultimately, the vaccine proved to be 100% effective with 25 cases occurring in the control group and zero cases in the vaccinated group. Similarly, SmithKline field tested their vaccine in multiple European countries where there was significant HAV disease (Andre et al. 1992), as well as a huge trial in Thailand involving 40,000 children (Innis et al. 1992).

All the HAV vaccine clinical trials concluded that these vaccines were safe and effective at preventing HAV infections in a community setting. Even before all the clinical trials were completed, formal licensing of both the inactivated and attenuated vaccines was beginning in many countries. The first to market was the SmithKline vaccine called Havrix™ which received European approval in 1992, while in the same year, the Chinese began using their live, attenuated vaccine in their country. Both Havrix™ and Merck's inactivated HAV vaccine (called Vaqta™) were licensed in the U.S. in 1995. Additional HAV vaccines soon followed as other companies developed their own proprietary versions of an inactivated vaccine. At last, there were excellent vaccines for hepatitis A and hepatitis B, and the incidence of both diseases could now be dramatically reduced through nationwide vaccination campaigns. Unfortunately, just as there was finally an effective means for thwarting these two hepatitis viruses, in 1989 hepatitis C (HCV) was discovered, and this virus has proven very refractory to vaccine development (see Chap. 15).

HAV Vaccines in the Twenty-First Century

We are now 30 years out from the first HAV vaccine, so there are three decades of experience to help us understand where we are with this virus and its vaccine. By the end of the twentieth century, several HAV vaccines were widely used in different nations, both the inactivated (HepA-I) and the attenuated, live (HepA-L) types of vaccines. Most of the world is using the HepA-I vaccine except China and India where HepA-L is the standard. Each type of vaccine has been administered to millions of people, and the cumulative evidence is that both vaccines are safe, highly immunogenic (i.e. they elicit anti-HAV antibodies), provide long-lasting immunity (>20 years), and give protection from clinical HAV disease in the 95–100% range in various studies (Zhang 2020). What remains evolving is the optimal way to utilize these vaccines to reduce or eliminate this disease cost-effectively. Currently, the WHO recommends three strategies depending on the incidence rate for HAV infections (The Immunological Basis for Immunization Series: Module 18 - Hepatitis A 2019). In developing countries where food and water sanitation are poor, HAV is endemic and most children become infected at an early age where infection rarely leads to disease but does produce immunity. Trying to vaccinate all children would be costly and would not provide much clinical benefit as the majority of children will become immune through natural infection. Instead, it is suggested that these countries use a targeted approach that focuses on high-risk groups. Among the high-risk groups are the immunosuppressed, IV drug users, and people with chronic liver disease, all individuals who might suffer dangerous cases of hepatitis from an HAV infection. In contrast, developed nations with very low endemic rates are encouraged to provide universal childhood vaccination. In these countries, infants and children are not typically exposed to HAV and generally reach adulthood with no immunity. This results in a large at-risk population since adults tend to have more severe disease from an HAV infection and often require more clinical intervention. Implementing widespread vaccination in early childhood eliminates this issue and produces a largely immune adult cohort. This universal strategy was first adopted by Israel in 1999 and has since been widely implemented by over 30 countries around the world, including the United States. A 2015 study found that the HAV vaccination program in the United States prevented nearly 95,000 infections each year which would have resulted in approximately 46,000 outpatient visits, 1300 hospitalizations, and 15 deaths, an impressive reduction in potential suffering and medical costs (Dhankhar et al. 2015).

The final and most complicated situation exists in transitioning nations where the public works infrastructure is improving but hasn't fully reached the standards of developed nations. These intermediate countries often have wide regional variations in endemic HAV rates. It is less clear which approach is the most effective, both in cost and in disease reduction, but combinational strategies include universal vaccination in low-endemic areas and targeted vaccination in regions with high endemic rates. Each country will need to evaluate what is feasible and financially sound to provide the most beneficial protection for their population, and their approach may need to be modified over time as endemic rates change. While improving living conditions in many countries coupled with appropriate vaccination programs have resulted in significant reductions in morbidity and mortality from this virus, it is unclear whether or not HAV could ever be eliminated with the current vaccine technology.

What might ultimately help in the battle with HAV is cheaper vaccines. To produce the current inactivated or live, attenuated vaccines requires growing large quantities of virus, an expensive and cumbersome process. In addition, there has to be extensive quality control and testing to ensure that production batches do not contain any wild-type viruses due to incomplete inactivation of HepA-I vaccines or reverting mutations in HepA-L vaccines. A subunit-type vaccine such as the HBV vaccine would be an attractive alternative yet very little progress has been made in developing such a vaccine. Unlike hepatitis B, producing a subunit vaccine consisting of HAV surface proteins has been challenging. For HBV, the major surface protein (HBsAg) that is recognized by neutralizing antibodies structurally folds into its native conformation when cloned and expressed in yeast cells. This means that the cloned protein retains the same shape that it has in the virion so the purified protein induces neutralizing antibodies just as it would during an infection. For HAV, the situation is more complicated as four viral proteins comprise the virion structure: VP1, VP2, VP3, and VP4. VP1, 2, and 3 form the capsid surface while VP4 is internal. With three surface proteins, studies were needed to determine which are important for eliciting neutralizing antibodies. Regions of both VP1 and VP3 were shown to contribute to the antibody binding site, suggesting that an interface between adjacent protein molecules presents the surface feature that is recognized by the antibody (Nainan et al. 1992). This potential dual protein requirement makes it unclear whether or not either single protein would be sufficient as a vaccine. Instead, a subunit vaccine might need both proteins that have assembled just as they do in the virion, not necessarily an easy feat to accomplish. Additionally, the expression of cloned versions of these proteins has been problematic due to solubility and aggregation issues (Gauss-Muller et al. 1990; Nain et al. 2022). These issues have dampened enthusiasm for the development of a subunit vaccine for HAV and only modest effort has been made to address this type of vaccine. While there are potential strategies to circumvent these problems such as cloning and expressing only the antigenically important regions of the proteins rather than the full proteins, these approaches are still in the early investigative stage. For now, the traditional HepA-I and HepA-L vaccines remain our major weapons for combating HAV infections and reducing the worldwide impact of this virus.

References

Almeida JD, Waterson AP (1969) The morphology of virus-antibody interaction. Adv Virus Res 15:307–338. https://doi.org/10.1016/s0065-3527(08)60878-7

Andre FE, D'Hondt E, Delem A, Safary A (1992) Clinical assessment of the safety and efficacy of an inactivated hepatitis A vaccine: rationale and summary of findings. Vaccine 10 Suppl 1:S160–S168. https://doi.org/10.1016/0264-410x(92)90576-6

Binn LN, Bancroft WH, Lemon SM, Marchwicki RH, LeDuc JW, Trahan CJ, Staley EC, Keenan CM (1986) Preparation of a prototype inactivated hepatitis A virus vaccine from infected cell cultures. J Infect Dis 153(4):749–756. https://doi.org/10.1093/infdis/153.4.749

Boggs JD, Melnick JL, Conrad ME, Felsher BF (1970) Viral hepatitis. Clinical and tissue culture studies. JAMA 214(6):1041–1046. https://doi.org/10.1001/jama.214.6.1041

Bradley DW, Maynard JE, Hindman SH, Hornbeck CL, Fields HA, McCaustland KA, Cook EH Jr (1977) Serodiagnosis of viral hepatitis A: detection of acute-phase immunoglobulin M anti-hepatitis A virus by radioimmunoassay. J Clin Microbiol 5(5):521–530. https://doi.org/10.1128/jcm.5.5.521-530.1977

Cui F, Liang X, Wang F, Zheng H, Hutin YJ, Yang W (2014) Development, production, and postmarketing surveillance of hepatitis A vaccines in China. J Epidemiol 24(3):169–177. https://doi.org/10.2188/jea.je20130022

Daemer RJ, Feinstone SM, Gust ID, Purcell RH (1981) Propagation of human hepatitis A virus in African green monkey kidney cell culture: primary isolation and serial passage. Infect Immun 32(1):388–393. https://doi.org/10.1128/iai.32.1.388-393.1981

Deinhardt F, Holmes AW, Capps RB, Popper H (1967) Studies on the transmission of human viral hepatitis to marmoset monkeys. I. Transmission of disease, serial passages, and description of liver lesions. J Exp Med 125(4):673–688. https://doi.org/10.1084/jem.125.4.673

Deinhardt F, Wolfe L, Holmes AW, Junge U (1972) Viral-hepatitis in nonhuman primates. Can Med Assoc J 106:468-+

Dhankhar P, Nwankwo C, Pillsbury M, Lauschke A, Goveia MG, Acosta CJ, Elbasha EH (2015) Public health impact and cost-effectiveness of hepatitis A vaccination in the United States: a disease transmission dynamic modeling approach. Value Health 18(4):358–367. https://doi.org/10.1016/j.jval.2015.02.004

Dienstag JL, Feinstone SM, Purcell RH, Hoofnagle JH, Barker LF, London WT, Popper H, Peterson JM, Kapikian AZ (1975) Experimental infection of chimpanzees with hepatitis A virus. J Infect Dis 132(5):532–545. https://doi.org/10.1093/infdis/132.5.532

Drexler JF, Corman VM, Lukashev AN, van den Brand JM, Gmyl AP, Brunink S, Rasche A, Seggewibeta N, Feng H, Leijten LM, Vallo P, Kuiken T, Dotzauer A, Ulrich RG, Lemon SM, Drosten C, Hepatovirus Ecology C (2015) Evolutionary origins of hepatitis A virus in small mammals. Proc Natl Acad Sci USA 112(49):15190–15195. https://doi.org/10.1073/pnas.1516992112

Ebert JW, Maynard JE, Bradley DW, Lorenz D, Krushak DH (1978) Experimental infection of marmosets with hepatitis A virus. Primates Med 10:295–299

Ellerbeck EF, Lewis JA, Nalin D, Gershman K, Miller WJ, Armstrong ME, Davide JP, Rhoad AE, McGuire B, Calandra G et al (1992) Safety profile and immunogenicity of an inactivated vaccine derived from an attenuated strain of hepatitis A. Vaccine 10(10):668–672. https://doi.org/10.1016/0264-410x(92)90087-z

Feinstone SM (2019) History of the discovery of hepatitis A virus. Cold Spring Harb Perspect Med 9(5). https://doi.org/10.1101/cshperspect.a031740

Feinstone SM, Kapikian AZ, Purceli RH (1973) Hepatitis A: detection by immune electron microscopy of a viruslike antigen associated with acute illness. Science 182(4116):1026–1028. https://doi.org/10.1126/science.182.4116.1026

Feinstone SM, Daemer RJ, Gust ID, Purcell RH (1983) Live attenuated vaccine for hepatitis A. Dev Biol Stand 54:429–432

Flehmig B, Vallbracht A, Wurster G (1981) Hepatitis A virus in cell culture. III. Propagation of hepatitis A virus in human embryo kidney cells and human embryo fibroblast strains. Med Microbiol Immunol 170(2):83–89. https://doi.org/10.1007/BF02122672

Flehmig B, Haage A, Pfisterer M (1987) Immunogenicity of a hepatitis A virus vaccine. J Med Virol 22(1):7–16. https://doi.org/10.1002/jmv.1890220103

Flehmig B, Heinricy U, Pfisterer M (1989) Immunogenicity of a killed hepatitis A vaccine in seronegative volunteers. Lancet 1(8646):1039–1041. https://doi.org/10.1016/s0140-6736(89)92443-4

Frosner GG, Deinhardt F, Scheid R, Gauss-Muller V, Holmes N, Messelberger V, Siegl G, Alexander JJ (1979) Propagation of human hepatitis A virus in a hepatoma cell line. Infection 7(6):303–305. https://doi.org/10.1007/BF01642154

Gauss-Muller V, Frosner GG, Deinhardt F (1981) Propagation of hepatitis A virus in human embryo fibroblasts. J Med Virol 7(3):233–239. https://doi.org/10.1002/jmv.1890070308

Gauss-Muller V, Zhou MQ, von der Helm K, Deinhardt F (1990) Recombinant proteins VP1 and VP3 of hepatitis A virus prime for neutralizing response. J Med Virol 31(4):277–283. https://doi.org/10.1002/jmv.1890310407

Gholizadeh O, Akbarzadeh S, Ghazanfari Hashemi M, Gholami M, Amini P, Yekanipour Z, Tabatabaie R, Yasamineh S, Hosseini P, Poortahmasebi V (2023) Hepatitis A: viral structure, classification, life cycle, clinical symptoms, diagnosis error, and vaccination. Can J Infect Dis Med Microbiol 2023:4263309. https://doi.org/10.1155/2023/4263309

Innis BL, Snitbhan R, Kunasol P, Laorakpongse T, Poopatanakool W, Suntayakorn S, Suknantapong T, Safary A, Boslego JW (1992) Field efficacy trial of inactivated hepatitis-a vaccine among children in Thailand. Vaccine 10:S159–S159. https://doi.org/10.1016/0264-410x(92)90575-5

Karron RA, Daemer R, Ticehurst J, D'Hondt E, Popper H, Mihalik K, Phillips J, Feinstone S, Purcell RH (1988) Studies of prototype live hepatitis A virus vaccines in primate models. J Infect Dis 157(2):338–345. https://doi.org/10.1093/infdis/157.2.338

Khot S, Park BS, Longstreth WT Jr (2011) The Vietnam war and medical research: untold legacy of the U.S. doctor draft and the NIH "yellow berets". Acad Med 86(4):502–508. https://doi.org/10.1097/ACM.0b013e31820f1ed7

Klein MK (1998) The legacy of the "yellow berets": the Vietnam war, the doctor draft, and the NIH associate training program. NIH History Office

Kojima H, Shibayama T, Sato A, Suzuki S, Ichida F, Hamada C (1981) Propagation of human hepatitis A virus in conventional cell lines. J Med Virol 7(4):273–286. https://doi.org/10.1002/jmv.1890070404

Krugman S, Giles JP, Hammond J (1967) Infectious hepatitis. Evidence for two distinctive clinical, epidemiological, and immunological types of infection. JAMA 200(5):365–373. https://doi.org/10.1001/jama.200.5.365

Lewis JA, Armstrong ME, Larson VM, Emini EA, Midthun K, Ellerbeck E, Nalin D, Provost PJ, Calandra GB (1991) Use of a live, attenuated hepatitis-a vaccine to prepare a highly purified, formalin-inactivated hepatitis-a vaccine. In: Viral hepatitis and liver disease, pp 94–97. https://www.webofscience.com/wos/woscc/full-record/WOS:A1991BY03U00024

Mao JS, Dong DX, Zhang HY, Chen NL, Zhang XY, Huang HY, Xie RY, Zhou TJ, Wan ZJ, Wang YZ et al (1989) Primary study of attenuated live hepatitis A vaccine (H2 strain) in humans. J Infect Dis 159(4):621–624. https://doi.org/10.1093/infdis/159.4.621

Mascoli CC, Ittensohn OL, Villarejos VM, Arguedas JA, Provost PJ, Hilleman MR (1973) Recovery of hepatitis agents in the marmoset from human cases occurring in Costa Rica. Proc Soc Exp Biol Med 142(1):276–282. https://doi.org/10.3181/00379727-142-37005

Midthun K, Ellerbeck E, Gershman K, Calandra G, Krah D, McCaughtry M, Nalin D, Provost P (1991) Safety and immunogenicity of a live attenuated hepatitis A virus vaccine in seronegative volunteers. J Infect Dis 163(4):735–739. https://doi.org/10.1093/infdis/163.4.735

Migueres M, Lhomme S, Izopet J (2021) Hepatitis A: epidemiology, high-risk groups, prevention and research on antiviral treatment. Viruses 13(10). https://doi.org/10.3390/v13101900

Nain A, Kumar M, Banerjee M (2022) Oligomers of hepatitis A virus (HAV) capsid protein VP1 generated in a heterologous expression system. Microb Cell Factories 21(1):53. https://doi.org/10.1186/s12934-022-01780-x

Nainan OV, Brinton MA, Margolis HS (1992) Identification of amino acids located in the antibody binding sites of human hepatitis A virus. Virology 191(2):984–987. https://doi.org/10.1016/0042-6822(92)90277-v

Provost PJ, Hilleman MR (1978) An inactivated hepatitis A virus vaccine prepared from infected marmoset liver. Proc Soc Exp Biol Med 159(2):201–203. https://doi.org/10.3181/00379727-159-40314

Provost PJ, Hilleman MR (1979) Propagation of human hepatitis A virus in cell culture in vitro. Proc Soc Exp Biol Med 160(2):213–221. https://doi.org/10.3181/00379727-160-40422

Provost PJ, Wolanski BS, Miller WJ, Ittensohn OL, McAleer WJ, Hilleman MR (1975) Physical, chemical and morphologic dimensions of human hepatitis A virus strain CR326 (38578). Proc Soc Exp Biol Med 148(2):532–539. https://doi.org/10.3181/00379727-148-38578

Provost PJ, Banker FS, Giesa PA, McAleer WJ, Buynak EB, Hilleman MR (1982) Progress toward a live, attenuated human hepatitis A vaccine. Proc Soc Exp Biol Med 170(1):8–14. https://doi.org/10.3181/00379727-170-41387

Provost PJ, Bishop RP, Gerety RJ, Hilleman MR, McAleer WJ, Scolnick EM, Stevens CE (1986a) New findings in live, attenuated hepatitis A vaccine development. J Med Virol 20(2):165–175. https://doi.org/10.1002/jmv.1890200208

Provost PJ, Hughes JV, Miller WJ, Giesa PA, Banker FS, Emini EA (1986b) An inactivated hepatitis A viral vaccine of cell culture origin. J Med Virol 19(1):23–31. https://doi.org/10.1002/jmv.1890190105

Sander AL, Corman VM, Lukashev AN, Drexler JF (2018) Evolutionary origins of enteric hepatitis viruses. Cold Spring Harb Perspect Med 8(12). https://doi.org/10.1101/cshperspect.a031690

Sattar SA, Jason T, Bidawid S, Farber J (2000) Foodborne spread of hepatitis A: recent studies on virus survival, transfer and inactivation. Can J Infect Dis 11(3):159–163. https://doi.org/10.1155/2000/805156

Sjogren MH, Hoke CH, Binn LN, Eckels KH, Dubois DR, Lyde L, Tsuchida A, Oaks S Jr, Marchwicki R, Lednar W et al (1991) Immunogenicity of an inactivated hepatitis A vaccine. Ann Intern Med 114(6):470–471. https://doi.org/10.7326/0003-4819-114-6-470

The Immunological Basis for Immunization Series: Module 18 - Hepatitis A. (2019). https://www.who.int/publications/i/item/97892516327

Van Damme P, Pinto RM, Feng Z, Cui F, Gentile A, Shouval D (2023) Hepatitis A virus infection. Nat Rev Dis Primers 9(1):51. https://doi.org/10.1038/s41572-023-00461-2

Wang X, Ren J, Gao Q, Hu Z, Sun Y, Li X, Rowlands DJ, Yin W, Wang J, Stuart DI, Rao Z, Fry EE (2015) Hepatitis A virus and the origins of picornaviruses. Nature 517(7532):85–88. https://doi.org/10.1038/nature13806

Werzberger A, Mensch B, Kuter B, Brown L, Lewis J, Sitrin R, Miller W, Shouval D, Wiens B, Calandra G et al (1992) A controlled trial of a formalin-inactivated hepatitis A vaccine in healthy children. N Engl J Med 327(7):453–457. https://doi.org/10.1056/NEJM199208133270702

WHO position paper on hepatitis A vaccines – October 2022 (2022) Weekly epidemiological record. 97(40):493–512

Wiedermann G, Ambrosch F, Kollaritsch H, Hofmann H, Kunz C, D'Hondt E, Delem A, Andre FE, Safary A, Stephenne J (1990) Safety and immunogenicity of an inactivated hepatitis A candidate vaccine in healthy adult volunteers. Vaccine 8(6):581–584. https://doi.org/10.1016/0264-410x(90)90013-c

Zhang L (2020) Hepatitis A vaccination. Hum Vaccin Immunother 16(7):1565–1573. https://doi.org/10.1080/21645515.2020.1769389

10

Herpes Varicella Zoster: The Other Pox (Chickenpox)

Keywords Congenital varicella syndrome (CVS) • Human herpesvirus-3 (HHV-3) • Latent infection • Oka strain • Restriction endonuclease mapping • Shingles • Michiaki Takahashi

Abbreviations

ACIP	Advisory Committee on Immunization Practices
CMV	Cytomegalovirus
CVS	Congenital varicella syndrome
EBV	Epstein-Barr virus
EM	Electron microscopy
FAMA	Fluorescent antibody to membrane antigen test
gE	Glycoprotein E
HHV	Human herpesvirus
HuEF	Human embryo fibroblast cells
KSHV	Kaposi' sarcoma herpesvirus
MMRV	Measles/mumps/rubella/varicella vaccine
PHN	Postherpetic neuralgia
RE	Restriction endonuclease
SPS	Shingle Prevention Study
SVV	Simian varicella virus
VZV	Varicella-zoster virus

Chickenpox, Creeping Among Us

Chickenpox is a word that doesn't invoke much terror. Unlike the highly fatal smallpox or the dangerously paralytic polio, chickenpox is generally a mild and mostly innocuous childhood illness. Even the multiple etymologies of the word chickenpox tend to reflect its temperate nature. No one knows for sure exactly how this moniker arose, but all the proposed origins have a common theme of describing a non-threatening disease. One widely accepted theory is that the "chicken" part of chickenpox was derived from chickpeas (garbanzo beans). Apparently, someone believed that half of a raw chickpea, being flesh-colored and translucent, resembled the vesicular chickenpox lesions, and coined the name chickenpox to reflect that resemblance. Another speculation is that the word "chicken" is a corruption of the Old English word "gican" which meant "itch". Thus, chickenpox may originally have been gicanpox or literally the "itching pox", an apt name for a disease whose rash can be horribly itchy.

Alternatively, the name may derive from some actual connection with chickens. At one point, some people believed that the disease was acquired from chickens, comparable to catching cowpox from cows, so calling it chickenpox would have been a reasonable designation. Another possible origin was that the chickenpox rash resembled the rash that occurs after chicken pecks, hence the combination of chicken with pox as a descriptor for the disease rash. A fifth conjecture is that the "chicken" designation simply reflected the weak or harmless nature of the disease. In fact, Samuel Johnson's Dictionary of the English Language published in 1755 even defined chickenpox as "an exanthematous (means a rash) distemper, so called from its being of no great danger" (Johnson 1755). Regardless of which of these possible origins for the word chickenpox are correct, none of them reflect a dangerous or feared disease.

While the origin of this disease's common name is lost in history, other aspects of the disease are better preserved in the historical record. The scientific name of chickenpox is varicella-zoster virus (VZV), and this virus is a member of the herpesvirus family, an ancient lineage that extends back roughly 400 million years (Grose 2012). The prefix "herpes" comes from the Greek word for "creeping" and refers to the slowing spreading nature of the rashes observed with many of these viruses. The first use of the word herpes as a descriptor is attributed to the Greek physician, Hippocrates, though likely being used to describe herpesvirus diseases other than chickenpox. Members of the herpes family infect birds, reptiles, and mammals (including humans), indicating that these viruses have existed since before those three animal lineages diverged. Among the human herpesviruses (HHV), there are nine known types grouped into three subfamilies called the alphaherpesviruses, the betaherperviruses, and the gammaherpesviruses. VZV (also known as HHV-3) is an alphaherpesvirus along with herpes simplex type 1 (HHV-1) and herpes simplex type 2 (HHV-2), and all three of these viruses are neurotropic as they can infect nerve cells as well as skin cells. The betaherpesviruses include the ubiquitous cytomegalovirus (CMV or HHV-5) plus HHV-6A, HHV-6B, and HHV-7; these latter three are more recently discovered and lack common names. Lastly, the gammaherpesviruses have two members, Epstein-Barr virus (EBV or HHV-4) which causes infectious mononucleosis, and Kaposi's sarcoma herpesvirus (KSHV or HHV-8) which is associated with a skin cancer called Kaposi's sarcoma.

The more direct ancestor of modern VZV likely arose in an African primate around 70 million years ago as the closest remaining relative to human VZV is the simian varicella virus (SVV) found only in Old World monkeys. From its African origins, the precursor viruses spread among the primate family tree and one predecessor co-evolved with the branch that became modern humans. This long co-history between humans and VZV means that the virus is highly adapted to its human host and is one of our oldest endemic viruses. As early humans migrated out of Africa the virus spread with them to populate human communities throughout the world, eventually evolving into the modern strains that exist today. For most of human history, the VZV infection was just another febrile disease with a skin rash, a combination of symptoms that can be caused by many different viruses. Jumping ahead to the ancient civilizations of more recent history, there are numerous descriptions of skin rashes and disease symptoms consistent with VZV (Galetta and Gilden 2015). However, in many instances, VZV infections were considered just mild cases of smallpox with no appreciation that this was a completely different disease. It wasn't until the mid-1700s that chickenpox and smallpox were convincingly differentiated by the English physician William Heberden. Various reports and studies over the next 100 years confirmed this distinction without having any definitive explanation for the cause of chickenpox. It wasn't until the late 1800s that direct transfer studies were done showing that fluid from lesions of chickenpox patients caused the disease when inoculated into healthy volunteers (Weller 1991). These results confirmed that chickenpox was caused by an infectious agent, but the identity of this agent remained elusive for decades. Work from various researchers in the early twentieth century was consistent with a viral cause, but animal studies and attempts to culture a virus were unsuccessful until the 1950s. Thomas Weller (the Nobel Prize winner along with John Enders for their development of cell culture techniques) and colleagues, in a series of elegant papers, finally established the viral nature of the disease and developed the techniques for isolation and propagation of the virus (Weller et al. 1958; Weller 1953; Weller and Stoddard 1952; Weller and Witton 1953). After millennia of infecting humans, the agent of chickenpox was at last in hand and tractable for scientific investigation, but there were more mysteries to solve.

The Duality of Herpes Viruses

The majority of the viruses in the previous chapters only cause acute infections and do not persist. For these acute viruses, once the initial infection is resolved the infecting virus is fully eliminated from the patient by the immune system and no virus remains in the body. In contrast, all human herpesviruses, including VZV can establish permanent, lifelong infections. During the initial infection, herpesviruses undergo the productive phase of their life cycle and reproduce new virions, just like any acute virus. As the initial infection is controlled by our immune system, herpes viruses sequester themselves in specific cell types and enter a latent phase. In the latent phase, viral gene expression is minimal and there is no production of new virions. Instead, the viral genomes remain in a quiescent state that you can think of as hibernation. In this stealthy state, the minimal viral activity avoids activating the immune response, and in the absence of an immune attack, the genomes can persist in the infected individual for life. In most people, the latent viruses stay latent and never re-enter the productive phase. Most individuals never know that they are harboring a fugitive from their immune system and there is no clinical impact. However, in some latently infected people, the virus is triggered to reactivate, like a bear coming out of hibernation, leading to another round of productive replication and clinical disease.

For the varicella-zoster virus, varicella is the acute chickenpox phase and zoster is the reactivation disease resulting from the quiescent virus coming out of the latent phase. VZV is a respiratory virus and infection typically starts with inhalation of the virus (Gershon et al. 2015). This may come from coughs or sneezes by an infected person but is more likely the result of shed skin cells (Tsolia et al. 1990). The virus is present in large quantities in the pox lesions, so as skin cells slough off into the environment they are easily inhaled or can be picked up on the hands by contact and transferred to the oral/nasal region of the recipient. Once introduced, the virus infects the tonsils where it proliferates and then infects responding T cells. A viremia develops which disseminates the virions throughout the body and carries viruses to the skin where they infect epithelial cells and replicate extensively. During this early dissemination phase, the patient may develop nonspecific symptoms such as headache, fever, sore throat, and malaise. This spreading process is fairly slow, and the classical rash doesn't develop until 10–21 days after the initial infection. The rash begins as reddish papules (raised bumps) on the chest, back, and face that progress quickly to fluid-filled, intensely itchy blisters in 10–12 hours. Over the next 24–48 hours the blisters will dry out and become scabby, and the scabs will heal over and fall off within 1–2 weeks. New papules continue to form for 3–5 days so this is known as an asynchronous rash since patients will have all three forms of lesions (papules, blisters, scabs) present at the same time. As these virus-filled lesions heal and slough off the skin they release new virions into the surrounding where they can infect household members and other close contacts.

In children, chickenpox is usually considered an innocuous disease requiring no treatment other than support care for the itching and fever. Complications were rare, but with approximately four million cases per year in the United States, mostly in young children, even the rare complications produced significant numbers. Before the vaccine, the United States reported about 11,000 hospitalizations and 100 deaths annually from chickenpox (Meyer et al. 2000). Bacterial skin infections, pneumonias, and encephalitis were the most common serious sequelae. Interestingly, while the number of cases in adults was much lower (because most adults had the disease as children and were immune), the disease is more problematic in adults. Adult infections tend to be more severe and more likely to develop life-threatening secondary issues. Pregnant women were at particular risk for two reasons (Lamont et al. 2011). First, pregnant women are much more susceptible to developing potentially fatal pneumonia that requires in-patient treatment including antiviral drugs and mechanical ventilation. Second, as for rubella (German Measles-Chap. 7), VZV can infect the fetus in utero and cause a variety of adverse events from low birth weight and premature delivery to fetal death. The second trimester is the critical target period, and fetuses infected during this time have a high risk of developing congenital varicella syndrome (CVS). CVS infants that survive present with a constellation of permanent physical and intellectual deficits. Common manifestations include limb defects, skin lesions, cataracts and eye dysfunctions, microencephaly, retardation, and seizures. A second critical infection period occurs around the time of birth. Mothers with active VZV in the last month of pregnancy up to the delivery can transmit the virus to the neonate. Neonates infected in this period show classical chickenpox symptoms but are at much higher risk for developing fatal complications such as pneumonia, meningitis/encephalitis, hepatitis, and liver failure. The horrendous effects of VZV on pregnant women and their unborn children once inspired "chickenpox parties" (akin to the rubella parties in Chap. 7) to expose and "immunize" young girls so that they would be immune adults with no fear of acquiring chickenpox while pregnant.

If the acute state by VZV was all there was to its life cycle it would be a fairly straightforward virus. The greater complexity comes from the ability of VZV to establish a lifelong latent infection (Laemmle et al. 2019). As the immune system controls the acute stage replication of the virus in the skin, VZV can enter the sensory nerves that provide sensation to the skin (Fig. 10.1). The viremia that occurs during infection can seed VZV to nerve cells throughout the body. In a still poorly understood process, VSV persists in these nerve cells indefinitely with only minimal expression of viral genes, allowing it to

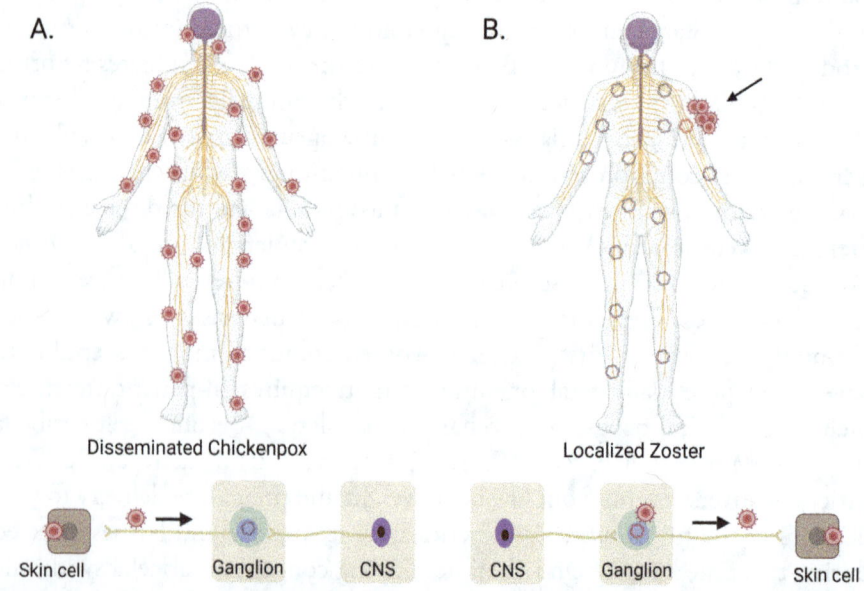

Fig. 10.1 The disease duality of varicella-zoster virus (VZV). (**a**) Chickenpox is a whole-body, disseminated disease with virions present in the skin lesions. At each lesion, viruses replicating in the skin cells can enter the peripheral nerves and move along the nerve fibers to the ganglia where the viral DNA remains latent in the cell nuclei. The result is that nerve ganglia throughout the body are harboring latent VZV. (**b**) In zoster, viral DNA from a single ganglion reactivates to produce new virions which move back down the nerve fiber to the original area of skin. The virions infect the skin cells and replicate to produce the localized lesions characteristic of shingles (arrow)

avoid immune elimination. Anyone infected with VSV will have the virus distributed widely throughout their body in this latent state. In about one-third of individuals who harbor VZV, there can be a reactivation event that triggers the virus to move back down the nerve fibers to the skin where it replicates again and causes the painful, blistering lesion of shingles (aka zoster). As the reactivation usually only occurs in a small subset of nerves, only the skin innervated by those nerves is affected. Consequently, unlike the original disseminated, whole-body lesions of chickenpox, shingles are a more localized condition. Nonetheless, zoster is not only painful, but it can be associated with a variety of significant sequelae (Gershon et al. 2015). The most common complication of zoster is postherpetic neuralgia (PHN) which occurs in about 15% of cases. PHN is long-term, sometimes permanent, nerve pain that persists after the skin lesions have healed. It can be excruciating to the point of disabling the patient. Other serious consequences that depend on the site of reactivation include muscle weakness, meningoencephalitis, vascular complications, and eye involvement. These conditions can be treated with antiviral drugs (acyclovir, valaciclovir, or famciclovir) though the success rate varies considerably from patient to patient.

The causes and mechanisms of reactivation are not well understood, but waning immunity is believed to be a factor. As we age, our overall immune function declines, and in some individuals, the anti-VZV immunity, particularly cell-mediated immunity, may drop below the threshold required to keep the virus in check. This may allow the virus to reactivate and reenter the productive phase in the skin. Importantly, the shingles lesions are full of viruses just as are the lesions of chickenpox. Therefore, individuals with active shingles are infectious and can pass the virus on to non-immune people. Evolutionarily, this is a clever scheme for a virus to maintain itself in human populations. When communities were small, chickenpox would have spread rapidly and soon there would have been no susceptible hosts left. Normally a viral population would die if there were no new hosts to infect, but latency provides a mechanism to persist. The virus could hide in all its victims for decades, biding its time until new generations were born. Then a reactivation event could produce viruses and now spread to all these susceptible hosts. This cycle of infection, latency, and reactivation is a process that has likely been going on since the dawn of herpes viruses and has enabled them to be highly successful human pathogens.

Interestingly, zoster was recognized as a distinct clinical entity long before the connection between varicella and zoster was discovered (Wood 2000). The ancient Greeks were familiar with the zoster and recognized the distinctive pattern of localized blisters that often formed a spreading line of lesions.

In fact, the word zoster derives from a Greek term that referred to a belt-like girdle covering one side of the wearer, a distribution similar to the localization on one side of the trunk that was often seen with zoster lesions (Sykes 1902). This distinctive presentation of zoster was much more obvious than the generalized rash associated with chickenpox which can resemble rashes seen with other diseases. Even after chickenpox was formally recognized and accepted as a specific disease entity, the relationship between varicella and zoster was not immediately understood due to the extremely long period between these two manifestations of VZV infection. With one being primarily a childhood illness and the other seen mostly in the elderly, connecting these two diseases across decades was not obvious. One of the first indications that chickenpox and shingle might have the same causative agent was described by the Viennese physician, Janos von Bókay, in the late 1880s though not formally reported until 1909 (Nogueira and Traynor 2004). He noted that children in a household with a zoster case developed chickenpox and that the lesions were visibly similar, suggesting that both presentations were manifestations of the same type of infection. Similar observations by other physicians accumulated over the next decades, but more definitive evidence would come in 1925 via direct transfer experiments conducted by Karl Kundratitz. Taking fluid from lesions of either chickenpox or zoster, he could induce chickenpox by injecting the fluid into naïve children who had no previous history of chickenpox. The ability to induce chickenpox from either type of lesion strongly supported that the causative agent was the same for both diseases. Subsequent immunological and histological studies from many labs would support this conclusion until the final confirmation came through electron microscopy (EM).

Developed in Germany during the 1930s, electron microscopy remained a local technique that was largely inaccessible to other countries due to the rise of Nazism and Germany's growing isolation from the rest of the world (Kruger and Mertens 2018). One of the inventors of the EM was Ernst Ruska who would eventually be awarded a Nobel Prize for his contributions to originating this device. Created to examine minute inanimate materials, there initially was skepticism about applying it to delicate biological samples. As fate would have it, Ernst's brother Helmut was a physician and biologist. Intrigued by his brother's device and the possibility of seeing the invisible, Helmut became an early adopter in the 1940s and applied the EM to view biological materials. One of his many observations with this new tool was the first visualization of the VZV virion and the demonstration that the virions present in chickenpox lesions were identical to the virions from zoster. At last, these two diseases were convincingly connected through one virus, though this work went long unrecognized because it was published in Germany during the World War II years and was slow to be appreciated by researchers in other nations. The final piece of confirmatory evidence was Thomas Weller's work in 1958 isolating and cultivating identical viruses from both chickenpox and zoster lesions (Weller et al. 1958). Finally confident that herpes zoster and herpes varicella were caused by the same virus, their names were combined to the now standard appellation of varicella-zoster virus.

With the virus in hand, a remaining mystery was why the virus sometimes caused chickenpox and at other times caused zoster, a conundrum that wouldn't be fully resolved until the development of molecular biological tools in the 1970s. There were various theories proposed to explain the alternative disease presentations ranging from where the initial site of infection was to the age of the patient when first exposed to the virus. In contrast, the concept of neural latency for herpes simplex virus had been known since the 1930s and there were hints throughout the 1900s that herpes zoster could affect nerve cells and might linger in these cells. With this burgeoning recognition of viral latency, the concept that zoster was a reactivation of latent varicella was a slowly growing idea that became firmly established with R. Edgar Hope-Simpson's landmark paper of 1965 (Hope-Simpson 1965). Summarizing years of observations and experiments, he refuted the argument that chickenpox and zoster were just different manifestations of an acute infection with VZV. Instead, he cogently presented the proposition that chickenpox is the acute form of a VZV infection, and that zoster results from the reactivation of a latent virus that has been quiescent for months to years in nerve cells. It was a powerfully presented hypothesis, but as persuasive as his arguments were, the technology still wasn't quite there to prove his contention. However, over the next two decades, irrefutable evidence accumulated that VZV did reside in nerve cells. Applying EM and molecular hybridization techniques to autopsied nerve tissue from recently deceased patients, with or without active VZV infections, VZV virions and protein (Esiri and Tomlinson 1972; Nagashima et al. 1975), DNA (Gilden et al. 1983), and RNA (Hyman et al. 1983) were all identified in the nerve ganglia. The presence of all three principal macromolecules (protein, DNA, and RNA) in nerve cells from multiple patients, including ones with no recent history of VZV infection, was convincing data that the virus truly infected and remained in nerve cells.

The one final piece of the varicella-zoster puzzle was the assertion that zoster resulted from the reactivation of a previous varicella infection. Finding the virus in nerve tissue confirmed that VZV could successfully infect this type of cell but didn't prove that the virus in the nerve cells was the same virus that previously caused chickenpox in that individual. It was still possible that these patients had two separate exposures to VZV, one resulting in disseminated chickenpox and the other leading to an infection of the nerve cells. What was needed was a direct molecular comparison of the chickenpox-producing virus and the shingles-producing virus from the same person, a difficult task since these two diseases often occur years to decades apart. This conundrum was solved by a research team led by Dr. Stephen Straus at the NIH (Straus et al. 1984). In 1983, they treated a 9-year-old boy with a genetic immunodeficiency (Wiskott-Aldrich syndrome) who presented with a severe case of chickenpox. They were able to collect a viral sample from his chickenpox lesions and store the sample for later analysis. After treatment with antiviral drugs, the child recovered fully and was released. Unexpectedly, the boy returned several months later with shingles. Apparently, due to his immunodeficiency, his body was unable to keep the latent virus in check and he suffered a rapid reactivation to give localized zoster. Gathering virus from the zoster lesions, Straus' group was able to compare the virus from the zoster lesions with the virus from the original chickenpox lesions. They examined the viral DNA from each sample using a technique called restriction endonuclease mapping. Restriction endonucleases (REs) are enzymes that cut DNA at specific sequences. When you take a large DNA like the VZV genome and cut it with an RE it will produce a unique series of fragments of different lengths that can be separated and visualized, a molecular fingerprint of sorts (Fig. 10.2). If the chickenpox virus and the shingles virus are the same their RE fingerprints will be identical and conversely, if the fingerprints don't match then two diseases resulted from two independent infections with different VZV strains. Their results showed identical RE patterns for both isolates, giving the first documentation that shingles truly resulted from the same virus that caused the original chickenpox. While many mechanistic details remained unknown, at least the basic principles of the VZV life cycle and disease process were finally settled by the early 1980s.

A Father's Love

What father wouldn't do anything for their child, especially a sick one? The story of the VZV vaccine starts not with a concerted effort by the scientific community or the marshaled resources of large pharma, but with one man's concern for his suffering son and other children who might experience severe chickenpox. While we typically think of chickenpox as a fairly mild illness in children, some kids have extreme cases with high fevers and massive outbreaks of horribly itching skin lesions. Such was the case for 3-year-old Teruyuki Takahashi, the son of Michiaki Takahashi. Dr. Takahashi received his medical

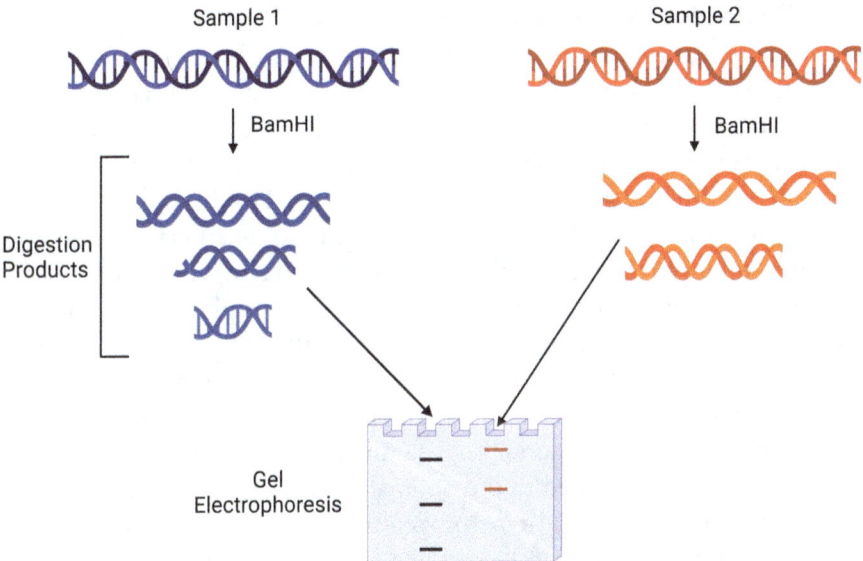

Fig. 10.2 Restriction endonuclease mapping. Restriction endonucleases (REs) are enzymes that cut DNA at specific sequences. Shown are two different DNA samples cut by a RE called BamHI. BamHI cuts sample 1 DNA into 3 fragments and sample 2 DNA into 2 fragments. These fragments can be separated by size and visualized using gel electrophoresis. If two DNA samples are identical then they will give the same fragment pattern

degree from Osaka University in 1954 and began a career as a physician and scientist. Working at the Research Institute for Microbial Disease at Osaka University, he was actively engaged in vaccine development for childhood illnesses, particularly measles, mumps, and polio. After 10 years at the Institute, he accepted a research fellowship at Baylor College of Medicine and moved his family to Houston, Texas. Soon after they arrived in Houston his son developed an aggressive case of chickenpox with extensive, whole-body lesions, high fever, and difficulty breathing. No antiviral drugs existed at that time so the anxious parents could only maintain a vigil and provide supportive care. Little Teruyuki survived and completely recovered, but this terrifying episode left a lasting impression on his father. As a vaccine researcher, Takahashi understood that a vaccine could be useful to prevent chickenpox, but at that time no vaccines had ever been developed against a herpes virus and there were significant challenges to this work.

After his fellowship ended, Takahashi returned to Osaka University and continued his work on other vaccines, although he never forgot his son's experience and the lack of a protective vaccine for chickenpox. In the early 1970s, he finally decided to embark on a VZV vaccine development project and received a fresh chickenpox sample from his colleague, Dr. Giichi Maruyama, a pediatrician. The sample came from a young boy with typical chickenpox who was otherwise completely healthy. VZV was still notoriously difficult to propagate in culture, but using this sample, Takahashi was able to establish an infection in primary human embryo fibroblast (HuEF) cells (Takahashi 1992). Using the standard approach for attenuation, Dr. Takahashi passaged the isolate multiple times in HuEF cells at a reduced temperature (34 °C) followed by multiple passages in non-human cells (guinea pig embryo fibroblast cells). Finally, the isolate was passaged briefly in WI-38 human cells to ensure it could still grow in human cells as human cells would be the host for actual vaccine production. The final virus strain after these passages was named the Oka strain in honor of the donor boy's family name. The final Oka strain was then analyzed and compared to the original wild-type VZV isolate. Oka was clearly different than its wild-type progenitor as it was slightly temperature-sensitive in cultured human cells and had an altered RE fingerprint pattern. (Subsequent work would show that Oka had at least 42 mutations although which of these are critical for attenuation is still unclear (Gershon et al. 2021)). Administration of the Oka strain to test animals, including primates, showed no evidence of adverse reactions, but now there was a conundrum. VZV is highly specific for human hosts and at that time there were no animal models that adequately recapitulated human chickenpox. The only way to test attenuation and efficacy as a vaccine to prevent chickenpox was to test the Oka strain in human volunteers. Small-scale studies in Japan, first with adults and then with 70 healthy children, produced no cases of chickenpox, confirming that the Oka strain was safely attenuated (Takahashi 1986). The participants also developed antibodies against VZV, a hopeful indicator that the vaccine might be inducing protective immunity.

With preliminary safety data in hand and the attenuation of Oka confirmed, normally the next step would be a phase II clinical trial that would further assess safety and begin to assess efficacy. However, for this potential chickenpox vaccine, there were reservations about conducting a standard phase II trial. First, the target population for this vaccine would be children. Using children in clinical trials is always avoided until safety concerns are resolved and there were serious questions about this vaccine, particularly regarding latency. While Oka was an attenuated strain that didn't appear to cause clinical chickenpox, it was still a live, replicating virus that potentially could establish latent infections in the vaccine recipients. Whether or not this would make the vaccinees susceptible to zoster attacks in the future was unknowable. A second concern was the potential transmission of the vaccine virus to contacts of the vaccine recipients. Other live attenuated vaccines can spread from the vaccinee to close contacts and the potential certainly existed for the Oka strain. What effect such secondary transmission might have on the receiving individuals was unpredictable and there was always the chance that the Oka strain might mutationally revert to a more virulent state as it spread to others. All of these issues imposed a need to move slowly and cautiously with this potential vaccine. Because of the uncertainties with this vaccine, the initial efficacy trials were targeted to at-risk individuals rather than a general community vaccine program. One experiment provided the vaccine to children in families where one child contracted chickenpox and was likely to spread it to the other siblings (Asano et al. 1977). In the vaccinated group, none of the 26 children receiving the Oka strain developed chickenpox after their household exposure. In contrast, in the control families, all 19 of the unvaccinated children developed chickenpox. A second targeted trial in 1974 tested the vaccine on children in a pediatric ward of Chukyu Hospital (Takahashi et al. 1974). A sick child in the ward developed chickenpox and potentially exposed all the other children to the virus. To prevent an outbreak, 39 other patients with no history of chickenpox were vaccinated after parental consent. None developed chickenpox and all developed antibodies. There was no specific control group as intentionally placing some sick children at risk by denying the vaccine would have been unacceptable. However, one child whose parents believed that he already had chickenpox was not vaccinated, and that boy did come down with a classic case of chickenpox. His case confirmed that the virus was transmitted in that setting and strongly suggested that the vaccine protected the other patients. A similar hospital situation arose a few years later in a different hospital pediatric ward housing several children with leukemias and other serious diseases. One leukemia patient developed chickenpox and potentially exposed seven other children, five of whom also had leukemias. The doctors and parents were divided about whether or not to vaccinate and ultimately only three of the seven received the Oka vaccine. None of the vaccinated children developed chickenpox while all four of the unvaccinated kids became ill with one dying. It was a small sample size, but it added to the accumulating data showing that the Oka strain was safe (at least in the short time range) and highly effective at preventing chickenpox. So at last there was a functional chickenpox vaccine, and one man's quest to vanquish this disease was nearer to fulfillment. Now the question was could the world be convinced to use this vaccine?

To Vaccinate or Not to Vaccinate

While the early studies of Takahashi and colleagues in Japan found the Oka strain to be an effective vaccine in targeted situations, there was still the need to prove its efficacy in the community context. Additionally, not all the original concerns had been addressed and new issues were being raised, both scientific and social. The important question of whether or not the Oka strain could become latent and its possible effect on the subsequent incidence of zoster remained unresolved. No one wanted to create a scenario where vaccinated individuals were at higher risk for zoster than people acquiring VZV through natural infection. Likewise, the potential reversion to virulence of the Oka strain was unknown. If the reversion rate was quite low, virulent cases were unlikely to show up in the small numbers of people vaccinated in the early studies, so there was no way to estimate this risk. Other investigators raised the possibility of oncogenic complications. During the 1970s, another type of herpes virus, the Epstein-Barr virus (EBV), was linked to a variety of oncogenic diseases including nasopharyngeal carcinoma and a type of lymphoma (Baumforth et al. 1999). While there was no known connection between VZV infection and cancer, this lack of evidence didn't exclude the possibility. If one herpes virus had oncogenic potential then perhaps other members of this viral family did also and the connection simply hadn't been made yet for VZV. The vaccine community had just gone through the scare with the SV40 virus contamination of the poliovirus vaccines (Chap. 6) and investigators were rightfully cautious about introducing any cancer-causing agent into the population through vaccination.

Another worrisome concern about the vaccine was its possible impact on zoster incidence in the millions of people who had already been infected with chickenpox and carried the VZV in their nerve cells. It was known that shingles are more prevalent in the elderly, and the viral reactivation was believed to be due to the dwindling immunity that naturally occurs with aging. In his seminal 1965 paper, Dr. Hope-Simpson speculated that one of the things that helped keep the virus latent was periodic exposures to VZV from chickenpox. During a person's lifetime, there was a good chance that they would be exposed to chickenpox multiple times. Essentially all the people we come in contact with are potential sources of exposure, especially our children, grandchildren, friends, neighbors, and work colleagues. Hope-Simpson proposed that these periodic family and community-based exposures acted like booster shots and stimulated anti-VSV immunity, helping to keep immune protection high enough to prevent viral reactivation and shingles. If this model was correct, then circulating VZV was beneficial for keeping zoster cases lower in adults. Therefore, if widespread adoption of the chickenpox vaccine lowered community levels of VZV then the decrease in chickenpox cases in children might produce a wave of increased zoster cases in adults. Given the potential complications and serious impact of zoster, this might not be an acceptable trade-off.

As well as these potential dangers and unintended consequences from the vaccine, there were critical technical issues that needed to be addressed. As with any new vaccine, the duration of protection was unknown. Short-term protection was evident from the early studies but whether this would persist for decades or merely a few months to years needed to be determined. An intrinsic component of the duration questions was the optimal dosing schedule—how many doses were needed and how much virus should be present in a dose to induce the maximum protective benefit? These are issues faced during the development of any vaccine, yet they were initially more problematic for VZV because the nature of the protective immunity was poorly defined. Yes, antibodies were induced by the vaccine, but the importance of antibodies in protecting from VZV infection was unclear. Live, attenuated vaccines induce both an antibody response (B cells) and a cellular immune response (T cells), and the contributions of each response to varicella vaccine efficacy were undefined. T-cell immunity is more difficult to assess so antibody levels are typically used as an indicator of immune induction by a vaccine. If a minimum antibody level that provides protection can be defined, then monitoring the rate of decline in antibody levels over time can be used to estimate how long vaccine immunity will persist. But in the early 1970s, no one knew precisely how important antibodies were to VZV immunity or what threshold level of antibody was needed to confer protection. Consequently, using antibody levels as a surrogate to monitor and project the duration of immunity was not initially feasible.

In addition to this multitude of scientific concerns and hurdles, there was the practical question of whether or not a vaccine for chickenpox was even needed or desirable. Dr. Takahashi was motivated by his personal experience to want this vaccine, but the use of vaccines must always be considered in the broader context of public health. For any vaccine, there is a risk-benefit analysis that must be examined because all vaccines pose at least some risk. The risk is generally small and includes minor side effects such as pain and redness at the site of injection (a localized inflammatory immune response), mild fever for a short duration, and some generalized symptoms that may include fatigue, malaise, and headaches. Serious adverse effects are rare but do happen in some unfortunate individuals. For example, people with undiagnosed immunodeficiencies may suffer serious illness from live, attenuated vaccines. Other rare individuals with no recognizable underlying conditions can suffer a variety of malign effects, generally due to some abnormal immune response to the vaccine material. We can't predict who is at risk and can only make estimates of the frequency of rare adverse effects once large numbers of people have received the vaccine. Note that this risk is not unique to vaccines and is present with any medical intervention from drugs to surgery. While the vast majority of patients will benefit from their medications or procedures, there will always be some who are harmed rather than helped.

For most viral vaccines the frequency of serious reactions is typically less than 1 in a million recipients. While this rate is quite low, it still means that among the millions of people who receive the vaccine, there will be people who are harmed. It is this risk that must be balanced with the vaccine benefits which are the prevention of disease and death in the community. For viruses with high rates of mortality and serious morbidity, widespread vaccination is demonstrably beneficial as the number of illnesses prevented and lives saved is far greater than the rare incidences of adverse vaccine reactions. In addition to the moral imperative for a society to reduce or eliminate diseases in its communities, there is also a financial component to public health decisions, the cost-benefit ratio. Disease imposes a significant monetary burden from direct treatment expenses such as doctor, hospital, and medication charges to the indirect cost of lost work time and productivity (this includes the time parents must take to care for sick children). Preventative measures that reduce disease incidence provide significant financial benefits to both the families and their greater community. However, all preventative measures have a cost. For example, building and operating sewage purification facilities to provide clean water is expensive yet this is balanced by the cost saved by preventing a wide range of water-borne diseases. Similarly, the production, distribution, and administration of vaccines is a complex and expensive process that has a discrete cost that can be defined for each vaccine dose. Consequently, the total vaccine cost to continuously inoculate generations of children must be assessed against the costs incurred by the disease itself. For serious viral diseases, most of us would agree that the prevention of disease is more important than cost, but for mild diseases like chickenpox, the cost-benefit analysis is an important consideration. Would the cost of vaccinating a significant portion of the population against chickenpox be worth the expense? Even though the VZV vaccine was created in 1974, this combination of scientific and economic issues prevented its acceptance and adoption until much later.

While much of the world waited and watched the VZV vaccine narrative, Dr. Takahashi and colleagues in Japan continued to test the vaccine to address the many issues and concerns raised. Additional trials were conducted over the next 10 years with increasing numbers of participants. Initially mostly in Japan but also beginning in other countries, the cumulative data from VZV vaccine trials continued to demonstrate both excellent safety and high protective efficacy of the Oka strain vaccine (Takahashi 1986). A recent meta-analysis examining VZV vaccine safety over the last 40 years confirmed these early results and found that serious adverse reactions were extremely rare with this vaccine (Ahern et al. 2023). Importantly, there is no evidence that VZV, acquired either through natural infections or through vaccinations, increases the risk for human cancer. Interestingly, there is even data indicating that having chickenpox may reduce the incidence of one type of brain cancer known as glioma (Amirian et al. 2016).

Other important information that came out of the early clinical trials addressed several of the concerning issues. First, there was no spread of the Oka vaccine strain to vaccinee contacts so inadvertent infections seemed to be unlikely (Asano et al. 1976), a finding that has been confirmed after decades of vaccination (Marin et al. 2019). Second, while the ability of the Oka strain virus to cause latency was established, possible reactivation to zoster was less frequent than for infections with wild-type VZV (Ha et al. 1980; Takahashi et al. 1985). One caveat to these studies was the type of patient used. Since the interval between chickenpox and zoster is usually years to decades in immunocompetent people, this risk could not easily or quickly be assessed in the general population. Instead, the initial information came from trials with leukemic children. As detailed above, there were several instances where leukemic children were vaccinated with Oka to protect them after a potential exposure to chickenpox. It was known already that leukemia patients who contracted chickenpox and recovered often had an occurrence of zoster shortly thereafter, presumably due to their immunocompromised status. This short interval between chickenpox and shingles in leukemic kids allowed a comparison of zoster incidence in young patients with natural chickenpox versus patients vaccinated with the Oka strain. Some vaccinated children did develop zoster, indicating that the vaccine strain could both establish latency and be reactivated. Fortunately, reactivation rates were lower with the vaccine strain than with the wild-type virus, so there was no increased zoster risk from the vaccine. Years later, much larger studies in healthy children likewise showed reduced zoster in vaccinated kids compared to kids who had chickenpox (Weinmann et al. 2013, 2019). Molecular experiments have also suggested that Oka establishes latency as effectively as the wild-type virus, but is impaired for reactivation, a result consistent with the clinical observations (Sadaoka et al. 2016). Together, these observations remove the concern that zoster might be more frequent or problematic in vaccine recipients compared to persons naturally infected with VZV.

Another unknown in the early years of using the Oka vaccine was the duration of protection. This question cannot be answered quickly and requires a significant number of vaccinees and a long follow-up period to collect reliable data. For duration studies, vaccinated individuals could be watched to determine how many ultimately get the target disease due to the waning of their vaccine-induced immunity. However, this could take many years, especially if community levels of the disease agent are low and most vaccinees simply never encounter the agent. Instead, antibody levels are typically used as a surrogate for following vaccine protection duration. First, a threshold level of antibody necessary to provide protection is defined. Then vaccinees can be tested periodically for their antibody levels. As antibody levels decline over time, a rate of decline can be calculated to provide a predicted estimate of how long their antibody levels will remain above the protection threshold. Problematically for the VZV infections, standard assays to measure antibody levels didn't accurately reflect protection, and duration studies were not possible with this approach. Fortuitously, in the same year the vaccine strain was produced, an American group developed a new immunological assay for VZV called the fluorescent antibody to membrane antigen test (FAMA) (Williams et al. 1974). FAMA quickly became the gold standard for measuring protective antibodies against VZV. With reliable measures available, numerous studies on duration and dosage showed that single-dose vaccinations had a reasonable duration and efficacy while two-dose regimens provided much longer protection in a higher percentage of the vaccine recipients (Flatt and Breuer 2012).

Similar to vaccine duration, determining the rate at which an attenuated vaccine strain reverts to virulence can require a large cohort of vaccinees to observe rare events. DNA viruses like VZV tend to have more stable genomes so reversion was not anticipated to be a major issue with this vaccine, nonetheless, experimental proof of the vaccine strain stability was lacking since the early trials involved too few vaccinees. Even after nearly 40 years of study, we still have only a limited understanding of the attenuation process for Oka due to the multitude of mutations in this virus and the existence of multiple substrains within the vaccine stock (Sadaoka and Mori 2018). However, there is no current clinical or scientific evidence of full reversion with the vaccine strain. Tens of millions of vaccine doses have been delivered without any indication of reversion, so this event, if it can even occur, must be so infrequent that it is not relevant to safety discussions about the VZV vaccine (Warren-Gash et al. 2017).

While not all the questions posed were answered completely, by the early 1980s there was sufficient confidence in the safety and efficacy of the Oka vaccine strain that commercial production and licensing began. Europe led the way with eight countries approving its use in 1984, though only in high-risk and certain vulnerable populations. Importantly, in 1985 the World Health Organization recognized the Oka strain as the best available stock for vaccines, and this strain has been widely adopted by vaccine manufacturers around the world ever since. Still, as important as the scientific data was for evaluating the risk-benefit of the VZV vaccine, the cost-benefit analysis was also important for deciding on vaccine adoption. Just because a safe and effective vaccine exists isn't necessarily sufficient justification to implement its usage. For a disease like chickenpox where the mortality and morbidity are low, if the cost of vaccine production and distribution exceeds the medical expenses incurred, it may not be worthwhile to the individual or the society to provide the vaccine. The related question is who should get the vaccine? Policymakers and regulatory agencies have to make decisions on vaccine usage that can range from recommending universal vaccination to targeting selected population groups that might see the most benefit from vaccine protection. Two studies in 1994 addressed the cost-benefit question and presented results indicating that there was a cost-benefit to preventing chickenpox through vaccination (Huse et al. 1994; Lieu et al. 1994). Coupled with the 20 years of scientific data, these cost analysis studies likely helped convince reluctant countries to consider implementing the chickenpox vaccine.

Surprisingly, 50 years after the vaccine was created and 40 years after its first approval for human use, the application of this vaccine still varies widely from country to country. For example, Japan approved the vaccine in 1986 as a voluntary vaccine for children one year and older, given as a single dose (Ozaki and Asano 2016). With only a voluntary status, varicella vaccination coverage never reached levels high enough in Japan to reduce annual chickenpox case numbers. Additionally, years of usage revealed that the one-dose regimen was not sufficient to provide long-lasting immunity, resulting in breakthrough cases of chickenpox in vaccinated individual long after their initial vaccination. Consequently, in 2014, Japan adopted a two-dose schedule and made the varicella vaccine part of the routine childhood vaccination schedule. This increased the vaccine coverage in the population and has slowly begun to reduce chickenpox cases.

The United States also started with the one-dose protocol, although much later than Japan (Gershon et al. 2021). Merck licensed the Oka strain in 1981 and then spent over 10 years developing and testing it for U.S. use. They did further attenuation, refined the manufacturing and production process several times, and conducted multiple clinical trials to evaluate their product and optimize dosing. Finally, the U.S. vaccine was licensed in 1995 as Varivax. Unlike Japan, the U.S. immediately recommended that Varivax be added as part of the routine childhood vaccination series rather than as a voluntary vaccine. This recommendation was strengthened in 1999 when the CDC's Advisory Committee on Immunization Practices (ACIP) recommended that all children entering daycare or elementary school should show proof of chickenpox vaccination. Due to the same breakthrough case problem seen in Japan, the U.S. changed to the two-dose regimen in 2007. In an attempt to increase vaccine usage, in 2005 Merck released a measles/mumps/rubella/varicella (MMRV) vaccine combining the chickenpox vaccine with the MMR vaccine. This is also a two-dose regimen vaccine and is quite effective, although the standalone varicella vaccine remains more widely used. With two vaccines available and strong guidelines for universal usage in children, vaccine coverage among U.S. children has increased significantly with a concomitant reduction in chickenpox cases and chickenpox-related serious illness. According to CDC estimates, before the vaccine, there were close to four million cases of chickenpox each year in the U.S. with around 2000 deaths. After the introduction of the vaccine in 1995, cases immediately began to decrease and over the next 20 years have declined by 97% with a corresponding decrease in morbidity and mortality (Marin et al. 2022).

The experiences in Japan and the United States are not unique and reflect the challenges of dealing with a vaccine against a disease that is not universally dangerous and feared. We have a vaccine, and it works, but the implementation of its usage is variable. Many countries have added the varicella vaccine to their stable of routine childhood vaccines while some only use it in targeted populations and some countries still don't use it all. There likely isn't a universal one-size-fits-all application to this vaccine. A country's development level, public health infrastructure, overall economic capability, and level of varicella circulation in the population are all factors that contribute to whether and how this vaccine should be utilized to best benefit the population (Papaloukas et al. 2014). This situation will likely continue for decades as countries evolve and evaluate their ongoing vaccination needs.

Saint Anthony's Fire (Zoster)

As cited above, the zoster risk to Oka vaccinees was shown to be less than for those individuals contracting wild-type varicella. What was more problematic was the possibility that reducing naturally circulating varicella would increase zoster among the enormous population of people already harboring the latent virus. The hypothesis was that latent varicella was kept in check by periodic exposures to chickenpox during a person's lifetime. Each random exposure acted like a booster shot, stimulating immunity and keeping protective immunity high enough to keep the virus latent. If a large proportion of children were vaccinated, then the incidence of varicella circulating in communities should be reduced, providing less opportunity for older individuals to get boosted by community exposures, resulting in more zoster. This concern was another variable in the cost-benefit equation that was part of the debate over this vaccine. If decreasing chickenpox, a relatively mild disease, led to an increase in zoster, a potentially much more severe disease, was a vaccination program justified? Differing opinions on this risk influenced some countries to reject or delay implementing varicella vaccination programs until more data was accumulated.

While there has been an increase in zoster prevalence over the last several decades, current evidence is not yet definitive about whether or not, or how much, varicella vaccination contributes to the rising case numbers (Harpaz 2019; Marinelli et al. 2017). Much of the epidemiological data fails to connect the zoster rise to the vaccine. First, zoster incidence began to increase long before the vaccine was widely used, and the increase is observed both in countries using the vaccine and countries not using the vaccine. Importantly, the rate of increase in zoster cases didn't change when countries implemented varicella vaccination, implying that the rise in zoster isn't precipitated by the vaccine. Second, the whole premise that circulating varicella is needed to stimulate immunity and prevent zoster has been challenged. One simple study compared zoster rates in cloistered monks (no exposure to children or circulating varicella) to a matched control cohort in the community (Gaillat et al. 2011). If repeated community exposures were necessary to prevent zoster then the monks should have higher rates than the controls, yet no difference was found. Consistent with this finding, a few studies have shown that "silent" reactivation of latent virus with no clinical symptoms can occur within individuals (Mourad et al. 2022; Mehta et al. 2014), although how frequent this is in normal populations is unknown. Since some virus is present during these silent reactivations, this newly produced endogenous virus likely acts as an internal booster to restimulate immunity. In contrast, another study found that external exposure to circulating varicella is a strong factor in restimulating immunity. It is quite possible that both endogenous and exogenous exposures can influence immunity and impact zoster development, and the relative contributions of each type of exposure are not yet well defined (Marinelli et al. 2017). Consequently, the effect of widespread varicella vaccination campaigns on future zoster cases in persons carrying latent varicella is still an unresolved issue and the reduction of chickenpox cases might increase cases of zoster. Fortunately, even without a complete understanding of the biological and epidemiological factors leading to zoster occurrences a solution exists, a vaccine for zoster itself.

The world is populated with tens of millions of people who have had chickenpox and are at risk for zoster as they age. Even if reducing cases of chickenpox through vaccination does not increase zoster incidence, there will still be an enormous burden of zoster each year among the people harboring the latent varicella virus. For example, it is estimated that there are a million cases of zoster each year in the United States alone. Furthermore, the growing number of individuals vaccinated against chickenpox will have latent Oka strain virus in their bodies and are also at some risk for zoster. Zoster can be a crippling and even fatal disease with symptoms ranging from agonizing neuropathy to life-threatening vascular issues or encephalitis (Galetta and Gilden 2015). Preventing zoster would have major benefits both for the individual and for reducing societal healthcare costs (Reddy 2017), and the impact of this reactivation disease was perceived early on during chickenpox vaccine development. Researchers postulated that if zoster results from waning immunity, then stimulating immunity in the elderly through additional inoculation with the chickenpox vaccine should reduce or prevent zoster occurrences. However, there was an important technical issue concerning protective immunity that had to be addressed. Prevention of zoster is more dependent upon T-cell mediated immunity than on neutralizing antibodies (Levin et al. 2003), and T-cell immunity is harder to induce in the elderly. Consequently, the standard Varivax dosage was not sufficient to produce an adequate response, and a much larger dosage was explored. Merck ultimately developed a potential zoster vaccine containing approximately 15 times more Oka virus than in Varivax and tested it during the 1990s, culminating in the Shingle Prevention Study (SPS) initiated in 1998 (Oxman et al. 2005). This was a large, double-blind study with nearly 40,000 elderly participants, half receiving the vaccine and half serving as controls. The study revealed that the vaccinated group had only half as many cases of zoster and the cases that did occur after vaccination were shorter in duration and of less severity. Similar studies, both in the United States and other countries, provided equivalent results demonstrating a good efficacy (typically between 60% and 80% depending on the age of the recipient) of the vaccine with no indication of serious adverse effects. This zoster vaccine was licensed by the FDA in 2006 and marketed as Zostavax which is given in a two-dose regimen. The CDC subsequently recommended the vaccine for people aged 60 and up. Continuing follow-up studies since the release of Zostavax confirmed its efficacy and safety though many noted that the protective effects declined with time.

Although Zostavax was reasonably effective at a 60–80% protective rate, greater protection was desirable. Additionally, with a live, attenuated vaccine, there is always a potential disease risk in recipients, especially immunocompromised individuals, a cohort more prevalent in the elderly. These limitations led to the development of a second shingles vaccine called Shingrix (GlaxoSmithKline Biologicals) which was approved by the FDA in 2017. Unlike the live Zostavax, Shingrix is a subunit vaccine consisting of a single viral protein called glycoprotein E (gE). The gE protein is abundant on the surface of the varicella virion and is known to induce both a strong T cell and B cell immune response. As was done for the S protein of the hepatitis B vaccine (Chap. 8), the gE gene from the varicella virus genome was cloned and the gE protein was expressed in cultured cells. The resultant protein was purified and mixed with an adjuvant to create the new vaccine. Being simply a protein, the Shingrix vaccine is noninfectious and cannot cause either chickenpox or zoster, making it suitable for both immunocompetent and immunocompromised persons. Additionally, Shingrix has a much higher protective efficacy reported at 97% in a large clinical trial (Lal et al. 2015). Because of these superior properties, Shingrix has replaced Zostavax and as of 2020, Zostavax is no longer on the market in the U.S. The one remaining issue is the low rate of utilization of this shingles vaccine. Whether because of cost, lack of awareness, or the perception that shingles is rare and not problematic, vaccine coverage in the target population (60 years and above) remains low. Expanding the usage of Shingrix may require greater effort by physicians and public health experts to spread the word about its safety and efficacy. With effective tools to prevent chickenpox (Varivax) and shingles (Shingrix), reducing the incidence of these two diseases is now a matter of public policy and personal choice rather than a scientific problem.

References

Ahern S, Walsh KA, Paone S, Browne J, Carrigan M, Harrington P, Murphy A, Teljeur C, Ryan M (2023) Safety of varicella vaccination strategies: an overview of reviews. Rev Med Virol 33(2):e2416. https://doi.org/10.1002/rmv.2416

Amirian ES, Scheurer ME, Zhou R, Wrensch MR, Armstrong GN, Lachance D, Olson SH, Lau CC, Claus EB, Barnholtz-Sloan JS, Il'yasova D, Schildkraut J, Ali-Osman F, Sadetzki S, Jenkins RB, Bernstein JL, Merrell RT, Davis FG, Lai R, Shete S, Amos CI, Melin BS, Bondy ML (2016) History of chickenpox in glioma risk: a report from the glioma international case-control study (GICC). Cancer Med 5(6):1352–1358. https://doi.org/10.1002/cam4.682

Asano Y, Yazaki T, Ito S, Isomura S, Takahashi M (1976) Contact infection from live varicella vaccine recipients. Lancet 1(7966):965–965

Asano Y, Nakayama H, Yazaki T, Kato R, Hirose S (1977) Protection against varicella in family contacts by immediate inoculation with live varicella vaccine. Pediatrics 59(1):3–7

Baumforth KR, Young LS, Flavell KJ, Constandinou C, Murray PG (1999) The Epstein-Barr virus and its association with human cancers. Mol Pathol 52(6):307–322. https://doi.org/10.1136/mp.52.6.307

Esiri MM, Tomlinson AH (1972) Herpes zoster. Demonstration of virus in trigeminal nerve and ganglion by immunofluorescence and electron microscopy. J Neurol Sci 15(1):35–48. https://doi.org/10.1016/0022-510x(72)90120-7

Flatt A, Breuer J (2012) Varicella vaccines. Br Med Bull 103(1):115–127. https://doi.org/10.1093/bmb/lds019

Gaillat J, Gajdos V, Launay O, Malvy D, Demoures B, Lewden L, Pinchinat S, Derrough T, Sana C, Caulin E, Soubeyrand B (2011) Does monastic life predispose to the risk of Saint Anthony's fire (herpes zoster)? Clin Infect Dis 53(5):405–410. https://doi.org/10.1093/cid/cir436

Galetta KM, Gilden D (2015) Zeroing in on zoster: a tale of many disorders produced by one virus. J Neurol Sci 358(1–2):38–45. https://doi.org/10.1016/j.jns.2015.10.004

Gershon AA, Breuer J, Cohen JI, Cohrs RJ, Gershon MD, Gilden D, Grose C, Hambleton S, Kennedy PG, Oxman MN, Seward JF, Yamanishi K (2015) Varicella zoster virus infection. Nat Rev Dis Primers 1:15016. https://doi.org/10.1038/nrdp.2015.16

Gershon AA, Gershon MD, Shapiro ED (2021) Live attenuated varicella vaccine: prevention of varicella and of zoster. J Infect Dis 224(12 Suppl 2):S387–S397. https://doi.org/10.1093/infdis/jiaa573

Gilden DH, Vafai A, Shtram Y, Becker Y, Devlin M, Wellish M (1983) Varicella-zoster virus DNA in human sensory ganglia. Nature 306(5942):478–480. https://doi.org/10.1038/306478a0

Grose C (2012) Pangaea and the out-of-Africa model of varicella-zoster virus evolution and phylogeography. J Virol 86(18):9558–9565. https://doi.org/10.1128/JVI.00357-12

Ha K, Baba K, Ikeda T, Nishida M, Yabuuchi H, Takahashi M (1980) Application of live varicella vaccine to children with acute leukemia or other malignancies without suspension of anticancer therapy. Pediatrics 65(2):346–350

Harpaz R (2019) Do varicella vaccination programs change the epidemiology of herpes zoster? A comprehensive review, with focus on the United States. Expert Rev Vaccines 18(8):793–811. https://doi.org/10.1080/14760584.2019.1646129

Hope-Simpson RE (1965) The nature of herpes zoster: a long-term study and a new hypothesis. Proc R Soc Med 58(1):9–20

Huse DM, Meissner HC, Lacey MJ, Oster G (1994) Childhood vaccination against chickenpox: an analysis of benefits and costs. J Pediatr 124(6):869–874. https://doi.org/10.1016/s0022-3476(05)83173-7

Hyman RW, Ecker JR, Tenser RB (1983) Varicella-zoster virus RNA in human trigeminal ganglia. Lancet 2(8354):814–816. https://doi.org/10.1016/s0140-6736(83)90736-5

Johnson S (1755) Dictionary of the English language. https://johnsonsdictionaryonline.com/

Kruger DH, Mertens T (2018) Classic paper: are the chickenpox virus and the zoster virus identical?: HELMUT RUSKA. Rev Med Virol 28(3):e1975. https://doi.org/10.1002/rmv.1975

Laemmle L, Goldstein RS, Kinchington PR (2019) Modeling varicella zoster virus persistence and reactivation - closer to resolving a perplexing persistent state. Front Microbiol 10:1634. https://doi.org/10.3389/fmicb.2019.01634

Lal H, Cunningham AL, Godeaux O, Chlibek R, Diez-Domingo J, Hwang SJ, Levin MJ, McElhaney JE, Poder A, Puig-Barbera J, Vesikari T, Watanabe D, Weckx L, Zahaf T, Heineman TC, Group ZOES (2015) Efficacy of an adjuvanted herpes zoster subunit vaccine in older adults. N Engl J Med 372(22):2087–2096. https://doi.org/10.1056/NEJMoa1501184

Lamont RF, Sobel JD, Carrington D, Mazaki-Tovi S, Kusanovic JP, Vaisbuch E, Romero R (2011) Varicella-zoster virus (chickenpox) infection in pregnancy. BJOG 118(10):1155–1162. https://doi.org/10.1111/j.1471-0528.2011.02983.x

Levin MJ, Smith JG, Kaufhold RM, Barber D, Hayward AR, Chan CY, Chan IS, Li DJ, Wang W, Keller PM, Shaw A, Silber JL, Schlienger K, Chalikonda I, Vessey SJ, Caulfield MJ (2003) Decline in varicella-zoster virus (VZV)-specific cell-mediated immunity with increasing age and boosting with a high-dose VZV vaccine. J Infect Dis 188(9):1336–1344. https://doi.org/10.1086/379048

Lieu TA, Cochi SL, Black SB, Halloran ME, Shinefield HR, Holmes SJ, Wharton M, Washington AE (1994) Cost-effectiveness of a routine varicella vaccination program for US children. JAMA 271(5):375–381

Marin M, Leung J, Gershon AA (2019) Transmission of vaccine-strain varicella-zoster virus: a systematic review. Pediatrics 144(3). https://doi.org/10.1542/peds.2019-1305

Marin M, Leung J, Anderson TC, Lopez AS (2022) Monitoring varicella vaccine impact on varicella incidence in the United States: surveillance challenges and changing epidemiology, 1995-2019. J Infect Dis 226(Suppl 4):S392–S399. https://doi.org/10.1093/infdis/jiac221

Marinelli I, van Lier A, de Melker H, Pugliese A, van Boven M (2017) Estimation of age-specific rates of reactivation and immune boosting of the varicella zoster virus. Epidemics 19:1–12. https://doi.org/10.1016/j.epidem.2016.11.001

Mehta SK, Laudenslager ML, Stowe RP, Crucian BE, Sams CF, Pierson DL (2014) Multiple latent viruses reactivate in astronauts during Space Shuttle missions. Brain Behav Immun 41:210–217. https://doi.org/10.1016/j.bbi.2014.05.014

Meyer PA, Seward JF, Jumaan AO, Wharton M (2000) Varicella mortality: trends before vaccine licensure in the United States, 1970–1994. J Infect Dis 182(2):383–390. https://doi.org/10.1086/315714

Mourad M, Gershon M, Mehta SK, Crucian BE, Hubbard N, Zhang J, Gershon A (2022) Silent reactivation of varicella zoster virus in pregnancy: implications for maintenance of immunity to varicella. Viruses 14(7). https://doi.org/10.3390/v14071438

Nagashima K, Nakazawa M, Endo H (1975) Pathology of the human spinal ganglia in varicella-zoster virus infection. Acta Neuropathol 33(2):105–117. https://doi.org/10.1007/BF00687537

Nogueira RG, Traynor BJ (2004) The neurology of varicella-zoster virus: a historical perspective. Arch Neurol 61(12):1974–1977. https://doi.org/10.1001/archneur.61.12.1974

Oxman MN, Levin MJ, Johnson GR, Schmader KE, Straus SE, Gelb LD, Arbeit RD, Simberkoff MS, Gershon AA, Davis LE, Weinberg A, Boardman KD, Williams HM, Zhang JH, Peduzzi PN, Beisel CE, Morrison VA, Guatelli JC, Brooks PA, Kauffman CA, Pachucki CT, Neuzil KM, Betts RF, Wright PF, Griffin MR, Brunell P, Soto NE, Marques AR, Keay SK, Goodman RP, Cotton DJ, Gnann JW Jr, Loutit J, Holodniy M, Keitel WA, Crawford GE, Yeh SS, Lobo Z, Toney JF, Greenberg RN, Keller PM, Harbecke R, Hayward AR, Irwin MR, Kyriakides TC, Chan CY, Chan IS, Wang WW, Annunziato PW, Silber JL, Shingles Prevention Study G (2005) A vaccine to prevent herpes zoster and postherpetic neuralgia in older adults. N Engl J Med 352(22):2271–2284. https://doi.org/10.1056/NEJMoa051016

Ozaki T, Asano Y (2016) Development of varicella vaccine in Japan and future prospects. Vaccine 34(29):3427–3433. https://doi.org/10.1016/j.vaccine.2016.04.059

Papaloukas O, Giannouli G, Papaevangelou V (2014) Successes and challenges in varicella vaccine. Ther Adv Vaccines 2(2):39–55. https://doi.org/10.1177/2051013613515621

Reddy KP (2017) Herpes zoster vaccine: time for a boost? J Gen Intern Med 32(2):145–147. https://doi.org/10.1007/s11606-016-3885-x

Sadaoka T, Mori Y (2018) Vaccine development for Varicella-Zoster virus. Adv Exp Med Biol 1045:123–142. https://doi.org/10.1007/978-981-10-7230-7_7

Sadaoka T, Depledge DP, Rajbhandari L, Venkatesan A, Breuer J, Cohen JI (2016) In vitro system using human neurons demonstrates that varicella-zoster vaccine virus is impaired for reactivation, but not latency. Proc Natl Acad Sci USA 113(17):E2403–E2412. https://doi.org/10.1073/pnas.1522575113

Straus SE, Reinhold W, Smith HA, Ruyechan WT, Henderson DK, Blaese RM, Hay J (1984) Endonuclease analysis of viral DNA from varicella and subsequent zoster infections in the same patient. N Engl J Med 311(21):1362–1364. https://doi.org/10.1056/NEJM198411223112107

Sykes W (1902) Origin of "zoster" in herpes zoster. Br Med J 1902:930–930

Takahashi M (1986) Clinical overview of varicella vaccine: development and early studies. Pediatrics 78(4 Pt 2):736–741

Takahashi M (1992) Current status and prospects of live varicella vaccine. Vaccine 10(14):1007–1014. https://doi.org/10.1016/0264-410x(92)90109-w

Takahashi M, Otsuka T, Okuno Y, Asano Y, Yazaki T (1974) Live vaccine used to prevent the spread of varicella in children in hospital. Lancet 2(7892):1288–1290. https://doi.org/10.1016/s0140-6736(74)90144-5

Takahashi M, Asano Y, Kamiya H, Baba K (1985) Varicella vaccine: case studies. Microbiol Sci 2(8):249–254

Tsolia M, Gershon AA, Steinberg SP, Gelb L (1990) Live attenuated varicella vaccine: evidence that the virus is attenuated and the importance of skin lesions in transmission of varicella-zoster virus. National Institute of Allergy and Infectious Diseases Varicella Vaccine Collaborative Study Group. J Pediatr 116(2):184–189. https://doi.org/10.1016/s0022-3476(05)82872-0

Warren-Gash C, Forbes H, Breuer J (2017) Varicella and herpes zoster vaccine development: lessons learned. Expert Rev Vaccines 16(12):1191–1201. https://doi.org/10.1080/14760584.2017.1394843

Weinmann S, Chun C, Schmid DS, Roberts M, Vandermeer M, Riedlinger K, Bialek SR, Marin M (2013) Incidence and clinical characteristics of herpes zoster among children in the varicella vaccine era, 2005–2009. J Infect Dis 208(11):1859–1868. https://doi.org/10.1093/infdis/jit405

Weinmann S, Naleway AL, Koppolu P, Baxter R, Belongia EA, Hambidge SJ, Irving SA, Jackson ML, Klein NP, Lewin B, Liles E, Marin M, Smith N, Weintraub E, Chun C (2019) Incidence of herpes zoster among children: 2003-2014. Pediatrics 144(1). https://doi.org/10.1542/peds.2018-2917

Weller TH (1953) Serial propagation in vitro of agents producing inclusion bodies derived from varicella and herpes zoster. Proc Soc Exp Biol Med 83(2):340–346. https://doi.org/10.3181/00379727-83-20354

Weller TH (1991) Historical-perspective. Transpl P 23(3):5–7

Weller TH, Stoddard MB (1952) Intranuclear inclusion bodies in tissue cultures inoculated with varicella vesicle fluid. AMA Am J Dis Child 83(1):75–76

Weller TH, Witton HM (1953) Propagation in tissue cultures of cytopathogenic agents apparently derived from varicella vesicle fluids. AMA Am J Dis Child 86(5):644–645; discussion, 615-646

Weller TH, Witton HM, Bell EJ (1958) The etiologic agents of varicella and herpes zoster; isolation, propagation, and cultural characteristics in vitro. J Exp Med 108(6):843–868. https://doi.org/10.1084/jem.108.6.843

Williams V, Gershon A, Brunell PA (1974) Serologic response to varicella-zoster membrane antigens measured by direct immunofluorescence. J Infect Dis 130(6):669–672. https://doi.org/10.1093/infdis/130.6.669

Wood MJ (2000) History of varicella zoster virus. Herpes 7(3):60–65

11

Rotavirus: The Democratic Virus

Keywords Intussusception • NSP4 • Reassortment • Rotarix • RotaShield • RotaTeq • Serotypes • Vaccine Adverse Events Reporting System (VAERS)

Abbreviations

ACIP	Advisory Committee on Immunization Practices
AGMK	African green monkey kidney cells
CD	Celiac disease
EM	Electron microscopy
GAVI	Global Alliance for Vaccines and Immunization
GI	Gastrointestinal system
GSK	GlaxoSmithKline
IgA	Immunoglobulin type A
IgG	Immunoglobulin type G
NIH	National Institutes of Health
PCV	Porcine circovirus
RCH	Royal Children's Hospital
TD1	Type 1 diabetes
VLP	Virus-like particle
VP	Virall protein
VRBPAC	Vaccine and Related Biological Products Advisory Committee

A Wheel of Misfortune

Diarrhea may be one of the most common human experiences, something that everyone suffers through multiple times in their life. Rich or poor, young or old, in developed nations or emerging countries, no one escapes bouts of diarrhea during their lifetime. Gastrointestinal infections leading to diarrhea can be caused by various microorganisms including protozoa, parasites, bacteria, and viruses. Among the viruses, a prominent culprit, especially in young children, is the rotavirus. Estimates suggest that by age five, almost all children, irrespective of country or socioeconomic status, will have had a rotavirus infection (Bishop 1996). It spares no one and infects everyone, hence its moniker as the democratic virus. Many children are infected in the first year or two of life when the disease can be most problematic. Symptoms begin after a short incubation period of 36–48 hours and start with fever, stomach pain, nausea, and vomiting. More concerning is the sometimes severe diarrhea that can persist from three to eight days. As adults, we typically find diarrhea vexing and inconvenient but not life-threatening. In contrast, extensive diarrhea in infants and toddlers can be a dangerous situation. With their small body mass and the difficulty in getting them to drink fluids, they can quickly become dehydrated from diarrhea and may need intravenous fluids to restore the electrolyte balance. In developing nations where medical intervention is unavailable and access to clean water is limited or nonexistent, rotavirus is a major cause of childhood mortality. Historically, rotavirus infections were associated with nearly a million deaths per year globally, predominantly in portions of Africa and Asia. This amounts to about 5% of all deaths in children under 5 years old each year, making rotavirus a major cause of early childhood mortality. Even in developed nations, rotavirus significantly impacts the healthcare system by causing hundreds of thousands of doctor visits and tens of thousands of hospitalizations yearly. Deaths, though rare, can occur when parents fail to recognize the seriousness of their child's dehydration until too late. More insidious than the immediate effects is the potential damage to the gut incurred by repeated infections (Kolling et al. 2012). Gut damage can alter the microbiome and have long-lasting health consequences, including deficits in cognitive development. Before the vaccine, the U.S. alone had about 50,000 hospitalizations and 50 deaths annually from rotavirus, so a significant number of children were at risk each year for the immediate and long-term effects of GI infections.

The story of human rotaviruses started with their discovery in the early 1970s (Bishop 2009). It may be surprising, but before the 1970s, there was little known about the specific causes of gastrointestinal diseases. It was presumed that many cases were due to infectious agents, but other factors such as diet, ingested toxins, malnutrition, and even genetic traits were also likely contributors. Teasing out the precise cause of any patient's diarrheal disease was complex and generally beyond the scope of the available clinical tools. Even for cases that seemed to have an infectious cause, identifying the pathogen was problematic. Our gastrointestinal (GI) system is rife with bacteria and bacteriophages, and sorting out any pathogenic ones from all the commensal organisms in stool samples was usually an unproductive exercise. You might think that finding viruses would be simpler as they should not normally be inhabitants of our GI system and would be more obvious. Yet the standard approach of culturing an unknown virus from a clinical sample was usually unsuccessful for stool. The massive amounts of bacteria in the stool would contaminate and overwhelm the cell cultures long before any viral isolation was possible, so another approach was needed.

A common occurrence in science is that there is a question or problem that intrigues the field for years until some innovative researcher puts the piece together to solve the puzzle. For rotavirus that individual was Ruth Bishop of the Royal Children's Hospital (RCH) in Melbourne, Australia. Educated at the University of Melbourne, Dr. Bishop received her microbiology Ph.D. in 1961. After additional training, she was hired by the Gastroenterology Research Unit at the RCH in 1968. While the main focus of this unit was on chronic GI illnesses like celiac disease, they also dealt with thousands of cases of acute diarrhea. Diarrheal disease was a major cause of infant mortality in Australia throughout much of the twentieth century, attracting a significant amount of research into causes and treatments. In particular, there was a severe diarrheal disease that was prevalent in children under 5 and often led to hospitalization. Like others before her, Bishop tried to identify a causative agent in stool without success. Her frustration inspired a search for a different tactic, and her solution was within her own institution. A serious impediment to much GI work in infants and small children was the inability to safely collect biopsy specimens from the upper intestines. Fortuitously, two gastroenterologists from RCH, Graeme Barnes and Rudge Townley, pioneered the development of a new procedure where these samples could be quickly and safely collected. Bishop wondered if it might be easier to detect a pathogen in these small biopsy samples rather than in the voluminous diarrheal stool. Receiving biopsy samples from the GI group, she enlisted Ian Holmes from the University of Melbourne to help with virus detection. Holmes was an expert virologist who was applying state-of-the-art electron microscopy to study viruses. Using a technique called negative staining, they almost immediately detected viral particles in the biopsy samples, describing them as 70-nanometer diameter spheres with a wheel-like appearance. Their 1974 publication describing these particles identified in four patients with severe, acute, GI disease finally introduced the world to this elusive pathogen that was so dangerous to young children (Bishop et al. 1974). Dr. Bishop remained a pioneer in this field for the remainder of her career and was dedicated to eliminating this insidious killer of infants and toddlers.

In a second bit of fortuitous science, this newly discovered virus wasn't so new after all. A diarrheal disease of infant mice was first described in the 1940s by none other than the esteemed cell culturist, John Enders (Chap. 6) (Pappenheimer and Enders 1947). By the early 1960s, this murine agent had been visualized by electron microscopy and was similar in appearance to viruses isolated from monkey rectal swaps and calves (Malherbe and Harwin 1963; Adams and Kraft 1963). With the publication of the human virus, studies quickly proved that these animal viruses and the human diarrheal virus were related members of the same family (Flewett et al. 1974). Because of their wheel-like appearance, this group of viruses was dubbed the rotaviruses since rota is the Latin word for wheel. We now know that this type of virus is extremely widespread among many mammalian and avian species, including most domestic animals, with each species having their own specific rotaviruses. Physical and molecular characterization of rotaviruses places them in the family *Reoviridae*. This family of viruses contains segmented genomes composed of double-stranded RNA (i.e. each virion contains multiple pieces of dsRNA). Human rotaviruses contain 11 dsRNA segments with most of the segment coding for one viral protein. Viruses with dsRNA genomes are less common than viruses with single-stranded genomes, and the reoviruses are the only known dsRNA viruses to infect humans. Because they have segmented genomes, reassortment can occur during co-infections (Gentsch et al. 2005), just as it does for the influenza virus (Chap. 5). Reassortment between different human strains and between human and animal rotaviruses can occur which enhances the genetic diversity of the human rotaviruses. These genetic exchanges, in conjunction with the usual mutational events such as replication errors, can translate into changes in the viral proteins. Altered protein sequences, especially for virion surface protein, can help the virus avoid immune detection and make the virus unrecognized by previously developed immune responses such as neutralizing antibodies. This complex and diverse population of rotavirus variants in circulation likely contributes to their ability to infect us multiple times in our lives.

Along with their uncommon genome structure, rotaviruses have an unusual virion structure with three distinct protein shells. The internal core surrounds the RNA segments and is made up primarily of a structural protein called viral protein 2 (VP2) but also contains VP1 and VP3. Outside of the core is a shell called the inner capsid that is comprised of VP6. The final outer capsid is formed from VP7 and is studded with stalks of VP4. As with all GI viruses, rotaviruses have no membrane envelope as these lipid structures would be easily destroyed by the harsh digestive conditions in the stomach. Consequently, VP7 and VP4 are the surface proteins of this virus and neutralizing antibodies are made primarily against these two proteins. In contrast to a membrane envelope, this tightly packed, multilayered capsid structure is highly resistant to acidic conditions and digestive enzymes. With an impervious capsid, the virus can easily pass through the stomach intact and into the intestines where it infects enterocytes, the cells that line our large and small intestines. Importantly for disease spread, the capsid structure that evolved to survive the stomach is also highly stable in the external environment. Numerous studies have investigated rotavirus viability under different conditions, and this virus is a champion at survival like the hepatitis A virus (Chap. 9). Rotavirus virions remain infectious for days to months in water, on some surfaces, and in fecal matter (Abad et al. 1994; Sattar et al. 1986; Raphael et al. 1985). This includes human hands where survival and transfer can occur for hours (Ansari et al. 1988). This means that infected people, or caregivers of infected young children, likely pick up viruses and deposit them in their surroundings. Additionally, they are also resistant to elevated temperatures for short exposures and can even survive the pasteurization of milk (Benkaddour et al. 1993). Unfortunately, many common cleaning agents and disinfectants have relatively little effect on rotaviruses so decontaminating hands and surfaces is not always effective at completely eliminating residual virus. The cumulative message from these studies is that rotaviruses are persistent in the environment and likely lurk in homes, hospitals, schools, daycare centers, offices, businesses, and anywhere else where infected individuals have been. This means that wherever people go there may be persistent rotavirus residing on surfaces, ready to be picked up on hands and introduced into mouths to start infections. No wonder nearly everyone becomes a victim of this virus in the first few years of life and can be re-infected multiple times in a lifetime.

A Virus Runs Through Them

Many GI infections cause similar symptoms of diarrhea and vomiting, and diagnosing the causative agent by clinical presentation is usually not possible. While rotavirus infections can be asymptomatic or very mild, especially in repeat cases, one characteristic that separates rotavirus from other agents is the prolonged and extensive diarrhea that occurs in some children. Infected individuals, even ones with few if any symptoms, shed viruses, and the shedding can begin a day or two before symptoms appear. In their feces, the amount of virus can reach greater than ten billion infectious viral particles/milliliter (Bishop 1996). With this enormous amount of viruses being produced and released by an infected person, environmental contamination inevitably occurs. In undeveloped countries, this virus can readily enter the food and water chain and spread widely through ingestion of these contaminated stables. Flies have also been implicated in transmission as they pick up viruses from contact with feces and then mechanically transfer them to whatever else they alight on (Tan et al. 1997). While food and water are less important sources of rotaviruses in developed nations, that doesn't stop viral spread. Flies are everywhere, and cleaning up diarrhea is messy. Minute amounts of diarrheal feces on the hands can contain millions of virions that can then be deposited on surfaces around the home or workplace. Handwashing alone may not completely remove all the viral particles, so routine hygiene is unlikely to prevent the spread of this virus (Ansari et al. 1988). Even flushing toilets can create aerosols that spread viral particles throughout the room. With all these mechanisms that spread viral particles, we just have to accept that there is little that can be done in our daily lives to halt the transmission of rotaviruses.

The virus's journey in a new victim begins with the ingestion of as few as ten viral particles (Ward et al. 1986). There are approximately eight billion people in the world, so it would take eighty billion rotavirus particles to infect the entire population of the world. That sounds like a lot, but eighty billion infectious virions are roughly the amount present in one-half tablespoon of diarrheal stool. Even if the infectious dose is ten times higher at 100 particles per person, this would still only require a few ounces of diarrheal stool to infect the entire world. The amount of rotavirus produced by one person in one day is hundreds of times more than needed to infect every man, woman, and child on the planet. The numbers are so staggering that avoiding the virus is highly unlikely, making rotavirus infection an inevitable part of the human experience.

With its highly stable and resistant capsid structure, ingested rotavirus successfully passes through the stomach and reaches the intestines where symptoms begin in one to two days after ingestion. The inner surface of the intestines is lined primarily with cells called enterocytes. This lining is not flat and instead forms fingerlike projections known as villi that greatly increase the surface area for the absorption of nutrients. The valleys between the villi are called the crypts, and within the crypts are stem cells that replenish the enterocytes as they age and die. Enterocytes are critical for digestion as they control the flow of nutrients from inside the intestines into the tissues and bloodstream. Additionally, they regulate the influx and efflux of water to maintain homeostasis; too little water results in constipation and too much causes diarrhea. It is these enterocytes that are the target cells for rotavirus. The viral surface VP4 protein is critical for binding receptor molecules on the enterocytes to start the viral reproductive cycle. After the viral particles bind to the surface of the enterocytes, digestive enzymes present in the intestine cleave the VP4 protein causing it to change shape which triggers viral penetration into the cell (Herrmann et al. 2021).

Once internalized, rotavirus replicates extensively inside the enterocyte. Many of the details of this process are undefined due to the lack of model systems that faithfully recapitulate human infection, but animal and cell culture studies provide a general picture of the events (Crawford et al. 2017; Omatola and Olaniran 2022). Enterocyte cell death occurs, releasing progeny viruses into the lumen of the intestines to infect more cells. The net destruction of enterocytes results in atrophy and shortening of the villi. This villi damage leads to reduced absorption function that leaves excess water in the lumen and contributes to diarrhea. Additionally, rotavirus is unusual in that one of its proteins, NSP4, is a toxin. NSP4 activates a signaling pathway that causes more sodium to enter the intestines, a situation that causes water to enter the intestines and directly leads to diarrhea. At the same time, NSP4 inhibits one of the systems that normally promotes reabsorption of water out of the intestines, and this may also contribute to the excess fluid in the intestines. Lastly, NSP4 induces the production of serotonin in the intestines which stimulates peristalsis to further promote explosive diarrheal. The combined result of these viral-induced effects can be profuse, watery diarrhea, especially on first exposure to rotavirus. Serotonin also activates the vagus nerves in the gut which causes a signal that elicits vomiting (Hellysaz and Hagbom 2021). It is this massive fluid and electrolyte loss from vomiting and diarrhea that can be perilous to infants and small children. And of course, this expelled fluid releases an enormous number of viral particles into the environment to find new victims to infect. Parenthetically, NSP4 also impairs the activity of lactase in the intestine, an enzyme required to break down lactose, the prominent sugar in milk (Beau et al. 2007). Consequently, children experiencing a rotavirus infection can become lactose intolerant, a state that can persist for several weeks post-infection.

Recovery from rotavirus infections is not completely understood (Latifi et al. 2024). Numerous studies, mostly in animal models, suggest a role for both humoral (antibodies) and cellular (T cells) immunity (Omatola and

Olaniran 2022). Two types of neutralizing antibodies, immunoglobulin G (IgG) and immunoglobulin A (IgA), form against both the surface proteins, VP4 and VP7. IgG is the typical systemic antibody that is elicited in response to most pathogens and vaccines. IgA is also systemic but in addition, is the primary antibody found on mucosal surfaces, primarily in the digestive and respiratory tracts. While there is evidence that IgA in the intestines facilitates clearance of rotavirus infections in mice (Blutt et al. 2012), a comparable role in humans has not been established. Antibodies also form against other viral proteins, but the contributions of these other antibodies to recovery and protection from future infection haven't been defined. Likewise, cellular immunity is activated during rotavirus infection and is crucial to rotavirus clearance in animal models (Desselberger and Huppertz 2011). Studies in humans show that rotavirus-specific T-cell populations do develop although the response is not as robust as with other viruses (Laban et al. 2022). Still, even with the limited data available, T-cells are believed to be important mediators of recovery from rotavirus infection and protection from future reinfections.

If the immunology that protects us from rotaviruses is complex and poorly understood, we at least have a much better picture of the natural history of this infection. Rotavirus is a worldwide agent with some moderate seasonality. In temperate climates cases rise in the cool, dry seasons of fall and winter. Less seasonality is seen in tropical regions where the cases are constant throughout the year. Regardless of geography, the peak age for the first symptomatic infection with rotavirus is from three months to two years of age. Children younger than three months are generally protected because they have maternal antibodies against rotavirus that have passed to the fetus during gestation (Bishop et al. 1983). Breastfeeding can also provide maternal antibodies that extend the protected period (Krawczyk et al. 2016). Once the maternal antibodies decline, infants lose this passive immunity and become susceptible. Because the virus is so plentiful and ubiquitous in the environment, most children will endure their first infection before age five. Lacking their own immunity to the virus, initial infections tend to be the most severe and most likely to produce significant morbidity and mortality. Unfortunately, recovery from the initial infection does not confer what researchers call "sterilizing immunity", which is complete protection from subsequent infections Instead, the initial infection confers a partial immunity that still allows symptomatic reinfection but does protect from severe clinical symptoms (Velazquez et al. 1996). Over time with repeated exposures to rotaviruses, immunity gradually increases until subsequent infections are usually asymptomatic. This natural history had important implications for vaccine development. If a viral infection doesn't produce sterilizing immunity, then achieving this level of protection with a vaccine may be difficult or impossible. Nonetheless, researchers hoped that it would be possible to develop a vaccine that provided the same level of immunity as an initial rotavirus infection. If such a vaccine just prevented serious infections requiring hospitalization and reduced fatalities, it would be a huge benefit to children worldwide.

RotaShield: Baby Steps and a Stumble

Before rotavirus and other GI viruses were identified, there was no way to tell what agent was causing a GI infection since the clinical presentation was similar for many different pathogens. The morbidity and mortality associated with the total burden of GI infections were well known, but without being able to ascribe a particular infection to a particular pathogen, the extent and importance of any one agent were uncertain. With rotavirus finally in hand thanks to Dr. Bishop, progress in understanding the virus and its contribution to global GI disease was inevitable. Bishop and her collaborators quickly developed an electron microscopy (EM) method to detect rotavirus in stool samples, obviating the need for the highly technical intestinal biopsies used in her original discovery of the virus (Bishop et al. 1974). Using this approach she conducted the first epidemiological studies that began to gather data on the prevalence of rotavirus in children and establish that this virus was truly the causative agent of certain GI infections (Davidson et al. 1975). Chemical and physical characterization of the virus soon followed along with the development of immunological reagents and assays for more rapid and convenient identification of rotavirus than with the cumbersome EM method (Kapikian et al. 1974; Ghose et al. 1978; Blacklow et al. 1976). For the next twenty years, these assays were applied in clinical studies across many countries to gather details about viral prevalence, distribution, strain types, transmission mechanisms, clinical manifestations, and the host immune response. By the early 1990s, it was abundantly clear that rotaviruses were a major public health problem causing significant morbidity and mortality, especially in underdeveloped nations, and that a vaccine was desperately needed. Importantly for potential vaccine development, the immune response to natural infection with rotavirus was being characterized. As described above, the protection induced by a case of rotavirus was not complete and individuals often suffered multiple rounds of infection from childhood into adulthood. However, the ability of a single infection to provide effective protection against further severe disease was established by several independent studies (Velazquez et al. 1996; Bishop et al. 1983; Ward and Bernstein 1994).

Subsequent infections typically became milder and shorter as broader immunity developed with each exposure to the virus. Therefore, even if sterilizing immunity was unachievable with a vaccine, at least ameliorating the most severe aspects of rotavirus infection seemed possible and should greatly reduce hospitalizations and deaths from rotavirus.

With a global need and a potentially huge market, several researchers and vaccine manufacturers began the development of potential rotavirus vaccines. As discussed for previous vaccines, there are multiple strategies for making a vaccine, and decisions about what might be the most effective approach must take into account viral and disease features. Another GI virus, poliovirus (Chap. 6), had a highly effective oral vaccine consisting of attenuated virus, and rotavirus seemed amenable to this same approach. The standard protocol for attenuation is to passage a human virus through animals and/or non-human cell cultures to adapt the virus to this new host environment. During this adaptation, the human virus often loses its natural virulence in humans and becomes a weakened strain. If this weakened strain retains sufficient antigenic identity with the parental strain so that it can still induce protective immunity, then a potential vaccine has been created. However, Albert Kapikian at the National Institutes of Health (NIH) chose to explore a different methodology to create a vaccine strain, a "Jennerian" strategy (Kapikian et al. 1986). Jennerian is a reference to Edward Jenner and his use of cowpox to vaccinate against smallpox (Chap. 2). In Jenner's case, there was an animal poxvirus that was related to human smallpox and exposure to cowpox worked like an attenuated strain to provide immunity to smallpox. For most human viruses there aren't suitable animal counterparts that can function as naturally attenuated versions of the human virus, so the Jennerian approach is not widely used. However, for human rotavirus, many animals harbored related rotaviruses that might be suitable as vaccine candidates. During the 1980s there were attempts to use bovine rotavirus as a vaccine, but ultimately clinical trials didn't demonstrate enough protection against human rotavirus for the bovine candidates to be a suitable vaccine (Georges-Courbot et al. 1991; Demol et al. 1986). The problem with any animal rotavirus is that its surface proteins, VP7 and VP4, may not be similar enough to the human rotavirus proteins to generate a protective immune response, and the bovine virus clearly wasn't working. It was time for a different strategy, and Kapikian adopted a modified Jennerian approach.

The concept of a modified Jennerian approach is to start with an animal rotavirus as the backbone and then modify its genome by replacing some of the animal genes with the corresponding human rotavirus genes. This gene replacement is relatively easy to do for rotaviruses because there is a simple genetic tool to mix human and animal rotavirus genes called reassortment (see Chap. 5). Since rotaviruses have segmented genomes, doing a co-infection with two different rotaviruses will produce progeny reassortants with all possible combinations of gene segments from each parent, and all you have to do is screen for the reassortants with the combination being sought. Wyeth started with a rhesus monkey rotavirus that induced some protection against human rotaviral disease but not enough to be sufficient for a commercial vaccine (Vesikari et al. 1990). An important shortcoming of the monkey virus used concerned the critical surface protein, VP7, that evokes neutralizing antibodies. Among human rotaviruses there are numerous natural variations of this protein, called serotypes, that can be distinguished with antibodies (O'Ryan 2009). These serotypes are designated sequentially as G1, G2, G3, etc. as they are discovered. While there are many VP7 serotypes, the most common rotaviruses in circulation carry serotypes G1, G2, G3, G4, or G9. Infection with a single rotavirus, for example, a G1 serotype, provides strong protection against reinfection with a G1 strain but lesser protection against other serotypes. The rhesus monkey rotavirus had a VP7 protein of the G3 serotype, so it gave only minimal protection against the other major serotypes. To circumvent this limitation, Kapikian created a tetravalent vaccine containing four rhesus rotaviruses each carrying a different VP7 protein serotype (Kapikian 2011). In the vaccine, there was the original rhesus virus with VP7 G3 plus three reassortants carrying VP7 G1, G2, and G4, respectively. Each reassortant was made by co-infecting cells with the rhesus and human viruses and selecting for reassortants where the gene segment encoding the rhesus VP7 protein was replaced by the equivalent human rotavirus segment (Fig. 11.1). This vaccine was designated RRV-TV for Rhesus Rotavirus Vaccine—Tetravalent and was licensed to the pharmaceutical company Wyeth who eventually marketed it under the trade name RotaShield.

With RRV-TV in hand, Kapikian and Wyeth conducted numerous small clinical trials, including studies in Venezuela (Perez-Schael et al. 1997), the United States (Rennels et al. 1996), and Europe (Joensuu et al. 1997). As expected, the vaccine was not sterilizing and did not prevent rotavirus infections but did reduce the most severe effects that normally led to hospitalizations and potential death. Collectively, these studies demonstrated good efficacy of the vaccine with no significant adverse events, although many recipients did experience fever, irritability, and some gastrointestinal symptoms. It was noted, however, that while it was not statistically significant, there were five cases of intussusception among the roughly 10,000 vaccine recipients with only one case among nearly 5000 control patients (Rennels et al. 1998). Intussusception is a medical condition where a portion of the

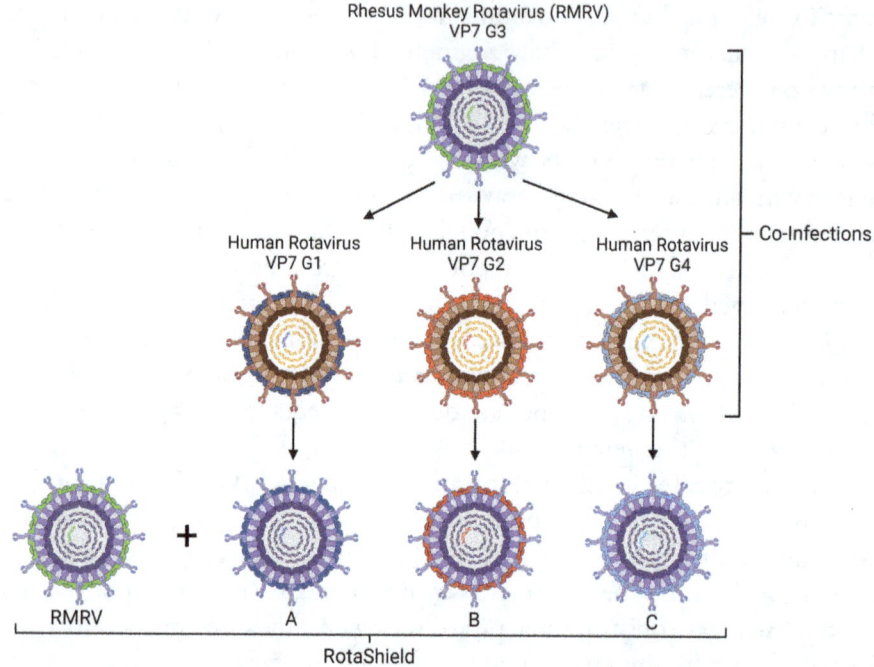

Fig. 11.1 RotaShieldRotaShield was a tetravalent vaccine based on a rhesus monkey rotavirus (RMRV). The parental RMRV expressed VP7 type G3 (green outer shell). To create reassortants between RMRV and human rotaviruses, cells were co-infected with the parental RMRV and human rotaviruses carrying a different VP7 type. The three human rotaviruses used expressed VP7 G1 (dark blue outer shell), G2 (red outer shell), or G4 (light blue outer shell). Three reassortants (A, B, and C) were isolated containing the complete RMRV genome with the RMRV VP7-encoding RNA segment (green) replaced with the human rotavirus equivalents (dark blue, red, or light blue). The RotaShield vaccine contained the parental RMRV plus the three reassortants

intestine slides within the adjacent region, like the collapsing tubes of a telescope. It is not uncommon in infancy and has a natural frequency of 1 in 2000 children. This condition can cause pain, bleeding, and bowel obstruction that may require surgery to correct, and it can be fatal if untreated. During the FDA approval process for vaccines, these intussusception cases were presented to the Vaccine and Related Biological Products Advisory Committee (VRBPAC) as part of the RRV-TV review but were viewed as normal background level cases (Schwartz 2012). Ultimately the FDA approved RotaShield for commercial use in August of 1998 with no requirement for post-distribution monitoring for intussusception cases. Similarly, the CDC's Advisory Committee on Immunization Practices (ACIP) had been following RotaShield's development and testing for several years, and they also concluded that the intussusception cases were not related to the vaccine. In March of 1999, the ACIP published its recommendation that RotaShield be part of the routine infant immunization schedule with oral doses delivered at two, four, and six months. By July of that year, over half a million children in the U.S. had received at least their first dose of this vaccine.

It was only after the vaccine rollout, often called the "phase IV clinical trial", that worries about intussusception began after fifteen cases were reported in VAERS by mid-1999. VAERS stands for Vaccine Adverse Events Reporting System, a database system jointly maintained by the FDA and CDC. Established in 1990, VAERS is an open system where anyone from healthcare providers to concerned citizens can report any or all suspected vaccine-related adverse effects. Since it is not a controlled environment, individual events reported aren't necessarily proof that they are caused by the vaccine. For example, someone could get vaccinated and while leaving the clinic they could slip on some ice and break their leg. If they wanted, they could report the broken leg in VAERS as a vaccine-related event, but a broken leg is not a direct vaccine side effect. Similarly, many things are temporally connected by are not causally related. Medical events that happen in the weeks following a specific vaccination may well have been destined to happen regardless of the vaccine. So again, if someone reports a particular medical issue after vaccination, that single event can't be directly attributed to the vaccine. What VAERS is useful for is detecting patterns where a symptom or condition shows up repeatedly at a frequency higher than expected in the vaccinee population, like this sudden rise in intussusception cases.

The VAERS data triggered a rapid review by the CDC into the relationship between RotaShield and intussusception. It was just a year earlier in 1998 that Andrew Wakefield published his now-discredited paper linking the MMR vaccine to autism (Chap. 7). Public skepticism and distrust of vaccines were pervasive and promoting another potentially dangerous vaccine was unacceptable. In July of 1999, less than a year after its approval by the FDA, the CDC issued a recommendation to suspend the use of Rotashield until a substantive investigation could be conducted. Combing through medical records of RotaShield vaccinated infants and matched controls, CDC staff used multiple methodologies to assess the risk of intussusception. As all the available data suggested a true increased risk of intussusception due to RotaShield, the ACIP and the CDC worked to prepare a report reversing its support for this vaccine. Anticipating a negative decision by the CDC, Wyeth voluntarily withdrew RotaShield from the market in mid-October of 1999, ending its usage and consigning this vaccine to history as a failure. A mere three weeks later the CDC issued its final announcement in the Morbidity and Mortality Weekly Report canceling its endorsement of RotaShield.

The story of RotaShield is complex and can be viewed as either a success or a failure of the system. Some would say that RotaShield should never have been approved originally as the intussusception risk was not sufficiently considered. A subsequent investigation of VRBPAC and ACIP by the U.S. Congress revealed conflicts of interest for members of both committees and procedural flaws in committee operations (Congress 2000). Critically, several members had disclosed or undisclosed ties to the pharmaceutical industry and/or were directly involved in the vaccine's development yet were allowed to participate in the discussion and/or vote on the RotaShield approval. Whether or not any of the concerns raised in the report impacted the RotaShield decision was not discerned. The data that VRBPAC and ACIP reviewed showed that the vaccine worked, there was a huge global demand for a rotavirus vaccine, and the intussusception cases were presented as not statistically significant, so maybe approval was the only possible outcome even if the process was flawed. Nonetheless, the congressional report did indicate that both the VRBPAC and the ACIP were not "fairly balanced" with a broad membership that represented all constituencies affected by vaccines, including community members. The report also issued 17 recommendations for corrective actions to improve the performance and integrity of both committees' decisions on future vaccines.

After RotaShield's approval, some would say that the system succeeded as designed to ensure public safety. The rapid identification of the intussusception issue and the almost immediate removal of the vaccine from the market highlighted that the VAERS safety net worked and the regulatory bodies acted appropriately. Yet even this contention is debatable as a slower and more cautious approach to RotaShield might have yielded a different decision. ACIP's review only lasted a few months before deciding to revoke their recommendation for RotaShield. Several subsequently published studies challenged the initial intussusception risk analysis and found that there was a significantly lower risk than was first calculated and maybe no risk at all (Simonsen et al. 2001; Kramarz et al. 2001). As the conflicting data were debated, the CDC and ACIP reappraised the vaccine in February 2002 but remained firm in their original withdrawal conclusion. Their decision that the vaccine should not be used in the United States doomed any future reimplementation of this vaccine. While the scientific results about intussusception risk had supporters on both sides, likely the public would never have accepted the vaccine regardless of CDC recommendations. In an era with heightened public anxiety about and scrutiny of vaccines, any risk, especially in infants, would have engendered widespread resistance to this vaccine.

The failure of RotaShield in the United States was not just a local setback but also had a global impact (Glass et al. 2004). Wyeth was counting on sales in the U.S. to provide the revenue for broader distribution and to encourage other countries to follow the U.S. lead and license the vaccine. Watching RotaShield's rejection in the U.S., other developed nations opted against adopting the vaccine and Wyeth's lucrative markets quickly disappeared. Without the projected income from developed nations, there was no sustainable way for Wyeth to offer the vaccine to developing nations at a reduced cost. Sadly, these developing countries were the ones that most needed the vaccine and where the most lives could have been saved. Even if there was a real risk from intussusception, this risk would have been quite small compared to the risk of dying from rotavirus infection, making the vaccine a viable option in many nations. However, this risk-to-benefit argument didn't sway other countries. Even as Wyeth tried to explore these other markets, the attitude in many nations was that if the vaccine isn't good enough for American children then it isn't good enough for our children. Furthermore, Wyeth had never tested RotaShield in developing countries where poor nutrition, exposure to different environmental pathogens, and generally lower levels of overall health were known to reduce vaccine efficacy. Without such data, convincing poorer nations to adopt this vaccine was unsuccessful. Consequently, RotaShield never gained a foothold outside its brief appearance in the U.S. and production stopped. It would be seven years and hundreds of thousands of children's deaths before an improved rotavirus vaccine was finally back on the market.

Back into the Fray

While Wyeth was the original frontrunner for a rotavirus vaccine, other companies also had rotavirus vaccines in development. When RotaShield failed, the market was wide open again but both the vaccine manufacturers and the regulatory agencies were highly apprehensive about the intussusception issue. Now much larger clinical trials and more stringent testing would be needed to convince the regulatory agencies that there was no intussusception risk. The first company to clear this hurdle was Merck whose rotavirus vaccine called RotaTeq was FDA-approved in early 2006 (Offit and Clark 2011). Like RotaShield, RotaTeq was a live, oral vaccine for infants. Sticking with the modified Jennerian approach, RotaTeq was designed as a pentavalent reassortant vaccine based on a bovine rotavirus designated WC3. Four of the reassortants are the WC3 virus with its VP7 gene replaced by the human rotavirus VP7 gene for serotypes G1, G2, G3, or G4. The final reassortant has the WC3 VP4 gene replaced by the human rotavirus equivalent. This cocktail of five viruses evokes an immune response against the four major VP7 serotypes of human rotaviruses and also against the human rotavirus VP4 which is the other major viral surface protein. With this potential vaccine in hand, what was needed next was a massive clinical trial with enough participants to detect any rare cases of intussusception. Merck ultimately invested $350 million to enroll 70,000 infants in the RotaTeq trial across eleven countries, although primarily in two developed nations, the U.S. and Finland (Vesikari et al. 2006). Then the largest and most expensive trial ever, RotaTeq tested remarkably well with 98% protection against severe disease and a 96% reduction in hospitalizations due to rotavirus. Critically, there was no indication of increased intussusception as case numbers were almost identical in the vaccinated and control groups. With this admirable performance and safety, the FDA approved it for use in the U.S. in early 2006 and the CDC ACIP committee quickly recommended it for universal use in infants.

GlaxoSmithKline (GSK) took a different approach both in the vaccine composition and the testing and marketing strategy. No one knew what caused spontaneous intussusception in kids or why the rhesus monkey rotavirus-based Rotashield caused cases in some vaccine recipients. This might be a phenomenon unique to the rhesus rotavirus or maybe any animal rotavirus could cause intussusception in humans. However, natural infections with human rotavirus were not associated with any intussusception risk (Velazquez et al. 2004; Bines et al. 2006; Mulcahy et al. 1982). Consequently, GSK focused on the more conventional strategy of creating an attenuated version of human rotavirus as their vaccine platform. In the late 1990s, investigators at the Children's Hospital in Cincinnati, Ohio developed and tested a candidate vaccine based on a human rotavirus isolated from the stool of an infected child (Ward and Bernstein 2009). This isolate, called 89-12, was chosen because children infected with this strain developed broad and robust immunity that gave excellent protection against subsequent rotavirus infections. To attenuate 89-12 it was passaged 33 times in culture, first in primary kidney cells directly isolated from African green monkeys (AGMK cells) and then in an AGMK cell line. The resulting strain was tested in 120 individuals from infants to adults and caused no significant illness indicating that it was well attenuated yet it induced high titers of neutralizing antibodies. This candidate strain was eventually licensed to GSK who refined it by purifying the virus and further attenuating it by additional passage in Vero cells (another AGMK line). Their final strain, known as RIX4411, became the vaccine called Rotarix (Vesikari et al. 2004).

Like RotaTeq, the hurdle for Rotarix was conducting a large enough clinical trial to ensure that there was no intussusception risk with this vaccine. Typically such trials had been conducted in developed nations because they had the infrastructure of physicians, hospitals, and research facilities that were needed to enroll and monitor patients. Furthermore, there were financial considerations that favored developed nations. The richer nations could help fund the trials through government grants to institutions and investigators involved in the testing of candidate vaccines, and these nations were the primary market since they could afford to pay the often significant price for vaccines newly brought to market. Only after the successful recovery of their investment could the vaccine manufacturers afford to explore subsidized vaccine distribution to less developed nations. However, in 1999 a new entity arose that changed the dynamics of testing and support. This entity was the Global Alliance for Vaccines and Immunization (GAVI), a private-public partnership devoted to increasing vaccine access for children around the world (Ashraf 2000). With a massive donation from the Bill and Melinda Gates Foundation, GAVI quickly became a powerful influence in the global fight against infectious diseases, with rotavirus as one of its targets. Funds were provided not only to support research and testing of new vaccines against important pathogens but also to purchase vaccines for poorer nations. Realizing that GAVI could support rotavirus vaccine sales to developing nations, GSK embraced a novel strategy of marketing Rotarix simultaneously in both developed and less developed nations. Initially, GSK field-tested Rotarix in Finland, Brazil, Mexico, and Venezuela to establish its efficacy, safety, and lack of intussusception (Ruiz-Palacios et al. 2006). These early studies were small phase I and phase II trials that didn't produce any contraindications for this vaccine. Subsequently, large phase III trials were conducted that involved more than 60,000 children. These trials demonstrated excellent vaccine efficacy with no indication of increased risk for intussusception. Importantly, administration of Rotarix in conjunction with other routine childhood vaccines showed no interference or reduction in immune response so concomitant use with other standard vaccines was feasible. With these excellent results, the vaccine was first licensed in Mexico in 2004 followed by Europe in 2006, and finally the U.S. in 2008. Subsequent trials in Africa and Asia showed similar safety but with lesser protection, but still enough to make the vaccine an important weapon against rotavirus morbidity and mortality (Phua et al. 2009; O'Ryan and Linhares 2009). With the success of RotaTeq and Rotarix, India (Rotavac and Rotasiil), China (Lanzhou lamb), and Vietnam (Rotavin-M1) have developed their versions of a rotavirus vaccine that are used primarily in their own countries. Globally there is now a diverse array of available rotavirus vaccines, and much of the current focus is on ensuring the distribution and utilization of these vaccines where they can do the most good for children's health. Parenthetically, there may even be unintended benefits of these vaccines. Both celiac disease (CD) and type 1 diabetes (T1D) have been linked to rotavirus infections, although these are complex diseases whose etiology is not well understood (Gomez-Rial et al. 2019, 2020). Some reports suggest the rotavirus vaccine may help prevent both CD and T1D (Rogers et al. 2019; Hemming-Harlo et al. 2019) but this issue is far from resolved as other studies have not supported this conclusion. It will be important to confirm or disprove the relationship between CD, T1D, rotavirus infection, and the vaccines, as well as search for additional gastrointestinal disorders that might be improved by preventing rotavirus infection.

Pitfalls, Questions, and Opportunities—The Push for the Future

Over the subsequent years since the licensing of RotaTeq and Rotarix, post-marketing assessment of adverse effects has detected a slightly increased intussusception risk with both vaccines (Tate et al. 2016), and this may be inherent in any of the rotavirus vaccines. However, the extent, if any, of this risk remains debated as some studies find no increased risk (Soares-Weiser et al. 2019). If there is a risk, it is in the range of 1–5 additional cases per 50,000–100,000 vaccine recipients, a very small rise compared to the huge reduction in serious rotavirus disease since these vaccines were introduced. According to WHO estimates, in the year 2000 there were more than 500,000 childhood deaths annually from rotavirus infections. By 2017, roughly ten years after the vaccines became available, this number had declined by somewhere between 60% and 80% with a corresponding decrease in hospitalizations and doctor visits (Clark et al. 2017). Given the low risk and the large impact the vaccines have had on preventing serious cases of rotavirus, there has been no suggestion to remove them from the market, and all the vaccines remain widely used in various countries. However, in 2010 an additional safety concern arose when porcine circovirus DNA was found in the vaccine. One of the major advances in biology over the last 50 years has been the dramatic improvements in DNA sequencing technology. Sequencing was initially a very slow, expensive, and laborious manual process that could take days to obtain short lengths of sequence on the order of a few hundred nucleotide bases. We now have automated equipment that can sequence millions of base pairs in just hours at a minimal cost. The explosive evolution of this technology allows investigators to interrogate complex samples and analyze all the DNA (or RNA) present in those samples. In 2010, a research group led by Eric Delwart examined the genomic sequences of ten live, attenuated vaccines, including Rotarix and RotaTeq, looking for genetic variations in the vaccine viruses (Victoria et al. 2010). An unanticipated finding was that the Rotarix vaccine also contained DNA of a pig virus called porcine circovirus 1 (PCV1). A similar identification of PCV1 in RotaTeq quickly followed along with evidence in RotaTeq for DNA from porcine circovirus 2 (PCV2) as well. One difference was that Rotarix didn't just have PCV1 DNA but also had infectious viral particles while RotaTeq did not appear to contain infectious virions for either PCV1 or PCV2 (McClenahan et al. 2011). The discovery of these contaminants caused confusion and consternation while the FDA reviewed the data and the safety implications.

Circoviruses are a family of DNA viruses with small (around 2000 base), circular, single-stranded genomes. In 2010, circoviruses were known to be viruses of pigs and birds but their existence in other animal species, including humans, was not known. PCV1 was harmless for pigs while PCV2 could cause fatal infections in baby pigs, but neither was believed to have any impact on humans. While any vaccine contaminant is potentially alarming, the FDA quickly decided that these PCV contaminants posed no risk to vaccine recipients based on data from the food industry (Li et al. 2010). Both PCV 1 and 2 were widely present in domestic swine herds and were found in virtually all pork products consumed by humans. There was plenty of evidence that humans were regularly exposed to these viruses both through food and working with pigs with no indication of any disease or adverse effects. Retrospective analysis of the vaccine lots found that PCV contamination was present in the vaccines since the early trials and again there was no indication that any of the vaccine recipients had an adverse effect due to either PCV1 or 2. Lacking any evidence of a dangerous infectious risk, the FDA allowed the continued production and distribution of both vaccines with the provision that the manufacturer work to develop PCV-free versions of their vaccines.

A second and more vexing question was the observation that vaccine efficacy was poorest in the developing countries, the nations that most needed protection from rotavirus disease due to inadequate access to clinical care for sick children (Burke et al. 2021). This effect is not unique to the rotavirus vaccines and occurs with several vaccines, likely reflecting that many individuals in these countries have less than fully robust immune systems. This phenomenon appears to be multifactorial involving the host, the particular pathogen, and environmental contributions (Omatola and Olaniran 2022). For example, general undernourishment and well as specific vitamin deficiencies adversely affect the ability of the immune system to respond to foreign antigens, including pathogens. In addition, growing evidence implicates the health of the gut microbiome as an important factor in rotavirus vaccine efficacy (Srivastava et al. 2020; Clark et al. 2019). Understanding how nutritional status and environmental conditions influence the gut microbiome may be key to improving the response to oral vaccines like the rotavirus vaccine.

Since each incremental increase in vaccine efficacy means more children saved, improving current vaccines or developing newer versions is an important goal for ongoing research. There are numerous new versions of rotavirus vaccines in development spanning the gamut of vaccine platform types (Carcamo-Calvo et al. 2021). In addition to new versions of the attenuated oral vaccine approach using different rotavirus isolates, several novel strategies are being investigated. If intussusception and mild gastrointestinal distress in some recipients is an unavoidable consequence of live, oral rotavirus vaccines then killed or subunit vaccines may be able to provide equivalent protection with fewer risks. Some groups are testing standard, inactivated whole virus that would be injected like other inactivated vaccines (Velasquez et al. 2015), while many others are exploring subunit vaccines such as the one developed for the hepatitis B virus (Chap. 9). The subunit vaccines typically contain one or more rotaviral proteins, usually VP7 or VP4, but sometimes additional viral proteins are included as well to induce broader immunity. What is particularly exciting about the subunit approach is that it lends itself to alternative delivery methods. For example, instead of expressing and purifying the viral proteins for injection, the viral proteins could be produced in plants that are then ingested to deliver the proteins (Feng et al. 2017). Other approaches include attaching viral proteins to nanoparticles for injection delivery (Liu et al. 2021) or combining VP7 and VP4 with core viral proteins to form virus-like particles (VLPs) (Xia et al. 2019). VLPs are empty particles that resemble the complete virion structure but they lack the viral genome so are completely noninfectious and unable to replicate. However, by presenting the relevant viral surface proteins in the natural context of the virion, VLPs can effectively mimic the actual virus and induce very effective immune responses. Lastly, a particularly exciting approach is mRNA vaccine technology which was first utilized in humans for the COVID-19 vaccine (Chap. 13). Instead of introducing viral proteins into the vaccine recipient, a mRNA that encodes the desired protein is injected. Once inside the person's cells, the mRNA utilizes the host cell machinery to translate the mRNA into protein that is then processed and presented to the immune system to evoke the immune response. Promising results in animal models have already been reported by multiple groups (Hensley et al. 2024; Roier et al. 2023; Lu et al. 2024), and this approach may herald a new era in rotavirus vaccine development. The future is exciting, and any further advances in efficacy, safety, or cost for rotavirus vaccines will only save more innocent lives.

References

Abad FX, Pinto RM, Bosch A (1994) Survival of enteric viruses on environmental fomites. Appl Environ Microbiol 60(10):3704–3710. https://doi.org/10.1128/aem.60.10.3704-3710.1994

Adams WR, Kraft LM (1963) Epizootic diarrhea of infant mice: identification of the etiologic agent. Science 141(3578):359–360. https://doi.org/10.1126/science.141.3578.359

Ansari SA, Sattar SA, Springthorpe VS, Wells GA, Tostowaryk W (1988) Rotavirus survival on human hands and transfer of infectious virus to animate and nonporous inanimate surfaces. J Clin Microbiol 26(8):1513–1518. https://doi.org/10.1128/jcm.26.8.1513-1518.1988

Ashraf H (2000) Public and private bodies unite to push for global immunisation. Lancet 355(9202):477. https://doi.org/10.1016/S0140-6736(00)82033-4

Beau I, Cotte-Laffitte J, Geniteau-Legendre M, Estes MK, Servin AL (2007) An NSP4-dependant mechanism by which rotavirus impairs lactase enzymatic activity in brush border of human enterocyte-like Caco-2 cells. Cell Microbiol 9(9):2254–2266. https://doi.org/10.1111/j.1462-5822.2007.00956.x

Benkaddour M, Tache S, Labie C, Bodin G, Eeckhoutte M (1993) Influence of temperature, acidity and lactic acid bacteria on the stability of rotavirus and coronavirus in milk. Revue de Médecine Vétérinaire 144(8/9):709–716

Bines JE, Liem NT, Justice FA, Son TN, Kirkwood CD, de Campo M, Barnett P, Bishop RF, Robins-Browne R, Carlin JB, Intussusception Study G (2006) Risk factors for intussusception in infants in Vietnam and Australia: adenovirus implicated, but not rotavirus. J Pediatr 149(4):452–460. https://doi.org/10.1016/j.jpeds.2006.04.010

Bishop RF (1996) Natural history of human rotavirus infection. Arch Virol Suppl 12:119–128. https://doi.org/10.1007/978-3-7091-6553-9_14

Bishop R (2009) Discovery of rotavirus: implications for child health. J Gastroenterol Hepatol 24 Suppl 3:S81–S85. https://doi.org/10.1111/j.1440-1746.2009.06076.x

Bishop RF, Davidson GP, Holmes IH, Ruck BJ (1974) Detection of a new virus by electron microscopy of faecal extracts from children with acute gastroenteritis. Lancet 1(7849):149–151. https://doi.org/10.1016/s0140-6736(74)92440-4

Bishop RF, Barnes GL, Cipriani E, Lund JS (1983) Clinical immunity after neonatal rotavirus infection. A prospective longitudinal study in young children. N Engl J Med 309(2):72–76. https://doi.org/10.1056/NEJM198307143090203

Blacklow NR, Echeverria P, Smith DH (1976) Serological studies with reovirus-like enteritis agent. Infect Immun 13(6):1563–1566. https://doi.org/10.1128/iai.13.6.1563-1566.1976

Blutt SE, Miller AD, Salmon SL, Metzger DW, Conner ME (2012) IgA is important for clearance and critical for protection from rotavirus infection. Mucosal Immunol 5(6):712–719. https://doi.org/10.1038/mi.2012.51

Burke RM, Tate JE, Parashar UD (2021) Global experience with rotavirus vaccines. J Infect Dis 224(12 Suppl 2):S792–S800. https://doi.org/10.1093/infdis/jiab399

Carcamo-Calvo R, Munoz C, Buesa J, Rodriguez-Diaz J, Gozalbo-Rovira R (2021) The rotavirus vaccine landscape, an update. Pathogens 10(5). https://doi.org/10.3390/pathogens10050520

Clark A, Black R, Tate J, Roose A, Kotloff K, Lam D, Blackwelder W, Parashar U, Lanata C, Kang G, Troeger C, Platts-Mills J, Mokdad A, Global Rotavirus Surveillance N, Sanderson C, Lamberti L, Levine M, Santosham M, Steele D (2017) Estimating global, regional and national rotavirus deaths in children aged <5 years: current approaches, new analyses and proposed improvements. PLoS One 12(9):e0183392. https://doi.org/10.1371/journal.pone.0183392

Clark A, van Zandvoort K, Flasche S, Sanderson C, Bines J, Tate J, Parashar U, Jit M (2019) Efficacy of live oral rotavirus vaccines by duration of follow-up: a meta-regression of randomised controlled trials. Lancet Infect Dis 19(7):717–727. https://doi.org/10.1016/S1473-3099(19)30126-4

Congress U (2000) Conflicts of interest in vaccine policy making. Majority Staff Report

Crawford SE, Ramani S, Tate JE, Parashar UD, Svensson L, Hagbom M, Franco MA, Greenberg HB, O'Ryan M, Kang G, Desselberger U, Estes MK (2017) Rotavirus infection. Nat Rev Dis Primers 3:17083. https://doi.org/10.1038/nrdp.2017.83

Davidson GP, Bishop RF, Townley RR, Holmes IH (1975) Importance of a new virus in acute sporadic enteritis in children. Lancet 1(7901):242–246. https://doi.org/10.1016/s0140-6736(75)91140-x

Demol P, Zissis G, Butzler JP, Mutwewingabo A, Andre FE (1986) Failure of live, attenuated oral rotavirus vaccine. Lancet 2(8498):108–108

Desselberger U, Huppertz HI (2011) Immune responses to rotavirus infection and vaccination and associated correlates of protection. J Infect Dis 203(2):188–195. https://doi.org/10.1093/infdis/jiq031

Feng H, Li X, Song W, Duan M, Chen H, Wang T, Dong J (2017) Oral administration of a seed-based bivalent rotavirus vaccine containing VP6 and NSP4 induces specific immune responses in mice. Front Plant Sci 8:910. https://doi.org/10.3389/fpls.2017.00910

Flewett TH, Bryden AS, Davies H, Woode GN, Bridger JC, Derrick JM (1974) Relation between viruses from acute gastroenteritis of children and newborn calves. Lancet 2(7872):61–63. https://doi.org/10.1016/s0140-6736(74)91631-6

Gentsch JR, Laird AR, Bielfelt B, Griffin DD, Banyai K, Ramachandran M, Jain V, Cunliffe NA, Nakagomi O, Kirkwood CD, Fischer TK, Parashar UD, Bresee JS, Jiang B, Glass RI (2005) Serotype diversity and reassortment between human and animal rotavirus strains: implications for rotavirus vaccine programs. J Infect Dis 192(Suppl 1):S146–S159. https://doi.org/10.1086/431499

Georges-Courbot MC, Monges J, Siopathis MR, Roungou JB, Gresenguet G, Bellec L, Bouquety JC, Lanckriet C, Cadoz M, Hessel L et al (1991) Evaluation of the efficacy of a low-passage bovine rotavirus (strain WC3) vaccine in children in Central Africa. Res Virol 142(5):405–411. https://doi.org/10.1016/0923-2516(91)90008-q

Ghose LH, Schnagl RD, Holmes IH (1978) Comparison of an enzyme-linked immunosorbent assay for quantitation of rotavirus antibodies with complement fixation in an epidemiological survey. J Clin Microbiol 8(3):268–276. https://doi.org/10.1128/jcm.8.3.268-276.1978

Glass RI, Bresee JS, Parashar UD, Jiang B, Gentsch J (2004) The future of rotavirus vaccines: a major setback leads to new opportunities. Lancet 363(9420):1547–1550. https://doi.org/10.1016/S0140-6736(04)16155-2

Gomez-Rial J, Sanchez-Batan S, Rivero-Calle I, Pardo-Seco J, Martinon-Martinez JM, Salas A, Martinon-Torres F (2019) Rotavirus infection beyond the gut. Infect Drug Resist 12:55–64. https://doi.org/10.2147/IDR.S186404

Gomez-Rial J, Rivero-Calle I, Salas A, Martinon-Torres F (2020) Rotavirus and autoimmunity. J Infect 81(2):183–189. https://doi.org/10.1016/j.jinf.2020.04.041

Hellysaz A, Hagbom M (2021) Understanding the central nervous system symptoms of rotavirus: a qualitative review. Viruses 13(4). https://doi.org/10.3390/v13040658

Hemming-Harlo M, Lahdeaho ML, Maki M, Vesikari T (2019) Rotavirus vaccination does not increase type 1 diabetes and may decrease celiac disease in children and adolescents. Pediatr Infect Dis J 38(5):539–541. https://doi.org/10.1097/INF.0000000000002281

Hensley C, Roier S, Zhou P, Schnur S, Nyblade C, Parreno V, Frazier A, Frazier M, Kiley K, O'Brien S, Liang Y, Mayer BT, Wu R, Mahoney C, McNeal MM, Petsch B, Rauch S, Yuan L (2024) mRNA-based vaccines are highly immunogenic and confer protection in the gnotobiotic pig model of human rotavirus diarrhea. Vaccines (Basel) 12(3). https://doi.org/10.3390/vaccines12030260

Herrmann T, Torres R, Salgado EN, Berciu C, Stoddard D, Nicastro D, Jenni S, Harrison SC (2021) Functional refolding of the penetration protein on a non-enveloped virus. Nature 590(7847):666–670. https://doi.org/10.1038/s41586-020-03124-4

Joensuu J, Koskenniemi E, Pang XL, Vesikari T (1997) Randomised placebo-controlled trial of rhesus-human reassortant rotavirus vaccine for prevention of severe rotavirus gastroenteritis. Lancet 350(9086):1205–1209. https://doi.org/10.1016/S0140-6736(97)05118-0

Kapikian AZ (2011) History of rotavirus vaccines part I: RotaShield. In: History of vaccine development, pp 285–314. https://doi.org/10.1007/978-1-4419-1339-5_28

Kapikian AZ, Kim HW, Wyatt RG, Rodriguez WJ, Ross S, Cline WL, Parrott RH, Chanock RM (1974) Reoviruslike agent in stools: association with infantile diarrhea and development of serologic tests. Science 185(4156):1049–1053. https://doi.org/10.1126/science.185.4156.1049

Kapikian AZ, Flores J, Hoshino Y, Glass RI, Midthun K, Gorziglia M, Chanock RM (1986) Rotavirus: the major etiologic agent of severe infantile diarrhea may be controllable by a "Jennerian" approach to vaccination. J Infect Dis 153(5):815–822. https://doi.org/10.1093/infdis/153.5.815

Kolling G, Wu M, Guerrant RL (2012) Enteric pathogens through life stages. Front Cell Infect Microbiol 2:114. https://doi.org/10.3389/fcimb.2012.00114

Kramarz P, France EK, Destefano F, Black SB, Shinefield H, Ward JI, Chang EJ, Chen RT, Shatin D, Hill J, Lieu T, Ogren JM (2001) Population-based study of rotavirus vaccination and intussusception. Pediatr Infect Dis J 20(4):410–416. https://doi.org/10.1097/00006454-200104000-00008

Krawczyk A, Lewis MG, Venkatesh BT, Nair SN (2016) Effect of exclusive breastfeeding on rotavirus infection among children. Indian J Pediatr 83(3):220–225. https://doi.org/10.1007/s12098-015-1854-8

Laban NM, Goodier MR, Bosomprah S, Simuyandi M, Chisenga C, Chilyabanyama ON, Chilengi R (2022) T-cell responses after rotavirus infection or vaccination in children: a systematic review. Viruses 14(3). https://doi.org/10.3390/v14030459

Latifi T, Kachooei A, Jalilvand S, Zafarian S, Roohvand F, Shoja Z (2024) Correlates of immune protection against human rotaviruses: natural infection and vaccination. Arch Virol 169(3):72. https://doi.org/10.1007/s00705-024-05975-y

Li L, Kapoor A, Slikas B, Bamidele OS, Wang C, Shaukat S, Masroor MA, Wilson ML, Ndjango JB, Peeters M, Gross-Camp ND, Muller MN, Hahn BH, Wolfe ND, Triki H, Bartkus J, Zaidi SZ, Delwart E (2010) Multiple diverse circoviruses infect farm animals and are commonly found in human and chimpanzee feces. J Virol 84(4):1674–1682. https://doi.org/10.1128/JVI.02109-09

Liu C, Huang P, Zhao D, Xia M, Zhong W, Jiang X, Tan M (2021) Effects of rotavirus NSP4 protein on the immune response and protection of the S(R69A)-VP8* nanoparticle rotavirus vaccine. Vaccine 39(2):263–271. https://doi.org/10.1016/j.vaccine.2020.12.005

Lu C, Li Y, Chen R, Hu X, Leng Q, Song X, Lin X, Ye J, Wang J, Li J, Yao L, Tang X, Kuang X, Zhang G, Sun M, Zhou Y, Li H (2024) Safety, immunogenicity, and mechanism of a rotavirus mRNA-LNP vaccine in mice. Viruses 16(2). https://doi.org/10.3390/v16020211

Malherbe H, Harwin R (1963) The cytopathic effects of vervet monkey viruses. S Afr Med J 37:407–411

McClenahan SD, Krause PR, Uhlenhaut C (2011) Molecular and infectivity studies of porcine circovirus in vaccines. Vaccine 29(29–30):4745–4753. https://doi.org/10.1016/j.vaccine.2011.04.087

Mulcahy DL, Kamath KR, de Silva LM, Hodges S, Carter IW, Cloonan MJ (1982) A two-part study of the aetiological role of rotavirus in intussusception. J Med Virol 9(1):51–55. https://doi.org/10.1002/jmv.1890090108

O'Ryan M (2009) The ever-changing landscape of rotavirus serotypes. Pediatr Infect Dis J 28(3 Suppl):S60–S62. https://doi.org/10.1097/INF.0b013e3181967c29

O'Ryan M, Linhares AC (2009) Update on Rotarix: an oral human rotavirus vaccine. Expert Rev Vaccines 8(12):1627–1641. https://doi.org/10.1586/erv.09.136

Offit PA, Clark HF (2011) Rotavirus vaccines part II: raising the bar for vaccine safety studies. In: History of vaccine development, pp 315–327. https://doi.org/10.1007/978-1-4419-1339-5_29

Omatola CA, Olaniran AO (2022) Rotaviruses: from pathogenesis to disease control-a critical review. Viruses-Basel 14(5):875. https://doi.org/10.3390/v14050875

Pappenheimer AM, Enders JF (1947) An epidemic diarrheal disease of suckling mice: II. Inclusions in the intestinal epithelial cells. J Exp Med 85(4):417–422. https://doi.org/10.1084/jem.85.4.417

Perez-Schael I, Guntinas MJ, Perez M, Pagone V, Rojas AM, Gonzalez R, Cunto W, Hoshino Y, Kapikian AZ (1997) Efficacy of the rhesus rotavirus-based quadrivalent vaccine in infants and young children in Venezuela. N Engl J Med 337(17):1181–1187. https://doi.org/10.1056/NEJM199710233371701

Phua KB, Lim FS, Lau YL, Nelson EA, Huang LM, Quak SH, Lee BW, Teoh YL, Tang H, Boudville I, Oostvogels LC, Suryakiran PV, Smolenov IV, Han HH, Bock HL (2009) Safety and efficacy of human rotavirus vaccine during the first 2 years of life in Asian infants: randomised, double-blind, controlled study. Vaccine 27(43):5936–5941. https://doi.org/10.1016/j.vaccine.2009.07.098

Raphael RA, Sattar SA, Springthorpe VS (1985) Long-term survival of human rotavirus in raw and treated river water. Can J Microbiol 31(2):124–128. https://doi.org/10.1139/m85-024

Rennels MB, Glass RI, Dennehy PH, Bernstein DI, Pichichero ME, Zito ET, Mack ME, Davidson BL, Kapikian AZ (1996) Safety and efficacy of high-dose rhesus-human reassortant rotavirus vaccines--report of the National Multicenter Trial. United States Rotavirus Vaccine Efficacy Group. Pediatrics 97(1):7–13

Rennels MB, Parashar UD, Holman RC, Le CT, Chang HG, Glass RI (1998) Lack of an apparent association between intussusception and wild or vaccine rotavirus infection. Pediatr Infect Dis J 17(10):924–925. https://doi.org/10.1097/00006454-199810000-00018

Rogers MAM, Basu T, Kim C (2019) Lower incidence rate of type 1 diabetes after receipt of the rotavirus vaccine in the United States, 2001–2017. Sci Rep 9(1):7727. https://doi.org/10.1038/s41598-019-44193-4

Roier S, Mangala Prasad V, McNeal MM, Lee KK, Petsch B, Rauch S (2023) mRNA-based VP8* nanoparticle vaccines against rotavirus are highly immunogenic in rodents. NPJ Vaccines 8(1):190. https://doi.org/10.1038/s41541-023-00790-z

Ruiz-Palacios GM, Perez-Schael I, Velazquez FR, Abate H, Breuer T, Clemens SC, Cheuvart B, Espinoza F, Gillard P, Innis BL, Cervantes Y, Linhares AC, Lopez P, Macias-Parra M, Ortega-Barria E, Richardson V, Rivera-Medina DM, Rivera L, Salinas B, Pavia-Ruz N, Salmeron J, Ruttimann R, Tinoco JC, Rubio P, Nunez E, Guerrero ML, Yarzabal JP, Damaso S, Tornieporth N, Saez-Llorens X, Vergara RF, Vesikari T, Bouckenooghe A, Clemens R, De Vos B, O'Ryan M, Human Rotavirus Vaccine Study G (2006) Safety and efficacy of an attenuated vaccine against severe rotavirus gastroenteritis. N Engl J Med 354(1):11–22. https://doi.org/10.1056/NEJMoa052434

Sattar SA, Lloyd-Evans N, Springthorpe VS, Nair RC (1986) Institutional outbreaks of rotavirus diarrhoea: potential role of fomites and environmental surfaces as vehicles for virus transmission. J Hyg (Lond) 96(2):277–289. https://doi.org/10.1017/s0022172400066055

Schwartz JL (2012) The first rotavirus vaccine and the politics of acceptable risk. Milbank Q 90(2):278–310. https://doi.org/10.1111/j.1468-0009.2012.00664.x

Simonsen L, Morens D, Elixhauser A, Gerber M, Van Raden M, Blackwelder W (2001) Effect of rotavirus vaccination programme on trends in admission of infants to hospital for intussusception. Lancet 358(9289):1224–1229. https://doi.org/10.1016/S0140-6736(01)06346-2

Soares-Weiser K, Bergman H, Henschke N, Pitan F, Cunliffe N (2019) Vaccines for preventing rotavirus diarrhoea: vaccines in use (Review). Cochrane Database Syst Rev 10:CD008521. https://doi.org/10.1002/14651858.CD008521.pub5

Srivastava V, Deblais L, Huang HC, Miyazaki A, Kandasamy S, Langel SN, Paim FC, Chepngeno J, Kathayat D, Vlasova AN, Saif LJ, Rajashekara G (2020) Reduced rotavirus vaccine efficacy in protein malnourished human-faecal-microbiota-transplanted gnotobiotic pig model is in part attributed to the gut microbiota. Benef Microbes 11(8):733–751. https://doi.org/10.3920/BM2019.0139

Tan SW, Yap KL, Lee HL (1997) Mechanical transport of rotavirus by the legs and wings of Musca domestica (Diptera: Muscidae). J Med Entomol 34(5):527–531. https://doi.org/10.1093/jmedent/34.5.527

Tate JE, Yen C, Steiner CA, Cortese MM, Parashar UD (2016) Intussusception rates before and after the introduction of rotavirus vaccine. Pediatrics 138(3). https://doi.org/10.1542/peds.2016-1082

Velasquez DE, Wang Y, Jiang B (2015) Inactivated human rotavirus vaccine induces heterotypic antibody response: correction and development of IgG avidity assay. Hum Vaccin Immunother 11(2):531–533. https://doi.org/10.4161/21645515.2014.988553

Velazquez FR, Matson DO, Calva JJ, Guerrero L, Morrow AL, Carter-Campbell S, Glass RI, Estes MK, Pickering LK, Ruiz-Palacios GM (1996) Rotavirus infection in infants as protection against subsequent infections. N Engl J Med 335(14):1022–1028. https://doi.org/10.1056/NEJM199610033351404

Velazquez FR, Luna G, Cedillo R, Torres J, Munoz O (2004) Natural rotavirus infection is not associated to intussusception in Mexican children. Pediatr Infect Dis J 23(10 Suppl):S173–S178. https://doi.org/10.1097/01.inf.0000142467.50724.de

Vesikari T, Rautanen T, Varis T, Beards GM, Kapikian AZ (1990) Rhesus Rotavirus candidate vaccine. Clinical trial in children vaccinated between 2 and 5 months of age. Am J Dis Child 144(3):285–289

Vesikari T, Karvonen A, Korhonen T, Espo M, Lebacq E, Forster J, Zepp F, Delem A, De Vos B (2004) Safety and immunogenicity of RIX4414 live attenuated human rotavirus vaccine in adults, toddlers and previously uninfected infants. Vaccine 22(21–22):2836–2842. https://doi.org/10.1016/j.vaccine.2004.01.044

Vesikari T, Matson DO, Dennehy P, Van Damme P, Santosham M, Rodriguez Z, Dallas MJ, Heyse JF, Goveia MG, Black SB, Shinefield HR, Christie CD, Ylitalo S, Itzler RF, Coia ML, Onorato MT, Adeyi BA, Marshall GS, Gothefors L, Campens D, Karvonen A, Watt JP, O'Brien KL, DiNubile MJ, Clark HF, Boslego JW, Offit PA, Heaton PM, Rotavirus E, Safety Trial Study T (2006) Safety and efficacy of a pentavalent human-bovine (WC3) reassortant rotavirus vaccine. N Engl J Med 354(1):23–33. https://doi.org/10.1056/NEJMoa052664

Victoria JG, Wang C, Jones MS, Jaing C, McLoughlin K, Gardner S, Delwart EL (2010) Viral nucleic acids in live-attenuated vaccines: detection of minority variants and an adventitious virus. J Virol 84(12):6033–6040. https://doi.org/10.1128/JVI.02690-09

Ward RL, Bernstein DI (1994) Protection against rotavirus disease after natural rotavirus infection. US Rotavirus Vaccine Efficacy Group. J Infect Dis 169(4):900–904. https://doi.org/10.1093/infdis/169.4.900

Ward RL, Bernstein DI (2009) Rotarix: a rotavirus vaccine for the world. Clin Infect Dis 48(2):222–228. https://doi.org/10.1086/595702

Ward RL, Bernstein DI, Young EC, Sherwood JR, Knowlton DR, Schiff GM (1986) Human rotavirus studies in volunteers: determination of infectious dose and serological response to infection. J Infect Dis 154(5):871–880. https://doi.org/10.1093/infdis/154.5.871

Xia M, Huang P, Jiang X, Tan M (2019) Immune response and protective efficacy of the S particle presented rotavirus VP8* vaccine in mice. Vaccine 37(30):4103–4110. https://doi.org/10.1016/j.vaccine.2019.05.075

12

Human Papillomaviruses: An Ancient Enemy

Keywords Cervical cancer • Gardasil • Oncoprotein • Sexually transmitted disease (STD) • Virus-like particle (VLP) • Harold zur Hausen • Warts

Abbreviations

ACIP	Advisory Committee on Immunization Practices
BPV	Bovine papillomavirus
CDC	Centers for Disease Control
CRPV	Cottontail rabbit papillomavirus
EBV	Epstein-Barr virus
EV	Epidermodysplasia verruciformis
HBsAg	Hepatitis B surface antigen
HPV	Human papillomavirus
HSV	Herpes simplex virus
PCR	Polymerase chain reaction
PV	Papillomavirus
SCC	Squamous cell carcinoma
STD	Sexually transmitted disease
VLP	Virus-like particle
WHO	World Health Organization

Our Silent Passengers

The human body is colonized by billions of bacteria and bacteriophages, comprising hundreds of species and genera that make up our microbiome (Walker and Hoyles 2023). These bacteria and their phages inhabit us from birth as babies pass through the vaginal canal and begin to acquire organisms from their mothers. Nearly all domains of our bodies from our skin to our digestive system team with this invisible multitude. By definition, these commensal organisms that make up our microbiome are mostly innocuous while many even contribute to our health and normal bodily functions. While the exact composition of individual microbiomes can differ, certain types of bacteria are universally found in humans and likely have a symbiotic relationship with us. We provide a host environment for the bacteria to thrive in and they provide some benefit to us. In contrast, viruses were not historically considered part of our endogenous flora. Instead, viruses were perceived as purely exogenous pathogens that we transiently acquired. This was certainly true for many common disease-causing viruses like influenza (Chap. 5), polio (Chap. 6), or hepatitis A (Chap. 9). Not everyone had these viruses, and after an infection, the virus was cleared and did not remain associated with the recovered individual. Even the discovery of viruses like hepatitis B (Chap. 8) or chickenpox (Chap. 10) that can establish permanent infections didn't change the basic paradigm. These viruses are still considered external pathogens found only in a subset of humans and are not considered part of the normal human microbiome. However, if any virus can be considered a commensal organism that is part of the normal human flora, it is the human papillomaviruses (HPVs). These viruses are the causative agents of papillomas (hence the name papillomaviruses), what we commonly call warts. More importantly than benign skin warts, HPVs are the major cause of cervical cancer and an important contributor to some forms of anal, oral, and skin cancers (Tornesello and Buonaguro 2020).

Many of the viral diseases in previous chapters entered human populations thousands of years ago, but that length of time is nothing on an evolutionary scale. For human papillomaviruses (HPVs), we have to look back hundreds of thousands of years to trace their ancestry in ancient humans, and even hundreds of millions of years to find the origin of the papillomavirus (PV) family. Members of the PV family are DNA viruses with small (approximately 8000 base pairs), circular, double-stranded genomes. We believe that they likely infect every type of vertebrate animal from fish to primates, but not invertebrates (Van Doorslaer 2013). This universal presence in vertebrate hosts indicates that the PV family is truly ancient and must have arisen soon after the vertebrate lineage developed from invertebrates around 470 million years ago. As the first vertebrate radiated out to form all of its diverse descendants, the primordial PV was spread among the newly evolving host species and became highly host-specific with no ability to infect anything other than their natural host species. Since that time, PVs have co-evolved along with their various hosts, including the PVs infecting prehistoric hominins, the ancestral lineages that gave rise to modern humans. Available data suggest that the last common ancestor of all modern HPVs existed somewhere between 50 and 60 million years ago. Because PVs co-existed with hominins long before *Homo sapiens* arose, modern HPVs are intimately and specifically adapted to their human hosts and do not infect other animals, even primates (McBride 2022). Within prehumans, PVs diversified and slowly evolved through mutation to generate an extensive panoply of genetically distinct viruses adapted to different anatomical niches in the human host. These distinct entities are designated types and are distinguished based on the DNA sequence encoding the major viral capsid protein, L1 (de Villiers et al. 2004). The types are numbered sequentially starting with the first HPV isolated, HPV1. When an HPV is isolated from a patient, its L1 gene is sequenced to assign it to a known type. If the isolate's L1 gene has greater than 10% sequence divergence from any known type then it is considered a new type and has the next consecutive number assigned to it. Currently, there are around 450 known HPV types with more continuing to be discovered, although roughly half of these types have yet to be stringently characterized. HPV types divide into five distinct major genera on the HPV phylogenetic tree designated Alpha, Beta, Gamma, Mu, and Nu (Van Doorslaer 2013). The types that are most important clinically, and thus became the target for vaccine development, fall in the Alpha genus.

HPVs exclusively infect and survive in our skin, both the cutaneous (think arms, legs, torso) and the mucosal (oral, genital) regions of our bodies. Because of their long co-evolution with humans, the HPV viral life cycle is inextricably tied to the biology of the human epidermis, the outer layer of our bodies (de Villiers et al. 2004). The skin has two major components, the lower dermis and the upper dermis, and they are separated by a structure called the basement membrane (Luxembourg and Moeller 2017). The epidermis is a multi-layered structure that starts with a bottom layer of cells called the basal layer which sits atop the basement membrane. The basal layer is only a single cell thick but is the critical replicative component that continually replenishes our skin. Basal cells can divide and grow to maintain the basal layer and provide cells for the upper layers of the epidermis. When a basal cell divides, both daughter cells can remain in the basal layer associated with the basement membrane or one daughter can move upward away from the membrane. As basal cells separate from the basement membrane they lose their ability to replicate and enter a process called terminal differentiation. Over about 40–50 days, the differentiating cells will move upwards and pass through several discrete morphological stages to form the upper epidermal layers (spinosum, granulosum, lucidum [only found on palms, digits, and the soles of the feet], and corneum) (Koster 2009). By the time they reach the uppermost layer (corneum), the differentiated cells are essentially inert and form the waterproof layer of dead skin that covers our bodies. These dead cells are continuously sloughed off and replenished by newly rising cells so that our entire epidermis is continuously replaced through this cycle. It is this complex process of epidermal maintenance, differentiation, and turnover that the HPVs have coopted for their own reproduction.

An HPV virion landing on the outer layer of the skin is not sufficient to start an infection, as the outer layer is a dead cell barrier. To establish an infection, a virion must first reach a replicative basal cell at the bottom layer of the epidermis (Moody 2017). This is not as difficult as it might seem because the epidermal layer is thin and easily damaged. Even minor bumps and scratches, too insignificant to produce any visible injury, can cause microabrasions that penetrate the epidermal layers and expose the basal cells to virions. Basal cells have appropriate receptors for HPV, and virions that reach this layer are taken up, uncoated, and their viral DNA transported to the cell nucleus. In the cell nucleus, the viral genome undergoes a brief amplification period to generate multiple copies of the viral DNA that remain in the nucleus as small plasmids. Once amplified, the number of viral genome copies remains stable, and the subsequent replication of these viral genomes becomes coordinated with the cellular division process. When the cell replicates its chromosomal DNA before cell division, the viral DNA is replicated in concert to double its copy number. As the cell divides into two daughter cells the viral DNA copies are partitioned between the two cells to roughly maintain the viral copy number. This so-called "maintenance state" allows the viral DNA to persist indefinitely in the basal cells unless the infected cells are eliminated by the immune response. As there is only minimal expression of viral proteins in this stage, these low levels of foreign proteins are hard to detect, so the immune response is not rapid. Eventually, most infections are cleared though it can take weeks to months. More problematically, some infections avoid immune surveillance and persist for years to decades. It is the persistent infections with certain HPV types that can lead to cancers and the discovery of this relationship spurred vaccine development.

During the maintenance state, the viral DNA replicates and persists to sustain the infection, but no new virions are produced in basal cells. Production of virions requires the infected cell to traverse the differentiating epidermal layers because different stages of the viral life cycle occur in specific layers of the epidermis. When an infected basal cell divides, one of the daughter cells containing HPV DNA may leave the basal layer. When this daughter cell begins to differentiate, this triggers the productive stage of the viral life cycle that leads to new virions. As the infected cell rises into the spinosum layer, another more extensive round of viral DNA amplification occurs to create many copies of the viral genome. Progressing into the granulosum layer triggers the expression of the two viral capsid proteins, called L1 and L2, that form the virion structure. With L1 and L2 available, virion assembly commences, and the newly replicated DNA molecules are packaged into capsids as the cell traverses through the granulosum and into the corneum layer. By the time the infected cell reaches the final corneum stage, it is full of new virions. These dead cells are shed and disintegrate releasing the HPV virions onto the skin surface where they can be spread to other people (Buck et al. 2005). Because they must reproduce in this dynamic, layered structure that we call our skin, HPVs cannot reproduce in any single type of cultured cell in the laboratory. Most of the viruses in previous chapters are typically grown in cell culture to produce large quantities of viruses for experimental study and vaccine production. However, the refractory nature of HPVs makes growing them in the laboratory through standard infection of cultured cells impossible. The inability to culture this family of viruses delayed and complicated their study until molecular technologies enabled their cloning in the 1970s. Even today, this family of viruses can't be grown in standard cell culture and human papillomaviruses only infect humans. Without a convenient culture system or an animal host for HPVs, the ability to do classical virologic studies is restricted, and there are areas of HPV biology and infection that remain poorly understood.

In addition to the experimental limitations of this virus family, it is important to note that the life cycle above often occurs with no visible sign of infection. Most people continuously harbor HPVs in their skin asymptomatically, particularly types falling in the Beta and Gamma genera. Infected individuals unknowingly shed virions as their outer skin corneum layer sloughs off (Antonsson et al. 2003a). It is these released virions that initiate infections in other individuals, usually through direct skin-to-skin contact (Gheit 2019). Like certain bacteria, some HPV types begin to be acquired at birth from the mother's vaginal flora (Antonsson et al. 2003b). As the neonate passes through the vaginal environment, maternally shed HPVs can easily contact and infect the newborn. Subsequently, contact with the mother, family members, and other persons handling the baby can transfer many HPV types to the baby's skin, and by age 5 children are usually positive for multiple HPVs. Other HPV types that infect cutaneous skin can be picked up throughout life by casual skin contact with infected persons. In contrast, the HPV types infecting mucosal skin are generally picked up at a later age through sexual contact (Plotzker et al. 2023). With hundreds of HPV types circulating in humans, we are likely constantly being exposed to and carrying silent HPV infections off and on throughout our lives, making these viruses a steady contributor to our microbial flora.

In contrast to the inapparent Beta and Gamma HPV infections, members of the Alpha, Mu, and Nu genera cause warts on the cutaneous and/or mucosal skin (Gheit 2019). Cutaneous warts have several varieties depending on their appearance or location on the body, but the type most people are familiar with is the common wart. This type can be anywhere on the body and presents as a hard, raised, and rough lump on the skin. Mucosal warts are found in the oral or anogenital regions, and the genital ones are called condyloma accuminata. Regardless of their type or anatomical location, the wart lesion is the site of active viral reproduction. Importantly, for this group of HPVs, the viruses are also stimulating excessive cell replication. Normally, these differentiating skin cells do not replicate, so the thickness of the epidermal layer never changes. However, the expression of certain viral genes in these cells causes the cells to reenter their replicative cycle. As these cells abnormally begin to divide and reproduce in the differentiating layers they pile up and cause a protuberance that extends out from the normal skin layer to form the visible wart, a tumor although a benign one. Additionally, the viral genes induce overexpression of cellular keratin proteins. At normal expression levels, keratin proteins provide structural support to skin and hair cells and contribute to the water resistance of the outer skin layer. It is the overabundance of these proteins promoted by HPV infection that causes the surface of warts to be rough and scaly, especially in cutaneous warts.

A Bumpy Past

The recognition that warts are an infectious disease has a long history extending back to the Greek and Roman eras (Syrjanen and Syrjanen 2008). Accurate descriptions of warts in different anatomical locations, including genital lesions, were present in the ancient writings with at least suggestions that these lesions could be transferred from person to person. However, these skin bumps took many forms, and it is unlikely that their common etiology was appreciated until modern times. After the fall of the ancient civilizations, warts, like many other diseases, spent centuries being attributed to various unfounded causes (Karamanou et al. 2010). It wasn't until the late 1800s that more systematic studies of patients and their contacts began to legitimize the concept of infectious transmission of these skin lesions. Although often limited in scope, not controlled, and sometimes anecdotal, increasing numbers of reports noted the connection between lesions in one individual and subsequent lesions in the patient's contacts. An important contribution to the discussion of these lesions was proposed by a French dermatologist in 1893. Historically, genital and cutaneous warts were thought to be different diseases, but his examination of the pathology of both lesions suggested that they were similar and might have a common causative agent. By this era, the concept of filterable agents, i.e. what we now know to be viruses, was widely known in scientific circles, and this new concept spurred more direct examination of these skin lesions. Several investigators demonstrated direct transmission of skin warts using extracts from lesions, and in 1907, Dr. Gieuseppe Ciuffo transmitted genital warts using a filtered extract from a patient's lesions. Based on these early results, it seemed plausible that warts, both skin and genital types, could be viral diseases. Continued transmission studies with different varieties of wart lesions supported this conclusion although the diversity of HPV types was not appreciated. Instead, the thinking was that there was a single wart virus that manifested as different forms of lesions depending on the bodily site of infection. The inability to culture and characterize the virus would hamper studies of the human agents for the next 50 years.

One important predecessor to the eventual discovery and isolation of the first human papillomavirus was the work with rabbit papillomas in the 1930s begun by Richard Shope, the same virologist who was instrumental in early influenza work (Chap. 5) (Andrewes 1979). Wild cottontail rabbits often have warty lesions on their skin, some of which can grow so long and protuberant on the head as to resemble horns, giving rise to the apocryphal "jackalope". Shope demonstrated that extracts from these lesions would transmit the warts to recipient wild or domestic rabbits and that this disease appeared to be due to a viral agent (Shope and Hurst 1933). This pathogen would eventually be designated the cottontail rabbit papillomavirus (CRPV) although it is commonly referred to as the eponymous Shope papillomavirus. He also found that the lesions, especially in domestic rabbits, would spontaneously regress while generating neutralizing antibodies that protected the animals from reinfection. The ability to induce antibodies and protection was a key observation that augured the possibility of vaccination against this type of virus, although this rabbit work was decades before future studies demonstrated that both human and rabbit warts were caused by viruses of the papillomavirus family.

With the rabbit as a somewhat tractable animal model system for studying warts and their presumptive viral agent, other investigators began to exploit this system. Most notable was Peyton Rous who would go on the win a Nobel Prize for his work on chicken cancers and become known as the father of tumor virology. Rous's contribution to the papillomavirus field was his observation that the clinical outcome of wart lesions differed in wild versus domestic rabbits. In wild cottontail rabbits, the warts were simply benign tumors, but when transferred to domestic rabbits the warts frequently progressed to malignant cancers (Rous and Beard 1935). In subsequent studies, he and others demonstrated that known carcinogens could enhance and accelerate the oncogenic transition of the wart lesions (Scherp and Syverton 1949; Rous and Friedewald 1944). The progression of some human warts to cancer, particularly vulval, anal, and some oral lesions, was well known by this time from observational accounts of lesions in individuals. However, these human cases were not amenable to experimental manipulation as human warts couldn't be transferred to any animal system and human-to-human studies were limited and not ideal. Now with the rabbit work, there was an animal model that paralleled the human situation and could be manipulated by investigators to address fundamental questions about transmission, pathology, and disease progression. For decades the rabbit model remained one of the most tractable systems for studying papillomavirus biology and the virus-host interaction. Yet as important as the rabbit-CRPV system was for many years, by the 1950s more and more attention was directed at human warts and their as-yet-uncharacterized viral pathogen. The early focus was on genital condylomas as these were established as a sexually transmitted disease likely of viral origin (Barrett et al. 1954). Also, there was a growing awareness that the abnormal cervical cells that were possible precursors to cervical cancer had morphological characteristics (called koilocytotic atypia) in common with the cells present in visible wart lesions (Koss and Durfee 1956). With both genital warts and cervical cancer as prevalent diseases, there was a clear need to understand the mechanisms of these illnesses with the hope of developing treatments or preventative measures. While the relationship between HPV infection and cervical cancer was still decades away from being postulated and proven, the convergence of human clinical observation and experimental work on animal papillomavirus was starting to expose a completely unanticipated world of human viral cancer. On a historical side note, abnormal cervical cells were detected in patients via the Pap smear, a procedure developed in 1928 (Vilos

1998). While we now know that the abnormal cells detected by this assay are HPV-infected cells, the "Pap" in Pap smear derives from the inventor's name, Papanicolaou, and has nothing to do with the words papilloma or papillomavirus.

Sex, Cancer, and HPVs

Skin warts can be annoying, unsightly, and sometimes painful, but are typically benign tumors that often spontaneously regress. The clinical significance of skin warts is modest enough that by themselves they are unlikely to have driven the HPV field forward at a rapid pace. Instead, the accelerating factor was the 1975 postulation by Professor Harold zur Hausen that HPV infection was a probable prerequisite to cervical cancer (zur Hausen et al. 1975). In the mid-twentieth century, cervical cancer was one of the most prevalent cancers worldwide and a leading cause of death in women, particularly in countries with limited or no regular Pap smear screening programs (zur Hausen et al. 1975). The serious and widespread nature of this disease made it medically urgent to understand its causes and possible prevention, a need that was frustrated by a long record of false leads. A possible contribution by a transmissible agent, i.e. a sexually transmitted disease (STD), in the development of cervical cancer was considered as far back as the early 1800s but a specific pathogen remained elusive. Consideration for the potential role of an STD was based primarily on epidemiological evidence from a variety of studies in the 1800s and 1900s that eventually correlated sexual activity with increased risk for cervical cancer. Some of the earliest observations anecdotally noted that the incidence of cervical cancer was very low to nonexistent in nuns and very high in prostitutes, two groups of women at the opposite extremes of sexual activity. Unfortunately, for nearly 100 years most of the attempts to correlate cervical cancer with specific risk factors were poorly designed and controlled. Many possibilities besides an STD were considered, such as a woman's overall age, age at first intercourse, age at marriage, number of births, history of other diseases, geographic location, and exposure to a variety of environmental substances, but the studies often gave confounding results. A variety of interesting but incorrect hypotheses were proposed over the years, including that the act of intercourse was physically irritating to the cervix and stimulated oncogenic changes or that sperm itself was oncogenic. It wasn't until the more rigorous epidemiological studies in the 1960s and 1970s that the evidence became overwhelming that an STD agent was highly likely to be a causative factor for cervical cancer (Beral 1974). An underlying connection between many diverse epidemiological reports was that behaviors that promoted catching an STD also increased the risk of developing cervical cancer. Specifically, having multiple sexual partners or having a single male partner who had multiple previous partners, behaviors that provided more opportunity for catching some agent from one's partner, were among women's highest risk factors for cervical cancer (Rotkin 1973). As this STD explanation gained power and acceptance, the hunt was on for the key pathogen.

Even before the evidence became overwhelming that an unidentified pathogen was critically important for the development of cervical cancer, various STDs had been examined for a causal role. Important bacterial STDs such as syphilis and gonorrhea were considered since these were well-known, prevalent diseases that could be easily diagnosed, but many other infectious agents were also examined such as chlamydia, mycoplasma, and trichomonas (a protozoan parasite) (Alexander 1973). When none of these pathogens showed a significant correlation with cervical cancer incidence, attention turned to viral agents, and by the late 1960s a prime suspect emerged, herpes simplex virus (HSV) (Rawls et al. 1968; Plummer and Masterson 1971). Herpes simplex, like the chickenpox virus (herpes varicella zoster) in Chap. 10, is a member of the herpes virus family. These are DNA viruses with large genomes and complex virions, and they can establish permanent infections in their hosts. HSV comes in two highly related types designated type 1 and type 2, both of which cause localized skin infections that form painful blisters. Historically, type 1 (HSV-1) was found orally while type 2 (HSV-2) caused genital lesions though this distinction is not absolute and both types can be found in either location. Being skin infections, these viruses are spread via direct contact, and HSV genital infections are considered an STD. Another member of the herpes family, the Epstein-Barr virus (EBV), had already been linked to cancer (Smith and Bausher 1972), so postulating that genital HSV could be the cervical cancer agent was a reasonable hypothesis. HSV-2 seemed like the perfect candidate as it fulfilled all the requirements: it was an STD, it established permanent infections, and it was potentially oncogenic.

Into this scientific probe about the origins of cervical cancer came a young German physician-scientist, Dr. Harald zur Hausen (Zur Hausen 2019). Trained as a physician in his home country, Dr. zur Hausen was drawn to scientific research and pursued post-doctoral training in the United States during the 1960s with Gertrude and Werner Henle who were EBV pioneers. Working with EBV and another transforming virus, adenovirus, zur Hausen became one of the early investigators in tumor virology and developed a lifelong passion for understanding the role of viruses in human cancers. It was during this post-doctoral experience that zur Hausen encountered scientific papers on warts that progressed to cervical cancer and became curious about the relationship between these genital lesions and cancer. Returning to Germany after his training period with EBV ended, he began his own studies with warts but also investigated HSV-2. With his background in EBV, it was a natural transition for him to investigate the related HSV-2 and its newly proposed role in cervical cancer. One criterion that would support a connection between a certain cancer and a specific virus is that there should be evidence of that virus in the cancer cells, for example, some or all of the viral genome should be found in the transformed cell. For roughly five years zur Hausen used state-of-the-art nucleic acid hybridization techniques to search for HSV-2 DNA in patient tumor samples (zur Hausen et al. 1974b). The hybridization approach takes advantage of the strong and specific binding of two complementary strands of DNA. To detect the presence of a specific DNA, such as HSV-2 DNA, the investigator must first create a probe sequence. The probe is a piece of single-stranded nucleic acid (RNA or DNA) whose sequence matches part of the sequence of the target DNA, in this case, the HSV-2 genome. In zur Hausen's era, the probe would be labeled with a radioactive compound for later detection. To test a patient's biopsy sample for HSV-2, the cells were lysed and the DNA isolated. This would include the cellular genomic DNA which is fragmented during this process and any HSV-2 DNA if it were present. The isolated DNA is then denatured to separate double-stranded DNA into single strands that are accessible for the probe. Once the probe is added, if HSV-2 DNA is present the probe will find it and bind to its complementary sequence (the "hybridization" step). If no HSV-2 DNA is present then the probe has nothing to bind to and is lost. After the probe binding step, the samples are examined to detect the presence of the radioactive probe. If the probe hybridized to HSV-2 DNA present in the sample then there will be a radioactive signal, while samples with no HSV-2 DNA will not be radioactive. This is a powerful and sensitive technique, but zur Hausen failed to find evidence of HSV-2 in most tumor samples and finally concluded that this herpes virus was not a contributing factor to cervical cancer. Even though other prominent research groups were still touting HSV-2 as the cervical cancer virus, zur Hausen was looking elsewhere. He knew that genital warts (condylomas) were caused by a virus that was sexually transmitted, that these lesions could occasionally transition to malignancy, and that related wart viruses found in animals had oncogenic capability, all features consistent with the possibility that the human wart virus was itself oncogenic and involved in cervical cancer. While all of these previous observations were circumstantial, he intuited that the poorly characterized virus found in human condylomas might be the actual tumor agent (zur Hausen

1976), a prescient and nearly correct hypothesis that would eventually win him a Nobel Prize in 2008.

Suspecting that a human virus causes cancer is easy but proving it can be extremely difficult. First, direct experimentation with pathogens in humans is limited if possible at all, and intentionally exposing people to a suspected carcinogenic virus would be a nonstarter. There was also the issue of the period between exposure to a carcinogen and the development of a tumor. No one knew what this period might be for cervical cancer though it was likely months to years since sexual activity begins in our youth and cervical cancer was typically a disease of middle age or older. With these limitations, testing a human virus in some susceptible animal species where tumor formation may occur more rapidly is the general approach, but human papillomaviruses do not infect any animal except humans so that line of experimentation was unavailable. The alternative approach is to look for consistent evidence of viral infection in the tumor samples. If a virus is truly a contributor to a particular type of cancer, then the virus should have infected the normal tissue that eventually transformed into the oncogenic cells, and the viral genome would be found in all the tumor samples. As for the HSV-2 work, zur Hausen approached this issue for the human papillomavirus by doing studies to detect HPV DNA in cervical cancer biopsies. While simple in concept, in practice these studies were problematic and confounded by the diversity of HPVs that was completely unknown to investigators in the early 1970s. The initial thinking was that the clinically different types of human warts were all caused by a single papillomavirus that was infecting anatomically different locations (Rowson and Mahy 1967). With no culture system, to isolate the virus investigators had to excise warts from patients and do biochemical extractions. Multiple investigators in the 1960s isolated HPV virions from human skin warts at different bodily locations and showed that virion sizes and shapes were identical when visualized by electron microscopy. Similarly, these independently isolated viruses all had DNA genomes of the same size and general composition, consistent with all these isolates being the same virus. With apparently only one virus causing a diverse set of lesions, a single hybridization probe should be sufficient to detect HPV DNA in any type of wart lesion, a completely erroneous assumption that took several years to sort out.

For many of the early investigations of HPV, the virus was isolated from plantar warts, so-called because they were lesions on the plantar surface (the sole) of the foot. These plantar lesions were commonplace, easily obtainable from patients, and were reputed to have high concentrations of viral particles making them a good source of starting material for virus purification. (Parenthetically, it was a plantar wart HPV that would become HPV1 when the numbering convention was eventually adopted.) When zur Hausen first attempted to demonstrate HPV in genital condylomas and cervical cancer biopsy samples in the early 1970s he used probes based on plantar wart HPV (zur Hausen et al. 1974a). Unexpectedly, neither the condylomas nor the cancer samples reacted with the probe, immediately suggesting that there must be a different type of HPV that was significantly different from the plantar HPV type. Luckily, the 1970s were the dawn of molecular biology with the emergence of many new techniques to distinguish related DNAs, including restriction endonuclease mapping (see Fig. 10.2) and ultimately DNA sequencing to read the nucleotide letters of entire genomes. Over the next 10 years, spearheaded by zur Hausen's group, the isolation and characterization of HPV DNAs from different clinical lesions began to reveal the amazing diversity of HPV types (zur Hausen 2002). HPVs detected in lesions from different anatomical sites were related but definitely not identical. The sequential numbering system was embraced to track the different isolates and eventually, taxonomic classification based on the complete genomic sequence was adopted and remains the basis for assigning type numbers to any new isolates.

With the ability to distinguish genital HPVs from cutaneous types, there was hope that the question of oncogenic potential and the relationship between genital HPV infections and cervical cancer could finally be clarified. However, there was an additional surprise that continued to obfuscate the field for several years, not all genital HPV infections were caused by the same HPV type. Depending on what type of probe was used, some studies easily found HPV in cervical cancer samples and other studies failed to detect the virus. Once again, Professor zur Hausen was instrumental in elucidating the solution to this puzzle. He and other labs continued to isolate and classify HPVs from genital lesions on different patients, eventually determining that classical condylomas could be caused by two different HPV types that were subsequently designated type 6 and type 11 (Gissmann et al. 1982; Gissmann and zur Hausen 1980). Even more importantly, when zur Hausen's group isolated the first HPV directly from a cervical tumor it proved to be distinct from types 6 and 11 and became type 16 (Durst et al. 1983). This was quickly followed by the identification of another distinct HPV from a cervical carcinoma that was named HPV 18 (Boshart et al. 1984). With these four genital HPV types in hand, condylomas and cervical tumors could now be tested for specific HPV types. Additionally, epidemiological studies could follow women with different HPV infections to determine the oncogenic risk of each of the four types. By the end of the 1980s, the conclusion was firmly established: infections with types 6 and 11 were benign and rarely if ever progressed to cancer while types 16 and 18 were highly associated with oncogenic development. Based on these observations, types 6 and 11 were designated low-risk papillomaviruses for cancer while types 16 and 18 were high risk. Widespread and extensive studies over the last 30 years have repeatedly confirmed this association and we now know that cervical cancer is almost exclusively due to infection with high-risk HPVs (Scott-Wittenborn and Fakhry 2021). The bad news is that you can literally catch cervical cancer, but the good news is that it can be prevented with the vaccine.

The oncogenic story of genital HPVs did not end with the correlation of types 16 and 18 with cervical cancer. Throughout the 1980s and 1990s, additional HPV types were discovered, and connections were made with cancers in other locations on the body (Milano et al. 2023). In addition to the cervix, high-risk HPV infection of the genital region is associated with anal cancer and with penile cancer, although both occur at a much lower frequency than cervical cancer. Not surprisingly, because the high-risk HPVs can also infect the oral mucosa, they contribute to cancers in this location as well. Current estimates suggest that HPV is a factor in 25–30% of head and neck cancers, specifically cancers of the oropharynx (the soft palate, the rear portion of the tongue, the throat, and the tonsils) and the larynx. Infections of the oral region appear to be acquired via two routes. Some infections occur at birth as the neonate passes through the vaginal canal and is exposed to maternal genital HPVs. More frequently, oral infections occur through oral sexual practices that have become more commonplace in recent decades. Along with the finding that HPVs have a role in cancers beyond the cervix, it became clear that types 16 and 18 are not the only high-risk HPVs. While types 16 and 18 are the most prevalent and account for about 70% of anogenital cancers, multiple other HPVs are now considered oncogenic. According to the National Cancer Institute, the current list of high-risk types includes numbers 31, 33, 35, 39, 45, 51, 52, 56, 58, and 59 with five or six additional types that have been suggested to have oncogenic potential. Fortunately, many of these types have a low incidence and their rarity suggests that they are not currently a major burden for cancer development. Instead, types 31, 33, 45, 52, and 58 are the most prevalent after 16 and 18, and these seven types collectively constitute the major risk for oral and anogenital cancer development.

While the role of high-risk mucosal HPVs in oral and anogenital cancer is well established, the contribution of cutaneous HPVs, particularly the Beta genus, to skin cancer is less certain (Neagu et al. 2023). The Beta HPVs include at least 50 types and constitute the majority of skin infections though they are generally asymptomatic. Nonetheless, sensitive detection methods such as PCR (polymerase chain reaction) can find these viruses in the skin of up to 90% of healthy individuals (de Koning et al. 2009). This high prevalence indicates that these viruses are ubiquitous and must be silently exchanged from person to person throughout our lives. However, because they don't typically cause visible lesions their presence was initially unsuspected and many of the Beta HPVs were first discovered in patients with a rare genetic disorder known as epidermodysplasia verruciformis (EV) (Burger and Itin 2014). Individuals with EV are dramatically impaired in their ability to control HPV infections. Beta HPVs that cause no lesions in normal individuals manifest as visible warts in EV patients, and gradually much of their cutaneous skin becomes covered by papillomatous lesions. Many HPV types are found in the EV lesions, but infections with types 5 and 8 have a high likelihood of developing into squamous cell carcinoma (SCC), particularly in portions of the skin exposed to UV radiation from sun exposure (McLaughlin-Drubin 2015). Likewise, patients who of immunocompromised due to disease or the immunosuppressive drugs required after organ transplants often manifest skin warts with a propensity for transformation into SCC (Reusser et al. 2015). So at least in the case of EV patients and the immunocompromised, the association between HPV infection and cancer development is strong. What remains less clear is the mechanism by which these Beta HPVs promote transformation and tumorigenesis. What has been equally difficult to establish is the role of Beta HPVs in SSC or basal cell skin cancers of normal individuals (Sichero et al. 2019). There are many suggestive studies implicating Beta HPVs in some skin cancers, but the precise role, if any, will require a better understanding of the virus properties and the other factors that may influence cancer formation. Even if the Beta HPVs turn out to be only a minor contributor to skin cancer in immunocompetent individuals, between the Betas and the Alphas, HPV infections constitute a significant source of human cancers worldwide.

Did Our Cousins Give Us an STD?

The ability of only certain HPVs to promote cancer development raises several intriguing questions. Where do these oncogenic HPVs come from, how do they subvert our normal cells into malignant ones, and what advantage does this ability give these viruses? Fifty years of HPV research has provided some insight into these questions even if many aspects of how these viruses interact with our cells are still being investigated. One surprising finding is the evolutionary heritage of HPV 16, the most prevalent of the high-risk HPVs and the one responsible for the greatest number of cervical cancer cases. Currently, there are four HPV 16 subtypes, A-D. Subtypes B and C are found mostly in natives of the African continent while A is the predominant type in most of the rest of the world along with a smattering of D. Analysis of genetic differences in HPV 16 samples from around the world suggests that all four types originated in ancient hominids in Africa hundreds of thousands of years ago (Chen et al. 2018; Pimenoff et al. 2017). When the first archaic humans (Neanderthals and Denisovans) left Africa roughly 500,000 years ago they carried with them a subset of the genetic variation in HPV 16, and the viruses they carried eventually became the HPV 16A lineage in Eurasia. Much later, between 60 and 120 thousand years ago, modern humans again left Africa for Eurasia and the HPV 16 they carried became type D. In our ancestors that remained in Africa, their HPV 16 infections eventually evolved into the modern B and C subtypes. The surprising result is that HPV 16A, a subtype carried by Neanderthals and Denisovans, became the predominant subtype carried by modern humans. We know that modern humans interbred with Neanderthals since we all carry Neanderthal genes in our genomes as a remnant of those trysts. As HPV 16 is a sexually transmitted disease (STD), the most likely scenario is that Neanderthals passed HPV 16A to our forebears during sexual encounters and it outcompeted HPV 16D to become the dominant strain in modern humans. Sadly, it appears that our Neanderthal relatives gave us an STD that became a cancer-causing virus that has afflicted humans ever since.

A second important question is what feature of high-risk HPVs differs from low-risk types and makes the high-risk HPVs drivers for cancer development. While the biology is complex, three viral proteins called E5, E6, and E7 are the primary culprits and are considered viral oncoproteins for high-risk HPVs (Basukala and Banks 2021). Each of these proteins plays multiple roles in the viral life cycle via their ability to interact with and influence the function of cellular proteins. The combined effect of these viral proteins is to alter the cellular environment to make it favorable for viral replication, basically converting the cell into a factor for HPV production. For example, HPV can drive the infected cell into a hyperproliferative state where all the cellular enzymes and raw materials for DNA synthesis are readily available for the virus to use in replicating its genome. Additionally, the oncoproteins disrupt cellular defense mechanisms, including immune systems, that would normally attack the invading virus and prevent its replication. While the low-risk genital HPVs also have these three proteins, their biological activities are somewhat different and/or more subdued in the low-risk types which eliminates their oncogenic potential. Even though there are functional differences, the goal of these three viral proteins is the same for both low- and high-risk HPVs: maintain the infection and facilitate the production of new virions that can spread to other individuals and sustain the virus in its host populations. Cancer only arises when individuals fail to control a high-risk infection. Typically, our immune systems eventually prevail over the oncoproteins, and these high-risk infections are usually cleared within 1–2 years with no subsequent cancer risk. Unfortunately, in around 5% of infections, the individual never clears the virus, and the infection persists for years to decades. Now the potent viral oncoproteins are continuously exerting their effects on the cells causing dysregulation of many critical cellular systems that normally keep our cells healthy and cancer-free. Furthermore, the high-risk E6 and E7 oncoproteins attack our DNA through several pathways leading to DNA damage. As these mutational changes in a cell's genome accumulate over time, transforming mutations may occur that finally convert the normal cell into a cancerous cell, leading to the eventual tumor. Since this mutational damage is essentially a random process, even individuals with persistent infections may never develop cancer if the critical mutations never occur. Regrettably, there is no way to predict which individuals with chronic infections will progress to cancer and currently, there are no antiviral drugs to cure these persistent infections. This leaves vaccination to prevent the initial infection as our primary defense against cervical cancer and other HPV-related cancers.

Lastly, it is important to note that cancer is bad news not only for the patient but also for the HPV. HPV-transformed cancerous cells typically no longer produce new viruses and are a dead end for HPV transmission, so there is no evolutionary advantage for high-risk HPVs to cause cancer. The viral oncoproteins have evolved to efficiently dysregulate our cells in ways that promote and enhance viral reproduction and transmission, not to cause cancer. Cancer is solely an accidental and inadvertent consequence of prolonged expression of these potent cell-altering proteins. Nonetheless, since so many people acquire high-risk infections during their lifetime, even a small percentage of persistent infections results in a large cancer burden that is now unnecessary and preventable through vaccination.

VLPs to the Rescue

In the 1980s and 90s, as the panoply of known HPVs expanded and the connection between infection with high-risk types and cancer development was established, a disappointing conclusion was reached: natural infection did not provide protective immunity. For many of the viruses in previous chapters, a single infection provides sufficient immunity that subsequent exposures to the virus produce no disease. Yet people often have multiple HPV infections at the same time and can be reinfected repeatedly throughout their lives. While infection does induce some immunity consisting of both T cells and antibodies, the level of immune response is highly variable among individuals (Carter et al. 2000) and does little to prevent subsequent infections. In particular, neutralizing antibodies that often play a primary role in preventing infections are only weakly if at all induced in many people. This is at least in part due to HPV infections being very superficial (i.e. confined to the epidermis rather than spread within the body), localized, and small in area so the amount of virus produced is minimal compared to a systemic infection. Importantly, expression of the virion capsid proteins, L1 and L2, is restricted to the upper layers of differentiating skin where there is limited immune surveillance. These capsid proteins, especially the major capsid protein, L1, are the target for neutralizing antibodies. With their restricted location and low levels of protein expression, the capsid proteins can avoid immune detection, and neutralizing anti-capsid antibody production can be meager. Unfortunately for vaccine production, if a natural infection with a wild-type virus doesn't generate protective immunity then trying to produce protection with a live attenuated virus vaccine is unlikely to succeed. Furthermore, there were two other major impediments to using a live HPV vaccine. First, even if a protective attenuated virus could be created, the inability to grow HPV in culture meant that producing vaccine-scale quantities of the virus was impractical. Second, since high-risk HPVs are oncogenic, even putting an attenuated form of the virus into children as a vaccine might be risky and would likely be unacceptable to most parents. With these restrictions, researchers quickly focused on inactivated vaccine approaches.

With L1 as the major target of neutralizing antibodies, several groups began to explore the expression and purification of this protein for use in a subunit-type vaccine. By the late 1980s, L1 alone or together with L2 had been expressed in *E. coli* and shown to be antigenic, although the expression and stability of these viral proteins in bacterial cells were limited (Tomita et al. 1987; Thompson and Roman 1987; Banks et al. 1987). These constraints were likely because eukaryotic cells (e.g. human cells) process and modify proteins differently than bacterial cells (prokaryotic). Consequently, subsequent investigators began to utilize eukaryotic cell systems to express the papillomavirus capsid proteins under conditions where they would more closely resemble their native state during an actual infection. Again, multiple groups were able to express cloned capsid proteins, but a seminal observation was made by Ian Frazier's group in Australia (Zhou et al. 1991). When they introduced expression vectors for HPV 16 L1 and L2 into monkey cells, not only were the proteins produced, but small, virus-like particles (VLPs) were detected in the cells. Like the hepatitis B surface antigen (HBsAg—Chap. 8), the HPV capsid proteins could spontaneously assemble into empty virions, potentially an ideal component for an HPV vaccine. However, their VLPs were smaller than the authentic virions suggesting that their assembly might be aberrant. Furthermore, Frazer's study did not examine whether or not the L1 protein in their VLPs had a native shape that could be recognized by neutralizing antibodies, a key necessity if recombinant capsid protein were going to be functional for a vaccine. Demonstration that cloned L1 protein expressed in mammalian cells did have the native conformation was quickly provided by a study from Richard Schlegel of Georgetown University (Ghim et al. 1992). While these studies were ongoing, Doug Lowy and John Schiller at the NIH were simultaneously conducting their own experiments on VLP production using bovine papillomavirus (BPV) L1 as their model system (Kirnbauer et al. 1992). Their work advanced the field in three important ways. First, they showed that L1 alone without L2 was sufficient for VLP production, simplifying the requirements for vaccine production. Second, their VLPs were larger and closer to the size of actual virions, suggesting a more authentic assembly process. And last, they demonstrated that their VLPs induced neutralizing antibodies, a critical requirement for VLPs to function as a vaccine. They also produced VLPs from HPV 16 L1 alone though at much lower levels than seen with BPV L1. In a subsequent paper, they showed that the HPV 16 L1 gene that was widely used by investigators around the world was a mutant form with impaired VLP assembly ability (Kirnbauer et al. 1993). By isolating and using a wild-type HPV 16 L1 they found the VLP production was robust and could likely be scaled up to vaccine production capacity. That same year, similar results were published by Robert Rose at the University of Rochester School of Medicine and Dentistry using HPV 11 L1 (Rose et al. 1993). Collectively, these four groups set the framework for producing and using HPV VLPs as a potential anti-HPV vaccine. Each group would file a patent on their discoveries, and it would take over a decade for the U.S. patent office and the appeals courts to resolve the conflicting claims but the patent for the HPV vaccine technology ultimately went to Ian Frazier and his university.

With the technology in hand, all that remained was to find a commercial partner to develop the VLP process into a viable vaccine candidate, a quest that proved more difficult than expected. A decade prior, Dr. zur Hausen had approached vaccine manufacturers about the development of an HPV vaccine and received rejection along with skepticism, both scientific and practical. Scientifically, there were no successful vaccines against sexually transmitted diseases and many failures, so the conventional thinking was that targeting the genital tract with a vaccine was difficult if not impossible. Practically, it wasn't clear if there would even be a market for an HPV vaccine for numerous reasons revealed in market surveys conducted by several companies. First, the goal of the vaccine would be to prevent cervical cancer so only girls would be vaccinated, immediately cutting the sales market and potential profit in half. Second, there was the general issue of vaccine overload. From the 1950s on there was a proliferation of more and more vaccines, leading some of the public to be overwhelmed and resistant to adding additional vaccines to the childhood repertoire. This hesitancy was insufficient for most parents to reject vaccines for serious and immediate diseases, but HPV was a completely different scenario. Unlike polio, hepatitis, influenza, or any of the other acute diseases, high-risk HPVs are mostly silent infections whose oncogenic consequences don't manifest for decades if at all, making them a distant and remote threat that didn't engender much urgency in parents. Furthermore, the high-risk genital HPVs are STDs and not communicable in casual public situations like respiratory diseases, so the risk of infection seemed avoidable. Many parents rationalized that only promiscuous girls/women caught STDs and that their precious daughter(s) would never be in this category, unaware that almost everyone acquires HPV infections once they become sexually active (Koutsky 1997). Other parents simply didn't want to have conversations about sexual issues with their daughters or even feared that immunizing young girls against an STD would provide a sense of security that could promote promiscuity. Given these attitudes and the typically enormous development costs for a new vaccine, pharmaceutical companies were reluctant to consider HPV vaccine development and many companies passed on the technology.

The first commercial entity to accept the HPV vaccine challenge was a small start-up biopharmaceutical company called MedImmune, Inc. which was founded in 1988. In 1993, they licensed Richard Schlegel's technology and developed it into a bivalent HPV vaccine that protected against the two major high-risk types, HPVs 16 and 18. While MedImmune was working with Schlegel, Lowy and Schiller approached Merck to develop their VLP technology. Maurice Hilleman, the vaccine czar at Merck, immediately saw the potential of their VLPs as a vaccine candidate and vowed to turn it into a functional vaccine. Like MedImmune, Merck hoped to develop a vaccine that would protect against HPVs 16 and 18, but with the added benefit of also protecting against HPVs 6 and 11. HPV types 6 and 11 are low-risk viruses that do not promote cervical cancer development, but they are the most common causes of benign genital warts. Since genital warts were much more common than cervical cancer, Merck hoped to make the vaccine more desirable to a wider population by targeting these two prevalent low-risk types.

While the technology for making VLPs was well established and their ability to induce high levels of neutralizing antibodies was quickly confirmed in phase I trials, there was still considerable uncertainty that injecting VLPs into someone's arm would generate immunity that protected against HPV infection. Intramuscular injection of protein vaccines, like the HPV VLPs, primarily induces IgG antibodies that circulate in the bloodstream. This is ideal for viruses that enter the bloodstream as they are exposed to the antibodies and neutralized. In contrast, HPV infection is localized to the epidermis, and the virus is confined to layers of the skin where circulating antibodies might not easily penetrate. It was possible that having high levels of neutralizing antibodies in the bloodstream would have no impact on HPV infection, and the only way to test this was in a phase II clinical trial. The first published phase II results used the Merck vaccine candidate and was a resounding success (Koutsky et al. 2002). In a double-blind, placebo-controlled study of over 2000 young women, the vaccine gave 100% protection from HPV 16 infection. Additional studies with similar positive results soon followed for both the Merck and MedImmune vaccines, confirming that HPV VLPs were highly effective at preventing infections by their target HPVs. Although the molecular details of how VLPs induce such a long-lived protective immunity are still being resolved, it is clear that this is primarily a B cell (the antibody-producing cell) phenomenon (Prabhu et al. 2022). Unlike a natural viral infection where the levels of L1 produced are low and mostly inaccessible to the immune system, direct injection of much larger quantities of L1 in the form of VLPs stimulates a high level of anti-L1 antibodies and induces a long-lived memory response that protects for years to decades, making VLPs an outstanding vaccine for HPV. Merck's tetravalent vaccine was approved by the FDA in 2006 and released under the trade name Gardasil. The MedImmune product was eventually sold to GlaxoSmithKline and after FDA approval in 2007, it was marketed as Cervavix. Both vaccines were initially approved only for use in women and were given as a series of three intramuscular injections. Importantly, both vaccines are intended to prevent new infections (i.e. they are prophylactic) and may have little or no effect on existing infections (i.e. they are not therapeutic), so vaccination is recommended for adolescent females before they become sexually active and may acquire genital HPV infections.

The decade following the introduction of Cervarix and Gardasil saw widespread confirmation of their safety and efficacy, as well as some significant changes in vaccine strategy. By 2012, one or both of the HPV vaccines had been licensed in over 100 countries, and 10s of millions of doses had been administered. Massive reviews of their country's vaccine data by the United States, the United Kingdom, and Australia, respectively, found no statistically significant evidence of unusual or concerning adverse effects (Markowitz et al. 2012). Similarly, studies by the World Health Organization and the Institute of Medicine in the United States likewise concluded that the vaccines were protective and safe. Many independent studies found 90–100% protection from infection, making these vaccines extremely efficacious (Herrero et al. 2015). Although both Cervarix and Gardasil performed equally well at protecting from HPV 16 and 18, Gardasil's additional protection against types 6 and 11 slowly gave it the market advantage in the U.S. This advantage was further advanced when Gardasil in the U.S. was licensed in 2009 for use in males followed in 2011 by the recommendation from the Advisory Committee on Immunization Practices (ACIP) that Gardasil be included in the routine vaccination schedule for boys age 11 or 12. While males don't get cervical cancer, they can contract HPV 16 and 18 infections that they can pass on to their partners, so immunizing males should reduce the amount of circulating virus, thus reducing potential female infections. Vaccinating males simply to prevent them from someday in the future potentially giving a high-risk HPV infection to a sexual partner might have been a hard sell, but the vaccine also has direct benefits for males. While much less frequent than cervical cancer, HPVs 16 and 18 can cause penile cancer, so vaccination should prevent this in men. Similarly, there was accumulating evidence that certain oral and anal cancers in both men and women were caused by HPV 16 or 18 infections, and both men and women would be protected from these cancers by HPV vaccination. Lastly, males do get genital warts, primarily from HPV 6 and 11, and these infections would be prevented by Gardasil, making it more immediately beneficial to male vaccine recipients. Without this added capability to prevent the common low-risk HPV infections, Cervarix usage declined.

The final demise of Cervarix in the U.S. came soon after the introduction of Gardasil 9 for women in 2014 and its approval for males in 2015. HPV types 16 and 18 together only account for about 70% of cervical cancer cases, with five other high-risk types accounting for most of the remaining 30% of cases. Once the original Gardasil was proven effective and released to the public, Merck investigated adding additional L1 VLPs to their formulation. Ultimately they formulated a Gardasil derivative with the original four HPV L1 types plus L1 VLPs from the next five most important high-risk types: 31, 33, 45, 52, and 58. Studies with this new nine-valent vaccine confirmed that recipients made robust and protective immune responses against all nine HPVs (Luxembourg and Moeller 2017). Due to antibody cross-reactivity, the antibodies induced by Gardasil 9 may even provide some protection against other HPV types not represented in the vaccine. With its broad coverage, Gardasil 9 quickly became the vaccine of choice, and by 2016 Cervarix was off the U.S. market.

Ten Years On—The HPV Vaccine Success Story

By the end of 2024, HPV vaccines have been in use for nearly twenty years, and the second-generation Gardasil 9 has been on the market for 10 years. Over this period a large body of data has accumulated on the safety and efficacy of all the HPV vaccines. As expected, Gardasil 9, like its predecessor, was shown to be highly effective at reducing HPV infections (Kamolratanakul and Pitisuttithum 2021) and reducing the incidence of genital warts (Lukacs et al. 2020). Just as importantly, there is no evidence that Gardasil or Gardasil 9 vaccines have any significant safety risks (Soliman et al. 2021; Yih et al. 2021). In addition to local studies conducted in various countries, both the Centers for Disease Control (CDC) and the World Health Organization (WHO) have examined the data on the hundreds of millions of doses delivered and have seen no statistically significant increase in serious adverse effects compared to control groups. While serious adverse effects such as life-threatening allergic reactions can occur with any vaccine, there is no indication that this is more frequent with HPV vaccines. Other serious adverse effects such as cardiac or neurological abnormalities have been attributed to the HPV vaccines in the media, but when scientifically evaluated, these events are occurring at the same rate in unvaccinated individuals so they cannot be attributed to the vaccines. Typical minor adverse effects do occur and include pain and redness at the injection site, nausea, headaches, and dizziness. Some adolescents may experience fainting so care should be taken immediately after vaccination. Overall, there is an extremely low risk to the HPV vaccines and their health benefits are significant as recent studies are confirming.

The last key question about the HPV vaccines was whether or not they would truly prevent cancer, and the answer is yes. Since cervical and other cancers can take years or decades to develop after HPV infection, cancer prevention was never tested in clinical trials. Instead, the reduction of HPV infection was used as the test criterion for vaccine effectiveness. Prevention of high-risk HPV infection was expected to reduce the incidence of cancer, but this assumption needed experimental verification.

Assessing cervical cancer development required a large cohort of vaccinated women followed for sufficient time so that any significant reduction in cancer incidence could be observed. By the early 2020s, these data were finally available for the first-generation bi- and quadrivalent HPV vaccines. Studies in England (Falcaro et al. 2021), Scotland (Palmer et al. 2024), and Sweden (Lei et al. 2020) all showed dramatic reductions in precancers and cervical cancers in HPV-vaccinated women, proving that the HPV vaccines are fulfilling their promise to help eliminate cervical cancer. As an additional benefit, other studies are finding evidence for reductions in high-risk HPV-associated oral and anal cancers among vaccinated men and women (Zhang et al. 2021). Eventually, similar and even better results should be seen with Gardasil 9 since it blocks infection with the seven most important high-risk HPVs. The ability to prevent an important class of human cancers through vaccination is a major step toward improving human health and a remarkable culmination of Harold zur Hausen's groundbreaking work on HPVs.

References

Alexander ER (1973) Possible etiologies of cancer of the cervix other than herpesvirus. Cancer Res 33(6):1485–1490

Andrewes C (1979) Richard Edwin Shope 1901–1966 A biographical memoir

Antonsson A, Erfurt C, Hazard K, Holmgren V, Simon M, Kataoka A, Hossain S, Hakangard C, Hansson BG (2003a) Prevalence and type spectrum of human papillomaviruses in healthy skin samples collected in three continents. J Gen Virol 84(Pt 7):1881–1886. https://doi.org/10.1099/vir.0.18836-0

Antonsson A, Karanfilovska S, Lindqvist PG, Hansson BG (2003b) General acquisition of human papillomavirus infections of skin occurs in early infancy. J Clin Microbiol 41(6):2509–2514. https://doi.org/10.1128/JCM.41.6.2509-2514.2003

Banks L, Matlashewski G, Pim D, Churcher M, Roberts C, Crawford L (1987) Expression of human papillomavirus type 6 and type 16 capsid proteins in bacteria and their antigenic characterization. J Gen Virol 68(Pt 12):3081–3089. https://doi.org/10.1099/0022-1317-68-12-3081

Barrett TJ, Silbar JD, Mc GJ (1954) Genital warts-a venereal disease. JAMA J Am Med Assoc 154(4):333–334. https://doi.org/10.1001/jama.1954.02940380043010c

Basukala O, Banks L (2021) The not-so-good, the bad and the ugly: HPV E5, E6 and E7 oncoproteins in the orchestration of carcinogenesis. Viruses 13(10). https://doi.org/10.3390/v13101892

Beral V (1974) Cancer of the cervix: a sexually transmitted infection? Lancet 1(7865):1037–1040. https://doi.org/10.1016/s0140-6736(74)90432-2

Boshart M, Gissmann L, Ikenberg H, Kleinheinz A, Scheurlen W, zur Hausen H (1984) A new type of papillomavirus DNA, its presence in genital cancer biopsies and in cell lines derived from cervical cancer. EMBO J 3(5):1151–1157. https://doi.org/10.1002/j.1460-2075.1984.tb01944.x

Buck CB, Thompson CD, Pang YY, Lowy DR, Schiller JT (2005) Maturation of papillomavirus capsids. J Virol 79(5):2839–2846. https://doi.org/10.1128/JVI.79.5.2839-2846.2005

Burger B, Itin PH (2014) Epidermodysplasia verruciformis. Curr Probl Dermatol 45:123–131. https://doi.org/10.1159/000356068

Carter JJ, Koutsky LA, Hughes JP, Lee SK, Kuypers J, Kiviat N, Galloway DA (2000) Comparison of human papillomavirus types 16, 18, and 6 capsid antibody responses following incident infection. J Infect Dis 181(6):1911–1919. https://doi.org/10.1086/315498

Chen Z, DeSalle R, Schiffman M, Herrero R, Wood CE, Ruiz JC, Clifford GM, Chan PKS, Burk RD (2018) Niche adaptation and viral transmission of human papillomaviruses from archaic hominins to modern humans. PLoS Pathog 14(11):e1007352. https://doi.org/10.1371/journal.ppat.1007352

de Koning MNC, Weissenborn SJ, Abeni D, Bouwes Bavinck JN, Euvrard S, Green AC, Harwood CA, Naldi L, Neale R, Nindl I, Proby CM, Quint WGV, Sampogna F, Ter Schegget J, Struijk L, Wieland U, Pfister HJ, Feltkamp MCW, The Epi-Hpv-Uv-Ca G (2009) Prevalence and associated factors of betapapillomavirus infections in individuals without cutaneous squamous cell carcinoma. J Gen Virol 90(Pt 7):1611–1621. https://doi.org/10.1099/vir.0.010017-0

de Villiers EM, Fauquet C, Broker TR, Bernard HU, zur Hausen H (2004) Classification of papillomaviruses. Virology 324(1):17–27. https://doi.org/10.1016/j.virol.2004.03.033

Durst M, Gissmann L, Ikenberg H, zur Hausen H (1983) A papillomavirus DNA from a cervical carcinoma and its prevalence in cancer biopsy samples from different geographic regions. Proc Natl Acad Sci USA 80(12):3812–3815. https://doi.org/10.1073/pnas.80.12.3812

Falcaro M, Castanon A, Ndlela B, Checchi M, Soldan K, Lopez-Bernal J, Elliss-Brookes L, Sasieni P (2021) The effects of the national HPV vaccination programme in England, UK, on cervical cancer and grade 3 cervical intraepithelial neoplasia incidence: a register-based observational study. Lancet 398(10316):2084–2092. https://doi.org/10.1016/S0140-6736(21)02178-4

Gheit T (2019) Mucosal and cutaneous human papillomavirus infections and cancer biology. Front Oncol 9:355. https://doi.org/10.3389/fonc.2019.00355

Ghim SJ, Jenson AB, Schlegel R (1992) HPV-1 L1 protein expressed in cos cells displays conformational epitopes found on intact virions. Virology 190(1):548–552. https://doi.org/10.1016/0042-6822(92)91251-o

Gissmann L, zur Hausen H (1980) Partial characterization of viral DNA from human genital warts (Condylomata acuminata). Int J Cancer 25(5):605–609. https://doi.org/10.1002/ijc.2910250509

Gissmann L, Diehl V, Schultz-Coulon HJ, zur Hausen H (1982) Molecular cloning and characterization of human papilloma virus DNA derived from a laryngeal papilloma. J Virol 44(1):393–400. https://doi.org/10.1128/JVI.44.1.393-400.1982

Herrero R, González P, Markowitz LE (2015) Present status of human papillomavirus vaccine development and implementation. Lancet Oncol 16(5):E206–E216. https://doi.org/10.1016/S1470-2045(14)70481-4

Kamolratanakul S, Pitisuttithum P (2021) Human papillomavirus vaccine efficacy and effectiveness against cancer. Vaccines (Basel) 9(12). https://doi.org/10.3390/vaccines9121413

Karamanou M, Agapitos E, Kousoulis A, Androutsos G (2010) From the humble wart to HPV: a fascinating story throughout centuries. Oncol Rev 4(3):133–135. https://doi.org/10.1007/s12156-010-0060-1

Kirnbauer R, Booy F, Cheng N, Lowy DR, Schiller JT (1992) Papillomavirus L1 major capsid protein self-assembles into virus-like particles that are highly immunogenic. Proc Natl Acad Sci USA 89(24):12180–12184. https://doi.org/10.1073/pnas.89.24.12180

Kirnbauer R, Taub J, Greenstone H, Roden R, Durst M, Gissmann L, Lowy DR, Schiller JT (1993) Efficient self-assembly of human papillomavirus type 16 L1 and L1-L2 into virus-like particles. J Virol 67(12):6929–6936. https://doi.org/10.1128/JVI.67.12.6929-6936.1993

Koss LG, Durfee GR (1956) Unusual patterns of squamous epithelium of the uterine cervix: cytologic and pathologic study of koilocytotic atypia. Ann N Y Acad Sci 63(6):1245–1261. https://doi.org/10.1111/j.1749-6632.1956.tb32134.x

Koster MI (2009) Making an epidermis. Ann N Y Acad Sci 1170:7–10. https://doi.org/10.1111/j.1749-6632.2009.04363.x

Koutsky L (1997) Epidemiology of genital human papillomavirus infection. Am J Med 102(5A):3–8. https://doi.org/10.1016/s0002-9343(97)00177-0

Koutsky LA, Ault KA, Wheeler CM, Brown DR, Barr E, Alvarez FB, Chiacchierini LM, Jansen KU, Proof of Principle Study I (2002) A controlled trial of a human papillomavirus type 16 vaccine. N Engl J Med 347(21):1645–1651. https://doi.org/10.1056/NEJMoa020586

Lei J, Ploner A, Elfstrom KM, Wang J, Roth A, Fang F, Sundstrom K, Dillner J, Sparen P (2020) HPV vaccination and the risk of invasive cervical cancer. N Engl J Med 383(14):1340–1348. https://doi.org/10.1056/NEJMoa1917338

Lukacs A, Mate Z, Farkas N, Miko A, Tenk J, Hegyi P, Nemeth B, Czumbel LM, Wuttapon S, Kiss I, Gyongyi Z, Varga G, Rumbus Z, Szabo A (2020) The quadrivalent HPV vaccine is protective against genital warts: a meta-analysis. BMC Public Health 20(1):691. https://doi.org/10.1186/s12889-020-08753-y

Luxembourg A, Moeller E (2017) 9-Valent human papillomavirus vaccine: a review of the clinical development program. Expert Rev Vaccines 16(11):1119–1139. https://doi.org/10.1080/14760584.2017.1383158

Markowitz LE, Tsu V, Deeks SL, Cubie H, Wang SA, Vicari AS, Brotherton JM (2012) Human papillomavirus vaccine introduction--the first five years. Vaccine 30 Suppl 5:F139–F148. https://doi.org/10.1016/j.vaccine.2012.05.039

McBride AA (2022) Human papillomaviruses: diversity, infection and host interactions. Nat Rev Microbiol 20(2):95–108. https://doi.org/10.1038/s41579-021-00617-5

McLaughlin-Drubin ME (2015) Human papillomaviruses and non-melanoma skin cancer. Semin Oncol 42(2):284–290. https://doi.org/10.1053/j.seminoncol.2014.12.032

Milano G, Guarducci G, Nante N, Montomoli E, Manini I (2023) Human papillomavirus epidemiology and prevention: is there still a gender gap? Vaccines (Basel) 11(6). https://doi.org/10.3390/vaccines11061060

Moody C (2017) Mechanisms by which HPV induces a replication competent environment in differentiating keratinocytes. Viruses 9(9). https://doi.org/10.3390/v9090261

Neagu N, Dianzani C, Venuti A, Bonin S, Voidazan S, Zalaudek I, Conforti C (2023) The role of HPV in keratinocyte skin cancer development: a systematic review. J Eur Acad Dermatol Venereol 37(1):40–46. https://doi.org/10.1111/jdv.18548

Palmer TJ, Kavanagh K, Cuschieri K, Cameron R, Graham C, Wilson A, Roy K (2024) Invasive cervical cancer incidence following bivalent human papillomavirus vaccination: a population-based observational study of age at immunization, dose, and deprivation. J Natl Cancer Inst. https://doi.org/10.1093/jnci/djad263

Pimenoff VN, de Oliveira CM, Bravo IG (2017) Transmission between archaic and modern human ancestors during the evolution of the oncogenic human papillomavirus 16. Mol Biol Evol 34(1):4–19. https://doi.org/10.1093/molbev/msw214

Plotzker RE, Vaidya A, Pokharel U, Stier EA (2023) Sexually transmitted human papillomavirus: update in epidemiology, prevention, and management. Infect Dis Clin N Am 37(2):289–310. https://doi.org/10.1016/j.idc.2023.02.008

Plummer G, Masterson JG (1971) Herpes simplex virus and cancer of the cervix. Am J Obstet Gynecol 111(1):81–84. https://doi.org/10.1016/0002-9378(71)90929-x

Prabhu PR, Carter JJ, Galloway DA (2022) B cell responses upon human papillomavirus (HPV) infection and vaccination. Vaccines (Basel) 10(6). https://doi.org/10.3390/vaccines10060837

Rawls WE, Tompkins WA, Figueroa ME, Melnick JL (1968) Herpesvirus type 2: association with carcinoma of the cervix. Science 161(3847):1255–1256. https://doi.org/10.1126/science.161.3847.1255

Reusser NM, Downing C, Guidry J, Tyring SK (2015) HPV carcinomas in immunocompromised patients. J Clin Med 4(2):260–281. https://doi.org/10.3390/jcm4020260

Rose RC, Bonnez W, Reichman RC, Garcea RL (1993) Expression of human papillomavirus type 11 L1 protein in insect cells: in vivo and in vitro assembly of viruslike particles. J Virol 67(4):1936–1944. https://doi.org/10.1128/JVI.67.4.1936-1944.1993

Rotkin ID (1973) A comparison review of key epidemiological studies in cervical cancer related to current searches for transmissible agents. Cancer Res 33(6):1353–1367

Rous P, Beard JW (1935) The progression to carcinoma of virus-induced rabbit papillomas (Shope). J Exp Med 62(4):523–548. https://doi.org/10.1084/jem.62.4.523

Rous P, Friedewald WF (1944) The effect of chemical carcinogens on virus-induced rabbit papillomas. J Exp Med 79(5):511–538. https://doi.org/10.1084/jem.79.5.511

Rowson KE, Mahy BW (1967) Human papova (wart) virus. Bacteriol Rev 31(2):110–131. https://doi.org/10.1128/br.31.2.110-131.1967

Scherp HW, Syverton JT (1949) A chemical investigation of keratin and carcinomas deriving from rabbit papillomas (Shope). Cancer Res 9(1):12–16

Scott-Wittenborn N, Fakhry C (2021) Epidemiology of HPV related malignancies. Semin Radiat Oncol 31(4):286–296. https://doi.org/10.1016/j.semradonc.2021.04.001

Shope RE, Hurst EW (1933) Infectious papillomatosis of rabbits : with a note on the histopathology. J Exp Med 58(5):607–624. https://doi.org/10.1084/jem.58.5.607

Sichero L, Rollison DE, Amorrortu RP, Tommasino M (2019) Beta human papillomavirus and associated diseases. Acta Cytol 63(2):100–108. https://doi.org/10.1159/000492659

Smith RT, Bausher JC (1972) Epstein-Barr virus infection in relation to infectious mononucleosis and Burkitt's lymphoma. Annu Rev Med 23:39–56. https://doi.org/10.1146/annurev.me.23.020172.000351

Soliman M, Oredein O, Dass CR (2021) Update on safety and efficacy of HPV vaccines: focus on Gardasil. Int J Mol Cell Med 10(2):101–113. https://doi.org/10.22088/IJMCM.BUMS.10.2.101

Syrjanen S, Syrjanen K (2008) The history of papillomavirus research. Cent Eur J Public Health 16(Suppl):S7–S13. https://doi.org/10.21101/cejph.b0040

Thompson GH, Roman A (1987) Expression of human papillomavirus type 6 E1, E2, L1 and L2 open reading frames in Escherichia coli. Gene 56(2–3):289–295. https://doi.org/10.1016/0378-1119(87)90146-6

Tomita Y, Shirasawa H, Simizu B (1987) Expression of human papillomavirus types 6b and 16 L1 open reading frames in Escherichia coli: detection of a 56,000-dalton polypeptide containing genus-specific (common) antigens. J Virol 61(8):2389–2394. https://doi.org/10.1128/JVI.61.8.2389-2394.1987

Tornesello ML, Buonaguro FM (2020) Human papillomavirus and cancers. Cancers (Basel) 12(12). https://doi.org/10.3390/cancers12123772

Van Doorslaer K (2013) Evolution of the papillomaviridae. Virology 445(1–2):11–20. https://doi.org/10.1016/j.virol.2013.05.012

Vilos GA (1998) The history of the Papanicolaou smear and the odyssey of George and Andromache Papanicolaou. Obstet Gynecol 91(3):479–483. https://doi.org/10.1016/s0029-7844(97)00695-9

Walker AW, Hoyles L (2023) Human microbiome myths and misconceptions. Nat Microbiol 8(8):1392–1396. https://doi.org/10.1038/s41564-023-01426-7

Yih WK, Kulldorff M, Dashevsky I, Maro JC (2021) A broad safety assessment of the 9-valent human papillomavirus vaccine. Am J Epidemiol 190(7):1253–1259. https://doi.org/10.1093/aje/kwab022

Zhang JY, Qin ZL, Lou CY, Huang J, Xiong YF (2021) The efficacy of vaccination to prevent human papilloma viruses infection at anal and oral: a systematic review and meta-analysis. Public Health 196:165–171. https://doi.org/10.1016/j.puhe.2021.05.012

Zhou J, Sun XY, Stenzel DJ, Frazer IH (1991) Expression of vaccinia recombinant HPV 16 L1 and L2 ORF proteins in epithelial cells is sufficient for assembly of HPV virion-like particles. Virology 185(1):251–257. https://doi.org/10.1016/0042-6822(91)90772-4

zur Hausen H (1976) Condylomata acuminata and human genital cancer. Cancer Res 36(2 pt 2):794

zur Hausen H (2002) Papillomaviruses and cancer: from basic studies to clinical application. Nat Rev Cancer 2(5):342–350. https://doi.org/10.1038/nrc798

Zur Hausen H (2019) Cancers in humans: a lifelong search for contributions of infectious agents, autobiographic notes. Annu Rev Virol 6(1):1–28. https://doi.org/10.1146/annurev-virology-092818-015907

zur Hausen H, Meinhof W, Scheiber W, Bornkamm GW (1974a) Attempts to detect virus-secific DNA in human tumors. I. Nucleic acid hybridizations with complementary RNA of human wart virus. Int J Cancer 13(5):650–656. https://doi.org/10.1002/ijc.2910130509

zur Hausen H, Schulte-Holthausen H, Wolf H, Dorries K, Egger H (1974b) Attempts to detect virus-specific DNA in human tumors. II. Nucleic acid hybridizations with complementary RNA of human herpes group viruses. Int J Cancer 13(5):657–664. https://doi.org/10.1002/ijc.2910130510

zur Hausen H, Gissmann L, Steiner W, Dippold W, Dreger I (1975) Human papilloma viruses and cancer. Bibl Haematol 43:569–571. https://doi.org/10.1159/000399220

13

SARS-Coronavirus-2: The Unexpected Plague

Keywords ACE2 receptor • BioNTech • Lipid nanoparticle (LNP) • MERS • Moderna • mRNA vaccine • SARS • Spike protein

Abbreviations

ACE2	Angiotensin-converting enzyme 2
BCoV	Bovine coronavirus
COVID-19	Coronavirus Disease-2019
DARPA	Defense Advanced Research Projects Agency
HCoV	Human coronavirus
HIV	Human immunodeficiency virus
HKU1	Hong Kong University sample 1
hMPV	Human metapneumovirus
IBV	Infectious bronchitis virus
LNP	Lipid nanoparticle
MERS	Middle East Respiratory Syndrome
MHV	Mouse hepatitis virus
mRNA	Messenger RNA
NL63	Netherlands sample 63
PEG	Polyethylene glycol
SARS	Severe acute respiratory syndrome
SARS-CoV-2	Severe acute respiratory syndrome-Coronavirus-2
TGEV	Transmissible gastroenteritis virus
tRNA	Transfer RNA

A New Threat

The twentieth century saw multiple viruses that spread around the world causing misery and death. The 1917 influenza pandemic (Chap. 5) killed tens of millions of men, women, and children and spared no part of the world. In terms of sheer numbers of cases and deaths, influenza was the top viral killer of the century. Of all the viruses with lethal capability, influenza was considered the most likely culprit to cause another pandemic but that worry never materialized. Next came poliovirus that rose from obscurity, panicking parents, and spreading terror throughout the developed nations (Chap. 6). Its ability to kill and cripple defenseless children made it a horrific disease even if the number of actual deaths was much smaller than for influenza. And just as many dangerous viruses were being controlled through vaccination, the AIDS virus (human immunodeficiency virus—HIV) became a global scourge that is still not conquered (Chap. 15). HIV spread silently for many years until it exploded with a surge of cases in the 1980s, presenting in young men with bizarre and unusual infections, often from organisms that normally weren't pathogens. As a sexually transmitted and blood-borne disease, HIV changed both social mores and medical practice as universal precautions and safe sex became a standard. Each of these viruses dramatically impacted both science and society, and any of them could lay claim to be the most important virus of the last century. The twenty-first century is still in its infancy, but we already have a strong frontrunner for the most impactful virus of this century, SARS-Coronavirus -2 (SARS-CoV-2).

The year 2019 ended quietly with most of the world unaware of the emerging threat. Holiday celebrations abounded with the usual gaiety and festive crowds. Tens of thousands of people crowded Times Square and similar venues around the world to welcome 2020 with hope and anticipation of a new year full of promise. While most of the world partied, the city of Wuhan in Hubei Province China was beginning to recognize a troubling outbreak of a respiratory illness. In December 2019, a cluster of pneumonia cases caused by an unknown agent caught the attention of local public health officials. Initial investigations revealed that many cases were in proximity to the Wuhan Seafood Wholesale Market. In contrast to its name, this market sold not only fish but also many other types of animals. These so-called "wet markets" where different live animal species are in close contact with humans are an ideal environment for interspecies transmission of viruses. Fearing that the market might be the source of this new infection, on January 1, 2020, the market was closed as a preventative measure. (Note the subsequent investigations provided evidence that the virus was likely circulating in Hubei province as early as October 2019 (Pekar et al. 2021) and the original source of the virus remains both unknown and controversial.) Initially, it was easy for the public to dismiss these cases as just another unusual influenza strain or infection by one of many other relatively innocuous viruses that can cause respiratory symptoms. Such respiratory disease outbreaks occur every year, some from common viruses that are endemic and seasonal, and others from novel viruses that briefly slip into human populations. While worrisome to the local population, these sporadic outbreaks had always been confined and eventually died out as human-to-human transmission failed to support the wider spread of the agent. In the past, after a brief flurry of news stories, both the coverage and the public's interest in these outbreaks quickly waned as cases ceased. There was no reason to suspect that this illness in China would be any different, but that casual attitude would soon be shattered.

By the first week of January, the WHO had been notified of the illnesses, and sequencing of samples from patients identified a novel coronavirus. Two previous outbreaks of novel coronaviruses had caused serious disease, immediately raising fears that this new coronavirus could be a significant public health threat. The original SARS (Severe Acute Respiratory Syndrome) coronavirus appeared in China in 2002–2003 and had a fatality rate of nearly 10% (Pustake et al. 2022). The subsequent MERS (Middle East Respiratory Syndrome) coronavirus that emerged in Saudi Arabia in 2012 was even more deadly with a fatality rate of 34%. Nobody knew how deadly this new coronavirus might be but deaths were already being reported, so this was not just another cold-causing virus.

With the previous outbreaks of SARS and MERS, both outbreaks were contained by patient isolation and quarantines. As January 2020 progressed, it grew alarmingly apparent that the world might not be so lucky with this new virus dubbed SARS Coronavirus 2 (SARS-CoV-2). Within days of China's public announcement of this new disease, cases were detected in Thailand, Japan, and the Republic of Korea, with the first case in the United States soon after. These widely dispersed new cases indicated that the virus had already eluded any chance of quarantine in China and must be spreading effectively from person to person. Still, most countries, including the U.S., attempted to limit viral spread by banning travel to China, screening incoming travelers for signs of infection (primarily fever), and imposing quarantines on new arrivals to the country. However, by mid-February, it was becoming painfully clear to public health officials that none of these actions aimed at stopping the virus at national borders was going to work. By the time these measures were implemented most countries had already been infiltrated and the virus was spreading among the domestic population. Additionally, many asymptomatic carriers were identified and these individuals would have escaped detection at national borders while bringing in the virus. By the end of February, the WHO had named the symptoms caused by SARS-CoV-2 the Coronavirus Disease-2019 (COVID-19) and was warning that more draconian measures might be needed to curtail viral spread and the resultant illnesses and deaths.

In the United States, once the New Year's merriments ended, January and February dragged on with their short days and often gloomy weather. By March, the anticipation of springtime was beginning to blossom with little portent of the devastating pandemic soon to follow. Yes, there were increasing news reports of COVID-19 in China and other countries, but for many people, these illnesses mostly seemed too distant and too inconsequential to arouse much concern. Cases in the U.S. were few and there was still optimism that containment efforts with travelers would prevent widespread outbreaks. In a few short weeks, these hopes would be dashed as the virus exploded everywhere causing a national medical emergency and forcing both the U.S. and the world into a massive government-imposed pandemic response. In a stunning reversal of the old adage, March came in like a lamb and went out like a lion.

A Spiking Disaster

By early March 2020, the reality of the growing and uncontrolled global spread of SARS-CoV-2 could no longer be denied. Over 100 countries were reporting cases and deaths were starting to mount. In response to this crisis, the WHO declared COVID-19 a pandemic on March 11. Within days the U.S. began to implement strategic measures designed to reduce the community spread of the virus. Travel bans were implemented, schools began to close, social distancing policies were enacted, and some states shut down restaurants and bars to prevent viral spread in crowded indoor areas. All of these measures were generally known to reduce the transmission of respiratory viruses, but their ability to impact the spread of SARS-CoV-2 was at best educated guesswork. While there was a long scientific history for the study of coronaviruses, there was considerable variation among this viral family in terms of disease capabilities. At the beginning of the outbreak, SARS-CoV-2 was an unknown entity with no specific knowledge about its transmission properties or pathogenic mechanisms. Usually, viruses are well studied, both in the community and in the lab, before public health guidelines are created. However, with the threat of a spreading and uncontrolled infection, immediate decisions were needed. Certain assumptions were made based on known coronavirus properties, but without data specific to SARS-CoV-2 some decisions were inevitably going to be wrong and would need refinement and correction as reliable information about the virus was obtained. For example, face masks were initially deemed unhelpful but were subsequently mandated as research established that they reduced transmission of SARS-CoV-2 (Howard et al. 2021). Even though this evolution of policies and procedures was a natural and appropriate process as data accumulated and our understanding of the virus increased, the changing rules caused confusion and loss of confidence by the public. An important lesson learned for future pandemics is that better and more transparent communication from public health and governmental agencies could forestall some of the pushback on policies and mandates. Officials need to be clear about the evidence, or lack thereof, that supports their decisions and fully explain how new data are used to alter previous guidelines.

By the end of March, the U.S. and most of the world were in full pandemic lockdown but the pandemic showed no signs of abating. Case numbers worldwide had risen from a few thousand at the beginning of March to over one million confirmed cases by the beginning of April and the numbers were still rapidly rising. Leading the way was the U.S. with roughly half the total cases in the world and nearly 20,000 deaths. These huge case numbers began to impact the healthcare system causing shortages of personal protective equipment (masks, gowns, eye shields) and overwhelming hospitals in some areas where patient numbers vastly exceeded available beds, ventilators, and physicians. By June the total number of cases in the U.S. alone was over two million and by July there was a record of over 75,000 new cases reported in a single day. Before the virus was finally controlled, the U.S. would see 111 million cases and over 1.2 million deaths, making this the single biggest disease outbreak in the history of the country. Worldwide the impact was even more staggering with over 700 million cases and 7 million deaths.

If the spread and impact of this virus were unique, what were the factors that made this virus so problematic? While much about the pathogenic mechanisms of the virus remains unclear, one important factor is the viral spike protein. The spike proteins project from the surface of the coronavirus virion and their function is to engage with receptors on cells to start the viral entry process (Li 2016). Somewhat surprisingly, there is considerable variation in the protein sequence of spike proteins among coronaviruses, and this leads to the utilization of different receptors on host cells for different types of coronaviruses. Only cells that express a suitable receptor can be infected and the type of cells susceptible to the virus contributes to the type of disease symptoms that occur. The original SARS virus used a cellular protein called ACE2 (angiotensin-converting enzyme 2) as its receptor and SARS-CoV-2 was quickly shown to use the same receptor (Letko et al. 2020). ACE2 is widely distributed, not only in human lung tissue but also in the heart, kidneys, gastrointestinal tract, and brain. The widespread presence of the receptor allows SARS-CoV-2 to infect organs throughout the body and cause disease symptoms in multiple organ systems, a critical feature often seen in serious infections (Iida et al. 2021). Two other features of the SARS-CoV-2 spike protein also appear to contribute to its successful emergence as a human pathogen. First, the SARS-CoV-2 spike protein was found to bind ACE2 with an even higher affinity than the spike protein from the original SARS virus, and this ensures more efficient binding to host cells to start the infection process (Wrapp et al. 2020; Giovanetti et al. 2023). Second, in contrast to the spike protein from the SARS virus, the SARS-CoV-2 spike protein has a mutation that created a new cleavage site in the protein (Peacock et al. 2021). As part of the entry process, spike protein bound to its receptor on the cell surface must next be cleaved by a cellular enzyme. Cleavage of the spike protein causes it to change shape (a conformational change in scientific parlance) and this change is requisite to initiate the penetration of the virion through the cell membrane and into the cell. Depending on the receptor and the specific cleavage enzyme there are different pathways that the virions use to enter the cell. The mutation in SARS-CoV-2 introduced a cleavage site for a cellular enzyme called furin. Utilization of this furin-dependent entry pathway facilitated infection of lung cells and helped the virus avoid certain host defense systems. The ability to recognize and bind strongly to the ACE2 receptors, and to be furin-cleaved, likely helped SARS-CoV-2 efficiently infect humans and transmit successfully from person to person, both key requirements for a zoonotic virus to establish itself in a new host species. With this calamitous spike protein, SARS-CoV-2 was unusually suitable for human infection and rapid spread was almost inevitable once the virus entered the human population. While the mutation rate of coronaviruses is relatively low compared to many RNA viruses, mutants are still plentiful during the replication cycle in individual hosts (Amicone et al. 2022). Once in humans, selective pressure favored those mutations that increased viral fitness, giving rise to wave after wave of variants and helping SARS-CoV-2 become a permanent addition to the cadre of human viruses (Tosta 2022; Markov et al. 2023).

From Colds to Killers

One of the surprising features of SARS, MERS, and SARS-CoV-2 is their pathogenicity. Before these three viruses attacked, human coronaviruses (HCoVs) were considered fairly innocuous pathogens (Kahn and McIntosh 2005). As cell culture and virological techniques advanced in the 1950s and 1960s, many investigators began to look for viruses in samples from patients with mild respiratory illnesses, i.e. colds. Originally, little was known about the cause of colds other than they were likely due to one or a few viruses. We now know that colds are not a disease that is specific to a particular virus or virus family. Instead, colds are a constellation of generally mild respiratory symptoms that can be caused by over 200 viruses from multiple virus families including adenoviruses, rhinoviruses, enteroviruses, and coronaviruses (Heikkinen and Jarvinen 2003). As part of these investigations into the etiology of colds, the first HCoVs were discovered in the 1960s and were isolated from individuals with common colds, although their actual classification wasn't established until later. In 1965, two independent groups, one in the U.S. and one in Great Britain, each isolated a virus from a patient with a cold (McIntosh et al. 1967; Hamre and Procknow 1966). The two isolates may well have been the same virus, but the British sample was lost, and the sample code number used by the American group, 229E, was adopted as the virus name. Within a year, researchers at the NIH also isolated a virus from the respiratory tracts of multiple cold patients. They were able to passage the isolate in an organ culture system and it was designated OC43 for organ culture sample 43 (Tyrrell and Bynoe 1966). Subsequent examination of 229E and OC43 by electron microscopy confirmed that these cold viruses were identical in appearance and resembled several known animal viruses including mouse hepatitis virus (MHV), infectious bronchitis virus (IBV) in chickens, and transmissible gastroenteritis virus (TGEV) in swine. All of these seemingly diverse viruses shared a common morphology with an enveloped virion that had distinct, club-shaped surface projections (the spike proteins) giving the virions a crown-like aspect. Since they appeared to be related, and in honor of this unique virion appearance, they were designated coronaviruses (corona means roughly crown-like in Latin). Subsequently, these viruses were all found to be genetically related and have positive-sense, single-strand RNA genomes that are among the largest genomes found in RNA viruses. Eventually, the term coronavirus was adopted as the official taxonomic genus for this group of viruses.

For the next 40 years, 229E and OC43 remained the only HCoVs specifically isolated and characterized. Both were in wide circulation worldwide, were readily cultivated in the lab, and served as the prototypes for studies on human coronaviruses. With their mild symptomology and limited pathology, coronaviruses were not considered major pathogens for humans until the unexpected emergence of the original SARS virus in 2002 (Cherry and Krogstad 2004). Technically, the SARS virus was considered a pandemic as cases were detected in 29 countries after its discovery in China. However, SARS had far less impact than SARS-CoV-2 as there were fewer than 10,000 total cases identified and less than 800 deaths attributed to the virus. We were fortunate that for this virus there were few asymptomatic cases and people were most contagious when they were exhibiting symptoms, so they were easy to detect and isolate. By rapid identification and quarantine of infected individuals, the pandemic was quickly contained and never became a major global outbreak. Still, the world was shocked that a type of virus considered mostly a nuisance and not a serious health concern could instead be incredibly dangerous to humans with a fatality rate of almost 10%. With the natural host of SARS believed to be bats, the scientific community had to recognize that the zoonotic transfer of animal coronaviruses into humans was a reality and a source of potentially deadly infections. This stimulated a flurry of new research interest in this virus family, but that interest waned as SARS was contained, human infections ceased, and that particular coronavirus never reappeared.

As part of this renewed interest in coronaviruses post-SARS, two additional human coronaviruses, NL63 (Netherlands sample 63) and HKU1 (Hong Kong University 1) were discovered in 2004 and 2005, respectively (Fouchier et al. 2004; Woo et al. 2005). Like the previously known HCoVs, NL63 and HKU1 are found globally and are generally associated with mild upper respiratory infections (Liu et al. 2021). Collectively, these four HCoVs account for between 15% and 30% of common cold cases, making them a significant contributor to our yearly bouts with minor respiratory infections. Since no other HCoV isolates have ever been found, researchers believe that these four are the only existing endemic human coronaviruses. Note that part of the limited severity of these HCoV infections is that they are primarily confined to the upper respiratory tract where they cause headaches, nasal congestion, and sore throats. In contrast, the more dangerously pathogenic coronaviruses can invade the lower respiratory tract (into the lungs) where pneumonia and other serious complications can develop.

Ten years after the SARS outbreak, a second novel and highly pathogenic coronavirus emerged in the Middle East. First appearing in Saudi Arabia in 2012, the new coronavirus was isolated from a 60-year-old man who eventually succumbed to pneumonia and renal failure (Zaki et al. 2012). This isolate was named the Middle East Respiratory Syndrome (MERS) coronavirus, and its existence started a search for additional cases. Subsequent examination of other respiratory disease cases in the region discovered a cluster of MERS cases in Jordon and a scattering of cases in other countries in the region. Over the next 10 years, sporadic MERS cases were observed in other parts of the world, mostly in travelers to the Middle East who brought the disease back home with them. With a fatality rate of over 30%, MERS is an extremely dangerous virus, particularly for the elderly, the immunocompromised, and people with other confounding medical issues. Fortunately, the total case numbers remain small, and by 2024, the WHO reported that the cumulative number of known cases worldwide was only about 2600. For the MERS virus, person-to-person transmission is poor, so community spread of the virus is limited. Instead, most cases result from direct contact with the natural host, the dromedary camel (Haagmans et al. 2014), or in the close confines of the hospital environment. In hospitals, infected individuals have spread the virus to healthcare workers before their disease was determined and appropriate protective measures implemented (Chowell et al. 2015). Additionally, symptomatic individuals are the most infectious, so it is again easy to identify and quarantine patients who are most likely to pass on the virus. These characteristics prevented a massive outbreak and allowed effective containment of this virus whenever cases arose. Importantly though, the MERS virus is still endemic in camel populations in the Middle East and continues to pose a danger for individuals coming in close contact with these animals. As with any animal virus, there is also the risk that random mutations might occur that confer increased fitness for human infection and transmission, possibly making MERS a much more threatening pathogen.

The experience with SARS and MERS confirmed that animal coronaviruses can jump from their animal hosts into human populations with potentially dangerous consequences, a harbinger of the SARS-CoV-2 pandemic. The zoonotic nature of SARS and MERS also piqued interest in the origins of the endemic HCoVs. Like most human viruses, how the endemic HCoVs became human diseases is uncertain although it is likely that they all arose as zoonotic diseases that entered the human population from some other host animal (Harrison et al. 2023). For HCoV OC43, the closest genetic relative is a bovine coronavirus (BCoV), and the molecular clock data suggest that this is a relatively new human virus that entered humans in the late 1800s (Vijgen et al. 2005, 2006). Similar studies indicate that HKU1 is likely derived from a mouse coronavirus (Otieno et al. 2022) while bats are likely the original host for 293E and NL63 (Tao et al. 2017). Whether or not the progenitor animal viruses that evolved into our current HCoVs were initially highly pathogenic in humans is unknown but certainly possible. Each of these animal viruses might have caused serious local epidemics upon entry into human populations before evolving into the more innocuous current versions (global pandemics were less likely than local epidemics before the modern era due to the arduous and slow nature of travel before air travel). Now that SARS-CoV-2 has established itself as a new endemic human virus, we can only hope that it follows this paradigm and eventually becomes more of an annoying cold virus and less of a potential killer. Lamentably, other animal coronaviruses are widespread and abundant, and they remain a fertile source of viruses with the potential capacity to infect and sicken humans. Hopefully, the last 20 years have sensitized us to the danger that zoonotic coronaviruses present and encouraged enhanced vigilance in the scientific, medical, and public health communities (Williams et al. 2023).

The Ascendency of RNA

Even with the knowledge of SARS and MERS, the world was completely unprepared for the perfect storm that was SARS-CoV-2. To the world's great misfortune, SARS-CoV-2 had all the features needed for a pandemic that were lacking in SARS and MERS. Like SARS and MERS, SARS-CoV-2 is highly pathogenic for the elderly and the infirmed, but it is also more readily transmissible in the community setting. SARS had an R_0 in the 2–3 range (Lipsitch et al. 2003) and was reasonably transmissible while MERS was less than 1 and not very effective at human-to-human spread (Breban et al. 2013). Estimates vary widely for SARS-CoV-2 as different variants have arisen, but the R_0 range in different studies is from roughly 2–8 (Karimizadeh et al. 2023), suggesting a much higher transmission rate than either of the two earlier zoonotic coronavirus outbreaks. Additionally, a significant fraction of SARS-CoV-2 infections are asymptomatic but still infectious (Wang et al. 2023), and even symptomatic patients can become infectious 2–3 days prior to the onset of symptoms (He et al. 2020; Tindale et al. 2020). The presence of high levels of the virus in people who don't yet know that they are infected allows the virus to be spread unknowingly, and this feature helped thwart early isolation and containment measures. Cars, buses, trains, boats, and planes carried healthy shedders to destinations around the globe. Within a few months of the initial reports the virus had spread worldwide and was circulating out of control. By August 2020, roughly 8 months after the discovery of cases associated with the Wuhan market, there were 22 million reported cases and 800,000 deaths with nearly every country and territory on the planet reporting cases. With no drugs, treatments, or vaccines, the case numbers soared, and healthcare systems teetered on overload. In many places, the massive influx of patients needing critical care exceeded the capacity of hospitals and healthcare workers. Not only were death rates from COVID mounting, but the ability of the healthcare system to deal with the usual business of routine surgeries, cancer treatments, accident victims, and health emergencies (e.g. heart attacks and strokes) was severely impaired. To try to interdict the viral spread and prevent the collapse of hospital systems and other critical societal infrastructures, many countries, the U.S. included, began massive lockdowns, travel restrictions, and school closings to reduce the opportunities for community transmission. While subsequent studies mostly indicated that these public health measures were beneficial in reducing cases and deaths (Murphy et al. 2023; Agyapon-Ntra and McSharry 2023), they were not without significant downsides. Loss of income, general economic decline, physical and psychological issues from social isolation, and reduced effectiveness of schooling were among the principal negative impacts resulting from these policies (Bardosh et al. 2022). This situation was necessary and endurable for the short term to protect the populace, but not sustainable indefinitely. What was desperately needed was a preventive vaccine, something that typically took years to decades to develop and test. Fortunately, the story of the SARS-CoV-2 vaccine didn't start with COVID but began decades earlier, so the science was poised for a remarkably swift response to this terrifying pandemic.

In the previous chapters, various viral diseases were controlled using three vaccine strategies: live attenuated viruses, inactivated whole viruses, and viral subunits (proteins). There is a rich scientific history about these technologies and they each have proven efficacy, excellent safety records, and well-understood strengths and weaknesses. However, choosing the most effective approach for a new disease is somewhat of a guess. Given this uncertainty, different countries and different manufacturers chose among these three standard methods, and ultimately SARS-CoV-2 vaccines of each traditional type were developed and utilized (Grana et al. 2022). Nonetheless, the first vaccines to be released and the ones proven to be most effective were an entirely new class, the messenger RNA (mRNA) vaccines.

Understanding the mRNA vaccines requires a brief review of cell biology and the conversion of genetic information into functional proteins. In our cells, our DNA genomes contain the instructions for everything the cell needs to produce. Among these instructions are the protein-coding genes that specify the tens of thousands of proteins that each type of cell needs to make. These genes are stretches of DNA whose nucleotide sequence will be converted to the amino acid sequence of a protein. To initiate this process, the DNA sequence is first read by a complex of proteins referred to as the transcriptional machinery. In the nucleus of the cell, this transcriptional complex copies the DNA sequence into a corresponding RNA sequence which is the messenger RNA (mRNA). This mRNA is then transported out of the nucleus and into the cytoplasm where it is recognized by another type of protein complex called the ribosome. The ribosome is a decoding device that reads the RNA sequence, and with the help of another type of RNA known as transfer RNA (tRNA), translates the mRNA sequence into the amino acid sequence of the protein. This is a continuous and ongoing process in most cells, and at any given time our cells are teeming with mRNAs, tRNAs, and a multitude of other RNA types that perform other functions. When viruses infect a cell, they also produce mRNAs to express their viral proteins. The viral mRNAs are decoded by the host cell ribosomes to produce viral proteins just as cellular mRNAs are decoded to produce the cellular proteins. Theoretically, any mRNA put into a cell will be recognized and translated by the ribosomes, and this is the basis of mRNA vaccines. First, synthesize a mRNA that encodes for a viral surface protein (e.g. the spike protein for SARS-CoV-2). Second, introduce that mRNA into the cell where it will be translated by ribosomes into the viral protein just as it would during an infection. As a foreign protein, the expressed viral protein should elicit an immune response that protects against the viral infection. In some ways, this is similar to the protein subunit vaccines. Both approaches target a viral surface protein that is an important target for protective immunity, but the difference is in how the protein is produced. In the traditional subunit vaccine, the viral protein is expressed and purified in the lab, often an arduous and expensive process that has to be individualized for every protein or change in a protein sequence. Then the purified viral protein is injected into the patient to induce the immune response. For the mRNA vaccine, the mRNA is synthesized and purified in the lab which is a quicker and cheaper process that can be modified for sequence changes more easily than changing a protein. The resultant mRNA is injected into the vaccinee, the mRNA is taken up by the surrounding cells, and the person's cells produce the protein. While simple in concept, numerous technical issues had to be overcome to turn mRNA vaccine theory into practice, and the world was fortunate that scientists had been developing mRNA expression and delivery methods for many years before the COVID-19 pandemic.

The scientific origins of the mRNA vaccine extend at least as far back as the late 1970s when researchers first showed that foreign mRNAs introduced into cells were functional and produced proteins (Ostro et al. 1978; Dimitriadis

1978). The possibility of using mRNAs to introduce beneficial proteins into patients was explored by multiple academic labs and biotech companies in the 1980s, but there were serious constraints that kept mRNA technology more as a laboratory tool than as a commercial product. First, no method for synthesizing RNA existed until 1984 when biologists at Harvard developed an enzymatic method using an RNA-synthesizing enzyme isolated from a bacteriophage (Melton et al. 1984). Their method suddenly made producing long mRNAs relatively easy and remains the basic approach still used today. Second, unlike DNA, mRNA is notoriously unstable and easily degrades. Biologically this makes sense in our cells because each mRNA is supposed to be a transient messenger, not a permanent fixture of the cell. If mRNAs were stable and long-lived, they would keep getting translated into protein over and over resulting in the accumulation of more protein than needed. Instead, when a cell needs more of a particular protein, signals are sent to the nucleus to start transcription of the gene that codes for that protein. The resultant mRNA molecules are translated into protein but also rapidly degraded so that only a limited amount of protein is produced. Balancing transcription and degradation of mRNAs is just one of the many clever regulatory tricks that our cells use to carefully titrate the amount of each protein, but it is problematic for clinical applications of mRNAs. Not only is it difficult to maintain the integrity of RNA while working with it in labs and production facilities, but once the mRNAs are introduced into a patient's cells they will start to be degraded. For therapeutic or vaccine purposes, high levels of the target protein are desirable, but the rapid degradation of the introduced mRNA would severely limit the amount of the target protein produced. Although there is no single solution to this degradation problem, over the last 30 years extensive research has developed a portfolio of tricks that can sufficiently stabilize mRNAs for clinical applications (Cheng et al. 2023).

In addition to the synthesis and stabilities issues with mRNA, two other major constraints limited mRNA applications, lack of an effective delivery vehicle and the cell's innate immune response. From the earliest days of mRNA research, it was apparent that simply exposing cells to RNA did not promote efficient survival and uptake of the RNA. Not only is this "naked" RNA quickly degraded, but it also doesn't successfully enter the cells. mRNAs are large molecules and do not readily cross the cell membrane whose function is to keep out extraneous biomolecules. Early RNA researchers used liposomes to surround and protect the mRNA and facilitate the uptake of the RNA into cells. Liposomes are small spherical particles made up of cholesterol and fatty lipids that resemble the lipids of the cell membrane. Liposomes carrying RNA (or other biomolecules) will fuse with the cell membrane and release their cargo inside the cell. While usable, liposomes are not the most efficient delivery system for RNA or DNA. Starting in the 1990s, as the concept of using RNA or DNA therapeutically gained momentum, better and more bespoke systems for nucleic acid delivery were investigated. Ultimately this led to the technology of lipid nanoparticles (LNPs) that was pioneered in the lab of Pieter Cullis at the University of British Columbia (Chonn and Cullis 1995). The LNPs can be customized using different combinations of lipid molecules to produce efficient carrier systems that have good shelf life, low toxicity, suitable stability inside the body, and highly efficient cargo delivery. Different biotech companies developed their own proprietary formulations for LNPs and by the mid-2000s an efficient manufacturing device for combining RNA and DNA into LNPs was created. Now any RNA or DNA could be packaged for clinical applications.

One final obstacle to the implementation of mRNA for clinical use was the tendency for exogenous RNAs to stimulate the host's innate defense system. Our bodies have three types of immune defense systems, intrinsic, adaptive, and innate. The intrinsic system is primarily an initial defense against viral infections and isn't relevant to the discussion of mRNA vaccines. Usually, when we mention immunity we are referring to adaptive immunity. Adaptive immunity consists of B cells (which produce antibodies) and T cells. These cells are activated and amplified by a first exposure to foreign antigens (from an infection or a vaccine), and once activated they remember these antigens for years to decades. However, it takes time (roughly 5–10 days) after first exposure to produce enough activated cells to fight off an infection. Until those immune cells reach a sufficient level we are vulnerable, and this is why we get sick the first time we encounter a new pathogen. On any subsequent exposure to that pathogen the immune cells "remember" it and respond within hours which thwarts the infection and prevents us from getting sick again. What the innate immune system does is try to slow down infections in the vulnerable period before adaptive immunity takes over. Without innate immunity, any pathogen could multiply unchecked for days. Even relatively innocuous microorganisms could overwhelm us the first time we encounter them, causing serious disease and fatalities, so this system is critical to our survival. Mechanistically, innate immunity works by using sensors (called pathogen recognition receptors—PRRs) that are present on the surface of and within cells. When a cell's sensors detect foreign biomolecules they initiate multiple biochemical processes that stop cellular activities, including mRNA translation into protein, and/or kill the cell. This might seem harsh, but we have millions of cells, and it is much better to sacrifice a few cells to slow down the pathogen spread and allow the adaptive immune system time to respond.

The problem for mRNA applications is that these laboratory-produced mRNAs trigger the innate system. Consequently, cells that take up the exogenous mRNA stop protein synthesis. These cells also have increased RNA degradation which destroys the incoming mRNA. This combination of degradation and inhibition of protein synthesis prevents the exogenous mRNA from producing the target protein, defeating the whole purpose of the mRNA treatment and stymieing clinical applications. The key breakthrough that circumvented this obstacle and enabled mRNA therapeutic and vaccine technologies to advance was the finding that certain chemical modifications of RNA protected it from degradation and shielded it from recognition by the innate immune system (Kariko et al. 2005, 2008). This seminal work came from the laboratory of two longtime pioneers in RNA research at the University of Pennsylvania, Katalin Karikó and Drew Weissman. Their observations stimulated a decade of research on optimizing mRNA stability and expression through chemical modification (Ye et al. 2023). Collectively, these studies formed the basis for transforming mRNA vaccines from a concept into a real-world product. Karikó and Weissman were recognized with the 2023 Nobel Prize in Physiology and Medicine for their critical contribution to the technology that enabled the SARS-CoV-2 vaccine.

The mRNA Vaccine Revolution

By the early 2010s, most of the technical hurdles for mRNA applications had been addressed and workable solutions had been found or were being developed. Both large pharma and small biotech were initially eager to exploit this technology for potential treatment of genetic diseases, cancer, and infectious diseases. The U.S. Defense Advanced Research Projects Agency (DARPA) was particularly interested in this technology, and DARPA funded projects at several large companies including Novartis, Pfizer, Sanofi-Pasteur, and AstraZeneca. However, as an entirely new vaccine platform, the mRNA vaccine approach still faced significant challenges in transforming a laboratory technique into a safe and accepted commercial product. This included potential difficulties with regulatory agencies like the FDA which would require even more extensive and stringent testing of this new platform before approving any vaccines. With no immediate need or specific market for mRNA vaccines, progress was slow and cautious throughout that decade. Attention focused mainly on diseases with no vaccines such as Nipah virus (a highly fatal zoonotic agent), Zika virus (a mosquito-borne agent causing birth defects), and cytomegalovirus (a herpes virus causing birth defects), or on diseases such as rabies and influenza where an improved vaccine would be useful.

A key player in the mRNA vaccine development realm was the German biotech company, BioNTech, created to exploit mRNA for the treatment and prevention of human disease. Founded in 2008, BioNTech's early focus was on cancer therapies, including anti-cancer vaccines, and by 2015 they were conducting initial tests in humans. In 2018 they partnered with Pfizer to develop an mRNA-based influenza vaccine, moving their technology into the realm of infectious disease vaccines. The work progressed at a slow and deliberate pace to ensure that this platform could produce vaccines that were at least as effective and safe as the other standard vaccine types. All that changed with the COVID-19 outbreak hit at the end of 2019. In conjunction with their Pfizer partner, BioNTech launched Project Lightspeed to create an mRNA vaccine for this new disease entity (Killeen et al. 2023). The beauty of the mRNA vaccine is that its development can be extremely rapid and doesn't require the individualized optimization needed for other vaccine platforms. Live vaccines need an attenuated strain, and it can take months or years to develop a safe version, killed vaccines have to be empirically developed to find the optimal conditions for inactivation while retaining immunogenicity, and protein subunit vaccines require significant time to develop the expression and purification parameters that produce a stable product that effectively induces immunity. In contrast, for an mRNA vaccine, all that is needed is the genetic sequence for the target protein, in this case, the spike protein. Once that sequence is known, the mRNA can be synthesized quickly and inserted into a company's pre-existing LNP carrier with no major modifications or optimizations needed. Because they had years of experience with mRNA and the associated delivery technologies, within days after the publication of the first SARS-CoV-2 sequence, BioNTech had a prototype vaccine ready for initial testing. To streamline the testing process, a small-scale combined phase I/phase II study was conducted to assess the vaccine's safety and ability to induce anti-spike antibodies (Mulligan et al. 2020). Highly positive results from this initial study led to a large placebo-controlled, observer-blinded, phase II/phase III trial involving 43,548 participants (Polack et al. 2020). Early results in November 2020 confirmed the safety and protective efficacy (95%) of this new vaccine, and in December the FDA gave the vaccine (trade name COMIRNATY) Emergency Use Authorization (with subsequent full use authorization the next year). In less than one year, a new pathogen had emerged, been identified, and a preventative vaccine was created, a remarkable scientific achievement built on 20 years of previous RNA and vaccine work.

Similar to BioNTech, the biotech startup company called Moderna was founded in 2010 in Cambridge, Massachusetts, and also focused on RNA. One of the early recipients of a DARPA grant, Moderna quickly embarked on a broad strategy for RNA therapeutics that encompassed oncology, physiological diseases (e.g. cardiovascular and renal diseases), and vaccines for infectious diseases. Establishing partnerships with multiple pharmaceutical and biotech companies provided funding and resources for rapid product development. By 2015 they had an mRNA-based influenza vaccine in testing and by 2017 had a bivalent vaccine for human metapneumovirus (hMPV) and parainfluenza virus in human trials. mRNA vaccines against several other viruses, including MERS and Nipah virus, were also in development, and by the end of 2019, Moderna had nine candidate vaccines against infectious diseases being tested in humans. But like BioNTech, Moderna was moving cautiously and deliberately with this new platform to avoid any unexpected pitfalls. When COVID-19 hit the U.S., NIH and the federal government immediately ramped up efforts to create diagnostics, therapeutics, and possible vaccines through Operation Warp Speed. Just as BioNTech did, Moderna immediately used the SARS-CoV-2 sequence to produce a spike protein mRNA that was incorporated into their mRNA vaccine LNP platform. Working with NIH, the Moderna prototype vaccine went through primate testing (Corbett et al. 2020) and then into human volunteers within ten weeks (Jackson et al. 2020), followed by a major phase III trial involving over 30,000 volunteers (Baden et al. 2021). With phase III results showing 94% protective efficacy, the FDA also gave the Moderna vaccine (trade name SPIKEVAX) Emergency Use Authorization in December 2020 (with full use authorization in January 2022).

Unlike some public perception, both the Moderna and BioNTech mRNA vaccines were not created using novel technology invented just for the COVID-19 pandemic. Instead, both vaccines were based on RNA production and delivery systems that had been researched and tested for over ten years as vaccine candidates for multiple infectious diseases. The pandemic emergency simply necessitated jumping the SARS-CoV-2 mRNA vaccines ahead of the other mRNA vaccines being developed. Still, the timing for the development, testing, and initial public distribution of the SARS-CoV-2 mRNA vaccines was not only scientifically outstanding but was also incredibly fortuitous. While lockdowns, masks, and other restrictions were helping restrict transmission, they were not capable alone of stopping the pandemic. By February 2021 there were 100 million identified cases worldwide with two million deaths and the pandemic was still peaking. The rapid availability of the two mRNA vaccines allowed massive production and vaccination campaigns to be implemented in the U.S. and around the world. Eventually, other standard vaccine platforms would also produce SARS-CoV-2 vaccines, including inactivated virus vaccines (e.g. CoronaVac and Covaxin), adenovirus vector vaccines (e.g. Ad26.COV2.S and Vaxzevria), and protein subunit vaccines (e.g. Novavax and Covovax), but none of these was available early in the pandemic. Within a year of the mRNA vaccines reaching the public, the pandemic was residing and one study estimated that the vaccination campaign in the U.S. saved $900 billion in healthcare costs and prevented 66 million infections, 17 million hospitalizations, and two million deaths (Scheider et al. 2021). As herd immunity grew through community-acquired infection and vaccinations, the WHO finally declared that the pandemic emergency nature of COVID-19 was over in May of 2023. After three and a half years of disruption and terror with hundreds of millions of infections and nearly seven million deaths, vaccine science finally tamed this virus. Sadly, SARS-CoV-2 appears to have established itself as a new endemic human virus that will circulate for generations to come. It may someday become as innocuous as its cold-causing cousins, but for now, periodic vaccination is our only mechanism for keeping case numbers down and protecting the most vulnerable populations.

But Are They Safe?

With any new vaccine, there is always some uncertainty about potential safety issues. At its best, phase III testing typically involves thousands to tens of thousands of volunteers followed for 1–4 years. Rare adverse effects or effects that take many years to manifest may not be detected during the phase III testing. Ultimately the release of a vaccine into general usage (phase IV) is the final test of a vaccine's safety. As millions to tens of millions of individuals take the vaccine the sample size becomes large enough to detect rare adverse effects. Depending on the nature of the effect this may simply require an alert warning for physicians and patients, or the vaccine may be withdrawn as was done for RotaShield (Chap. 11). As a new therapeutic and vaccine platform, applications of mRNA to clinical problems proceeded slowly and thoughtfully in the 2010s. Researchers and clinicians wanted to ensure the safety and efficacy of mRNA products with a careful eye toward any unanticipated adverse effects. The general belief was that this approach should be quite safe as mRNA is abundant in our cells and the added exogenous mRNA wasn't fundamentally different from the natural cellular mRNAs. Additionally, even the stabilized mRNAs needed for clinical applications would eventually degrade in the cell, stopping the production of the protein they encoded. The introduced protein would also eventually be degraded, so once the target protein and its mRNA were gone there should be no permanent changes to the cells that transiently contained these biomolecules. Specifically for vaccines, the protein expressed from the vaccine mRNA is the same protein expressed by the pathogen, so having the cell express this protein from the vaccine mRNA should be fundamentally the same as the cell expressing it from the pathogen's mRNA. Nonetheless, every new technology, therapeutic, and vaccine has the potential to evoke unforeseen off-target effects that can only be identified through testing in cell culture, animal models, and ultimately in human test subjects. The mRNA technology was still in the early stages of human testing when the COVID-19 pandemic galvanized the field into explosive human testing. The global pandemic emergency demanded rapid action and condensed what would have been years of additional small-scale testing of mRNA vaccines into less than twelve months. Over the subsequent three years after the mRNA vaccines were released, more than five billion people were injected with them in a massive phase IV experiment. We know that the vaccine protected people and helped stop the pandemic, but the adverse effects are still being evaluated. As a new vaccine technology, LNP-mRNA formulations have not been extensively studied in large-scale populations and many unanswered questions remain about potential risks, off-target effects, and the regulatory framework needed to ensure and monitor their safety (Banoun 2023).

Like nearly all vaccines, a constellation of minor adverse effects has been observed for the COVID-19 mRNA vaccines including fever, pain at the injection site, fatigue, and headache (SeyedAlinaghi et al. 2022). Typical of such minor vaccine-associated issues, these symptoms typically resolve within a few days. A more consequential adverse effect is the severe and potentially life-threatening allergic reaction known as anaphylaxis. The rate of anaphylaxis of the COVID mRNA vaccines is 2.5–5 cases per million doses (Shimabukuro et al. 2021). While efforts are underway to identify the root cause of this response and reduce its incidence, this rate is similar to or better than what is seen for many conventional vaccines, so it is not specific to nor a contraindication for mRNA vaccines. A concern that is specific to the mRNA vaccines is the use of a compound called polyethylene glycol (PEG) in most of the LNP formulations. LNP vaccine recipients or individuals exposed to PEG in other contexts can develop antibodies against this compound (Estape Senti et al. 2022). Upon repeated exposure to PEG-containing LNP vaccines, the resulting antibody response can not only attack the LNP vesicles and reduce their effectiveness but may also cause increased systemic side effects. Again, while this is not a serious health concern, developing better LNP formulations that don't include PEG remains an area of active investigation.

One rare, but potentially serious adverse reaction associated with the SARS-CoV-2 mRNA vaccines is inflammation of the heart, particularly in young males after a second dose of the vaccine (Fatima et al. 2022). This inflammation can occur in the outer tissue that surrounds the heart (pericarditis) or in the heart muscle itself (myocarditis). Pericarditis is relatively mild and usually spontaneously resolves without the need for medical intervention. Myocarditis also is generally mild and self-limited but can progress to serious and even fatal outcomes, so it needs medical attention. The mechanism behind this response is unknown but may not be specific to mRNA as both pericarditis and myocarditis have been observed with other non-mRNA SARS-CoV-2 vaccines (Buoninfante et al. 2024). Until this response is understood, physicians and vaccine recipients, particularly males under the age of 30, need to be aware of this complication so that medical intervention can be implemented quickly if needed.

In addition to cardiac issues, the other major class of reported adverse effects is neurological complications (Chatterjee and Chakravarty 2023). This group encompasses a wide constellation of problems that range from minor issues such as headaches to serious effects that include seizures, Bell's palsy, Guillain-Barré Syndrome, spinal cord inflammation (transverse myelitis), and encephalopathy (brain dysfunctions such as memory loss, personality changes, and disorientation). As with the cardiac events, most of the neurological effects are transient and resolve with little or no treatment, although some have resulted in long-term problems and even deaths. Importantly though, many of these reported effects have not conclusively been linked to the mRNA vaccines. Many of these neurological events are observed with other non-mRNA vaccines so they may not be specific to mRNA vaccines. Furthermore, it is still unclear if the mRNA vaccines have a causal relationship with most of these rare but serious issues. Large studies with appropriate control populations are needed to assess whether or not each type of adverse event is vaccine-specific or merely a random occurrence that is no more frequent in the vaccinated population than the controls.

At the current time, the risks associated with mRNA vaccines against SARS-CoV-2 appear minimal and similar to what is seen with other conventional vaccines. As for all vaccines it comes down to a risk-benefit assessment. As the pandemic raged in 2020 and 2021, the risk from the virus was extreme both to the individual and the societal structure. Consequently, the protection and return to normalcy offered by vaccination far outweighed any small degree of vaccine-associated adverse effects. Now that the outbreak has been controlled and managed, continued evaluation of mRNA vaccine safety is needed to identify the full range of vaccine-associated complications and develop improved versions that do not induce these complications. The pandemic and the immediate success of the SARS-CoV-2 mRNA vaccines launched this platform and spurred an explosive acceleration of research development into new applications for other infectious diseases, cancer treatments, and physiological conditions (Parhiz et al. 2024). This technology is unlikely to be discarded so we must study and understand it fully so that its application creates safe and effective vaccines and therapies that improve human health.

References

Agyapon-Ntra K, McSharry PE (2023) A global analysis of the effectiveness of policy responses to COVID-19. Sci Rep 13(1):5629. https://doi.org/10.1038/s41598-023-31709-2

Amicone M, Borges V, Alves MJ, Isidro J, Ze-Ze L, Duarte S, Vieira L, Guiomar R, Gomes JP, Gordo I (2022) Mutation rate of SARS-CoV-2 and emergence of mutators during experimental evolution. Evol Med Public Health 10(1):142–155. https://doi.org/10.1093/emph/eoac010

Baden LR, El Sahly HM, Essink B, Kotloff K, Frey S, Novak R, Diemert D, Spector SA, Rouphael N, Creech CB, McGettigan J, Khetan S, Segall N, Solis J, Brosz A, Fierro C, Schwartz H, Neuzil K, Corey L, Gilbert P, Janes H, Follmann D, Marovich M, Mascola J, Polakowski L, Ledgerwood J, Graham BS, Bennett H, Pajon R, Knightly C, Leav B, Deng W, Zhou H, Han S, Ivarsson M, Miller J, Zaks T, Group CS (2021) Efficacy and safety of the mRNA-1273 SARS-CoV-2 vaccine. N Engl J Med 384(5):403–416. https://doi.org/10.1056/NEJMoa2035389

Banoun H (2023) mRNA: vaccine or gene therapy? The safety regulatory issues. Int J Mol Sci 24(13). https://doi.org/10.3390/ijms241310514

Bardosh K, de Figueiredo A, Gur-Arie R, Jamrozik E, Doidge J, Lemmens T, Keshavjee S, Graham JE, Baral S (2022) The unintended consequences of COVID-19 vaccine policy: why mandates, passports and restrictions may cause more harm than good. BMJ Glob Health 7(5). https://doi.org/10.1136/bmjgh-2022-008684

Breban R, Riou J, Fontanet A (2013) Interhuman transmissibility of Middle East respiratory syndrome coronavirus: estimation of pandemic risk. Lancet 382(9893):694–699. https://doi.org/10.1016/S0140-6736(13)61492-0

Buoninfante A, Andeweg A, Genov G, Cavaleri M (2024) Myocarditis associated with COVID-19 vaccination. NPJ Vaccines 9(1):122. https://doi.org/10.1038/s41541-024-00893-1

Chatterjee A, Chakravarty A (2023) Neurological complications following COVID-19 vaccination. Curr Neurol Neurosci Rep 23(1):1–14. https://doi.org/10.1007/s11910-022-01247-x

Cheng F, Wang Y, Bai Y, Liang Z, Mao Q, Liu D, Wu X, Xu M (2023) Research advances on the stability of mRNA vaccines. Viruses 15(3). https://doi.org/10.3390/v15030668

Cherry JD, Krogstad P (2004) SARS: the first pandemic of the 21st century. Pediatr Res 56(1):1–5. https://doi.org/10.1203/01.PDR.0000129184.87042.FC

Chonn A, Cullis PR (1995) Recent advances in liposomal drug-delivery systems. Curr Opin Biotechnol 6(6):698–708. https://doi.org/10.1016/0958-1669(95)80115-4

Chowell G, Abdirizak F, Lee S, Lee J, Jung E, Nishiura H, Viboud C (2015) Transmission characteristics of MERS and SARS in the healthcare setting: a comparative study. BMC Med 13:210. https://doi.org/10.1186/s12916-015-0450-0

Corbett KS, Flynn B, Foulds KE, Francica JR, Boyoglu-Barnum S, Werner AP, Flach B, O'Connell S, Bock KW, Minai M, Nagata BM, Andersen H, Martinez DR, Noe AT, Douek N, Donaldson MM, Nji NN, Alvarado GS, Edwards DK, Flebbe DR, Lamb E, Doria-Rose NA, Lin BC, Louder MK, O'Dell S, Schmidt SD, Phung E, Chang LA, Yap C, Todd JM, Pessaint L, Van Ry A, Browne S, Greenhouse J, Putman-Taylor T, Strasbaugh A, Campbell TA, Cook A, Dodson A, Steingrebe K, Shi W, Zhang Y, Abiona OM, Wang L, Pegu A, Yang ES, Leung K, Zhou T, Teng IT, Widge A, Gordon I, Novik L, Gillespie RA, Loomis RJ, Moliva JI, Stewart-Jones G, Himansu S, Kong WP, Nason MC, Morabito KM, Ruckwardt TJ, Ledgerwood JE, Gaudinski MR, Kwong PD, Mascola JR, Carfi A, Lewis MG, Baric RS, McDermott A, Moore IN, Sullivan NJ, Roederer M, Seder RA, Graham BS (2020) Evaluation of the mRNA-1273 vaccine against SARS-CoV-2 in non-human primates. N Engl J Med 383(16):1544–1555. https://doi.org/10.1056/NEJMoa2024671

Dimitriadis GJ (1978) Translation of rabbit globin mRNA introduced by liposomes into mouse lymphocytes. Nature 274(5674):923–924. https://doi.org/10.1038/274923a0

Estape Senti M, de Jongh CA, Dijkxhoorn K, Verhoef JJF, Szebeni J, Storm G, Hack CE, Schiffelers RM, Fens MH, Boross P (2022) Anti-PEG antibodies compromise the integrity of PEGylated lipid-based nanoparticles via complement. J Control Release 341:475–486. https://doi.org/10.1016/j.jconrel.2021.11.042

Fatima M, Ahmad Cheema H, Ahmed Khan MH, Shahid H, Saad Ali M, Hassan U, Wahaj Murad M, Aemaz Ur Rehman M, Farooq H (2022) Development of myocarditis and pericarditis after COVID-19 vaccination in adult population: a systematic review. Ann Med Surg (Lond) 76:103486. https://doi.org/10.1016/j.amsu.2022.103486

Fouchier RA, Hartwig NG, Bestebroer TM, Niemeyer B, de Jong JC, Simon JH, Osterhaus AD (2004) A previously undescribed coronavirus associated with respiratory disease in humans. Proc Natl Acad Sci USA 101(16):6212–6216. https://doi.org/10.1073/pnas.0400762101

Giovanetti M, Branda F, Cella E, Scarpa F, Bazzani L, Ciccozzi A, Slavov SN, Benvenuto D, Sanna D, Casu M, Santos LA, Lai A, Zehender G, Caccuri F, Ianni A, Caruso A, Maroutti A, Pascarella S, Borsetti A, Ciccozzi M (2023) Epidemic history and evolution of an emerging threat of international concern, the severe acute respiratory syndrome coronavirus 2. J Med Virol 95(8):e29012. https://doi.org/10.1002/jmv.29012

Grana C, Ghosn L, Evrenoglou T, Jarde A, Minozzi S, Bergman H, Buckley BS, Probyn K, Villanueva G, Henschke N, Bonnet H, Assi R, Menon S, Marti M, Devane D, Mallon P, Lelievre JD, Askie LM, Kredo T, Ferrand G, Davidson M, Riveros C, Tovey D, Meerpohl JJ, Grasselli G, Rada G, Hrobjartsson A, Ravaud P, Chaimani A, Boutron I (2022) Efficacy and safety of COVID-19 vaccines. Cochrane Database Syst Rev 12(12):CD015477. https://doi.org/10.1002/14651858.CD015477

Haagmans BL, Al Dhahiry SH, Reusken CB, Raj VS, Galiano M, Myers R, Godeke GJ, Jonges M, Farag E, Diab A, Ghobashy H, Alhajri F, Al-Thani M, Al-Marri SA, Al Romaihi HE, Al Khal A, Bermingham A, Osterhaus AD, AlHajri MM, Koopmans MP (2014) Middle East respiratory syndrome coronavirus in dromedary camels: an outbreak investigation. Lancet Infect Dis 14(2):140–145. https://doi.org/10.1016/S1473-3099(13)70690-X

Hamre D, Procknow JJ (1966) A new virus isolated from the human respiratory tract. Proc Soc Exp Biol Med 121(1):190–193. https://doi.org/10.3181/00379727-121-30734

Harrison CM, Doster JM, Landwehr EH, Kumar NP, White EJ, Beachboard DC, Stobart CC (2023) Evaluating the virology and evolution of seasonal human coronaviruses associated with the common cold in the COVID-19 era. Microorganisms 11(2). https://doi.org/10.3390/microorganisms11020445

He X, Lau EHY, Wu P, Deng X, Wang J, Hao X, Lau YC, Wong JY, Guan Y, Tan X, Mo X, Chen Y, Liao B, Chen W, Hu F, Zhang Q, Zhong M, Wu Y, Zhao L, Zhang F, Cowling BJ, Li F, Leung GM (2020) Temporal dynamics in viral shedding and transmissibility of COVID-19. Nat Med 26(5):672–675. https://doi.org/10.1038/s41591-020-0869-5

Heikkinen T, Jarvinen A (2003) The common cold. Lancet 361(9351):51–59. https://doi.org/10.1016/S0140-6736(03)12162-9

Howard J, Huang A, Li Z, Tufekci Z, Zdimal V, van der Westhuizen HM, von Delft A, Price A, Fridman L, Tang LH, Tang V, Watson GL, Bax CE, Shaikh R, Questier F, Hernandez D, Chu LF, Ramirez CM, Rimoin AW (2021) An evidence review of face masks against COVID-19. Proc Natl Acad Sci USA 118(4). https://doi.org/10.1073/pnas.2014564118

Iida S, Arashiro T, Suzuki T (2021) Insights into pathology and pathogenesis of coronavirus disease 2019 from a histopathological and immunological perspective. JMA J 4(3):179–186. https://doi.org/10.31662/jmaj.2021-0041

Jackson LA, Anderson EJ, Rouphael NG, Roberts PC, Makhene M, Coler RN, McCullough MP, Chappell JD, Denison MR, Stevens LJ, Pruijssers AJ, McDermott A, Flach B, Doria-Rose NA, Corbett KS, Morabito KM, O'Dell S, Schmidt SD, Swanson PA 2nd, Padilla M, Mascola JR, Neuzil KM, Bennett H, Sun W, Peters E, Makowski M, Albert J, Cross K, Buchanan W, Pikaart-Tautges R, Ledgerwood JE, Graham BS, Beigel JH, m RNASG (2020) An mRNA vaccine against SARS-CoV-2 - preliminary report. N Engl J Med 383(20):1920–1931. https://doi.org/10.1056/NEJMoa2022483

Kahn JS, McIntosh K (2005) History and recent advances in coronavirus discovery. Pediatr Infect Dis J 24(11 Suppl):S223–S227, discussion S226. https://doi.org/10.1097/01.inf.0000188166.17324.60

Kariko K, Buckstein M, Ni H, Weissman D (2005) Suppression of RNA recognition by Toll-like receptors: the impact of nucleoside modification and the evolutionary origin of RNA. Immunity 23(2):165–175. https://doi.org/10.1016/j.immuni.2005.06.008

Kariko K, Muramatsu H, Welsh FA, Ludwig J, Kato H, Akira S, Weissman D (2008) Incorporation of pseudouridine into mRNA yields superior nonimmunogenic vector with increased translational capacity and biological stability. Mol Ther 16(11):1833–1840. https://doi.org/10.1038/mt.2008.200

Karimizadeh Z, Dowran R, Mokhtari-Azad T, Shafiei-Jandaghi NZ (2023) The reproduction rate of severe acute respiratory syndrome coronavirus 2 different variants recently circulated in human: a narrative review. Eur J Med Res 28(1):94. https://doi.org/10.1186/s40001-023-01047-0

Killeen T, Kermer V, Troxler Saxer R (2023) mRNA vaccine development during the COVID-19 pandemic: a retrospective review from the perspective of the Swiss affiliate of a global biopharmaceutical company. J Pharm Policy Pract 16(1):158. https://doi.org/10.1186/s40545-023-00652-y

Letko M, Marzi A, Munster V (2020) Functional assessment of cell entry and receptor usage for SARS-CoV-2 and other lineage B betacoronaviruses. Nat Microbiol 5(4):562–569. https://doi.org/10.1038/s41564-020-0688-y

Li F (2016) Structure, function, and evolution of coronavirus spike proteins. Annu Rev Virol 3(1):237–261. https://doi.org/10.1146/annurev-virology-110615-042301

Lipsitch M, Cohen T, Cooper B, Robins JM, Ma S, James L, Gopalakrishna G, Chew SK, Tan CC, Samore MH, Fisman D, Murray M (2003) Transmission dynamics and control of severe acute respiratory syndrome. Science 300(5627):1966–1970. https://doi.org/10.1126/science.1086616

Liu DX, Liang JQ, Fung TS (2021) Human coronavirus-229E,-OC43,-NL63, and-HKU1 (Coronaviridae). In: Encyclopedia of virology, Academic Press, San Diego, California, USA, p 428

Markov PV, Ghafari M, Beer M, Lythgoe K, Simmonds P, Stilianakis NI, Katzourakis A (2023) The evolution of SARS-CoV-2. Nat Rev Microbiol 21(6):361–379. https://doi.org/10.1038/s41579-023-00878-2

McIntosh K, Dees JH, Becker WB, Kapikian AZ, Chanock RM (1967) Recovery in tracheal organ cultures of novel viruses from patients with respiratory disease. Proc Natl Acad Sci USA 57(4):933–940. https://doi.org/10.1073/pnas.57.4.933

Melton DA, Krieg PA, Rebagliati MR, Maniatis T, Zinn K, Green MR (1984) Efficient in vitro synthesis of biologically active RNA and RNA hybridization probes from plasmids containing a bacteriophage SP6 promoter. Nucleic Acids Res 12(18):7035–7056. https://doi.org/10.1093/nar/12.18.7035

Mulligan MJ, Lyke KE, Kitchin N, Absalon J, Gurtman A, Lockhart S, Neuzil K, Raabe V, Bailey R, Swanson KA, Li P, Koury K, Kalina W, Cooper D, Fontes-Garfias C, Shi PY, Tureci O, Tompkins KR, Walsh EE, Frenck R, Falsey AR, Dormitzer PR, Gruber WC, Sahin U, Jansen KU (2020) Phase I/II study of COVID-19 RNA vaccine BNT162b1 in adults. Nature 586(7830):589–593. https://doi.org/10.1038/s41586-020-2639-4

Murphy C, Lim WW, Mills C, Wong JY, Chen D, Xie Y, Li M, Gould S, Xin H, Cheung JK, Bhatt S, Cowling BJ, Donnelly CA (2023) Effectiveness of social distancing measures and lockdowns for reducing transmission of COVID-19 in non-healthcare, community-based settings. Philos Trans A Math Phys Eng Sci 381(2257):20230132. https://doi.org/10.1098/rsta.2023.0132

Ostro MJ, Giacomoni D, Lavelle D, Paxton W, Dray S (1978) Evidence for translation of rabbit globin mRNA after liposome-mediated insertion into a human cell line. Nature 274(5674):921–923. https://doi.org/10.1038/274921a0

Otieno JR, Cherry JL, Spiro DJ, Nelson MI, Trovao NS (2022) Origins and evolution of seasonal human coronaviruses. Viruses 14(7). https://doi.org/10.3390/v14071551

Parhiz H, Atochina-Vasserman EN, Weissman D (2024) mRNA-based therapeutics: looking beyond COVID-19 vaccines. Lancet 403(10432):1192–1204. https://doi.org/10.1016/S0140-6736(23)02444-3

Peacock TP, Goldhill DH, Zhou J, Baillon L, Frise R, Swann OC, Kugathasan R, Penn R, Brown JC, Sanchez-David RY, Braga L, Williamson MK, Hassard JA, Staller E, Hanley B, Osborn M, Giacca M, Davidson AD, Matthews DA, Barclay WS (2021) The furin cleavage site in the SARS-CoV-2 spike protein is required for transmission in ferrets. Nat Microbiol 6(7):899–909. https://doi.org/10.1038/s41564-021-00908-w

Pekar J, Worobey M, Moshiri N, Scheffler K, Wertheim JO (2021) Timing the SARS-CoV-2 index case in Hubei province. Science 372(6540):412–417. https://doi.org/10.1126/science.abf8003

Polack FP, Thomas SJ, Kitchin N, Absalon J, Gurtman A, Lockhart S, Perez JL, Perez Marc G, Moreira ED, Zerbini C, Bailey R, Swanson KA, Roychoudhury S, Koury K, Li P, Kalina WV, Cooper D, Frenck RW Jr, Hammitt LL, Tureci O, Nell H, Schaefer A, Unal S, Tresnan DB, Mather S, Dormitzer PR, Sahin U, Jansen KU,

Gruber WC, Group CCT (2020) Safety and efficacy of the BNT162b2 mRNA Covid-19 vaccine. N Engl J Med 383(27):2603–2615. https://doi.org/10.1056/NEJMoa2034577

Pustake M, Tambolkar I, Giri P, Gandhi C (2022) SARS, MERS and CoVID-19: an overview and comparison of clinical, laboratory and radiological features. J Family Med Prim Care 11(1):10–17. https://doi.org/10.4103/jfmpc.jfmpc_839_21

Scheider EC, Shah A, Sah P, Moghadas SM, Vilches T, Galvani AP (2021) The U.S. COVID-19 vaccination program at one year: how many deaths and hospitalizations were averted? Issue Briefs

SeyedAlinaghi S, Karimi A, Pashaei Z, Afzalian A, Mirzapour P, Ghorbanzadeh K, Ghasemzadeh A, Dashti M, Nazarian N, Vahedi F, Tantuoyir MM, Shamsabadi A, Dadras O, Mehraeen E (2022) Safety and adverse events related to COVID-19 mRNA vaccines; a systematic review. Arch Acad Emerg Med 10(1):e41. https://doi.org/10.22037/aaem.v10i1.1597

Shimabukuro TT, Cole M, Su JR (2021) Reports of anaphylaxis after receipt of mRNA COVID-19 vaccines in the US-December 14, 2020-January 18, 2021. JAMA 325(11):1101–1102. https://doi.org/10.1001/jama.2021.1967

Tao Y, Shi M, Chommanard C, Queen K, Zhang J, Markotter W, Kuzmin IV, Holmes EC, Tong S (2017) Surveillance of bat coronaviruses in Kenya identifies relatives of human coronaviruses NL63 and 229E and their recombination history. J Virol 91(5). https://doi.org/10.1128/JVI.01953-16

Tindale LC, Stockdale JE, Coombe M, Garlock ES, Lau WYV, Saraswat M, Zhang L, Chen D, Wallinga J, Colijn C (2020) Evidence for transmission of COVID-19 prior to symptom onset. elife 9. https://doi.org/10.7554/eLife.57149

Tosta E (2022) The adaptation of SARS-CoV-2 to humans. Mem Inst Oswaldo Cruz 116:e210127. https://doi.org/10.1590/0074-02760210127

Tyrrell DA, Bynoe ML (1966) Cultivation of viruses from a high proportion of patients with colds. Lancet 1(7428):76–77. https://doi.org/10.1016/s0140-6736(66)92364-6

Vijgen L, Keyaerts E, Moes E, Thoelen I, Wollants E, Lemey P, Vandamme AM, Van Ranst M (2005) Complete genomic sequence of human coronavirus OC43: molecular clock analysis suggests a relatively recent zoonotic coronavirus transmission event. J Virol 79(3):1595–1604. https://doi.org/10.1128/JVI.79.3.1595-1604.2005

Vijgen L, Keyaerts E, Lemey P, Maes P, Van Reeth K, Nauwynck H, Pensaert M, Van Ranst M (2006) Evolutionary history of the closely related group 2 coronaviruses: porcine hemagglutinating encephalomyelitis virus, bovine coronavirus, and human coronavirus OC43. J Virol 80(14):7270–7274. https://doi.org/10.1128/JVI.02675-05

Wang B, Andraweera P, Elliott S, Mohammed H, Lassi Z, Twigger A, Borgas C, Gunasekera S, Ladhani S, Marshall HS (2023) Asymptomatic SARS-CoV-2 infection by age: a global systematic review and meta-analysis. Pediatr Infect Dis J 42(3):232–239. https://doi.org/10.1097/INF.0000000000003791

Williams BA, Jones CH, Welch V, True JM (2023) Outlook of pandemic preparedness in a post-COVID-19 world. NPJ Vaccines 8(1):178. https://doi.org/10.1038/s41541-023-00773-0

Woo PC, Lau SK, Chu CM, Chan KH, Tsoi HW, Huang Y, Wong BH, Poon RW, Cai JJ, Luk WK, Poon LL, Wong SS, Guan Y, Peiris JS, Yuen KY (2005) Characterization and complete genome sequence of a novel coronavirus, coronavirus HKU1, from patients with pneumonia. J Virol 79(2):884–895. https://doi.org/10.1128/JVI.79.2.884-895.2005

Wrapp D, Wang N, Corbett KS, Goldsmith JA, Hsieh CL, Abiona O, Graham BS, McLellan JS (2020) Cryo-EM structure of the 2019-nCoV spike in the prefusion conformation. Science 367(6483):1260–1263. https://doi.org/10.1126/science.abb2507

Ye Z, Harmon J, Ni W, Li Y, Wich D, Xu Q (2023) The mRNA vaccine revolution: COVID-19 has launched the future of vaccinology. ACS Nano 17(16):15231–15253. https://doi.org/10.1021/acsnano.2c12584

Zaki AM, van Boheemen S, Bestebroer TM, Osterhaus AD, Fouchier RA (2012) Isolation of a novel coronavirus from a man with pneumonia in Saudi Arabia. N Engl J Med 367(19):1814–1820. https://doi.org/10.1056/NEJMoa1211721

14

Respiratory Syncytial Virus: A Shape-Shifting Adversary

Keywords Chimpanzee coryza agent (CCA) • F protein • G protein • Syncytia • Vaccine Associated Enhanced Disease (VAED)

Abbreviations

CCA	Chimpanzee coryza agent
CDC	Centers for Disease Control
GSK	GlaxoSmithKline
HBsAg	Hepatitis B surface antigen
RSV	Respiratory syncytial virus
VAED	Vaccine Associated Enhanced Disease

The Silent Pandemic

Like human coronaviruses before SARS-CoV-2, respiratory syncytial virus (RSV) is another of those viruses causing common colds (Nam and Ison 2021). Found worldwide, this ubiquitous virus circulates continuously with cases peaking in the winter months. Typically, the virus produces a mild and self-limiting upper respiratory tract infection. The clinical presentation includes the usual cold symptoms of headache, slight fever, congested or runny nose, a dry cough, and sometimes a sore throat. Recovery is within a week or two, although throat irritation and coughing can persist longer. Once recovered there are no post-infection after effects and some immunity to reinfection is generated. While the immunity may not completely prevent subsequent infections it does reduce their severity. Being just an occasional nuisance for most people, RSV generally doesn't receive much public attention and lacks the name recognition of more serious viral diseases. However, this inconspicuous virus is not without clinical significance. Two population groups are particularly susceptible to more serious RSV infections, infants (<1 year of age) and the elderly (over 65 years of age). In both these cohorts, RSV can progress to serious and potentially fatal lower respiratory tract infections. In the United States, RSV is the leading cause of infant hospitalizations (Suh et al. 2022), and this trend extends globally with an estimated 3.2 million infant hospitalizations each year from RSV (Shi et al. 2017). Importantly, RSV causes approximately 100,000 infant deaths each year worldwide, with the heaviest burden in countries with limited healthcare capabilities (Cohen and Zar 2022). Similar to the clinical risk in infants, the elderly are also prone to more severe cases of RSV, and the risk is likely underestimated (McLaughlin et al. 2022). The Centers for Disease Control (CDC) estimates that there are 60,000–160,000 RSV hospitalizations of the elderly each year in the U.S. with 6000–10,000 deaths annually. It is the combined impact on these two vulnerable populations that spurred vaccine development for this virus, although it would take nearly 70 years after discovery to finally have a successful vaccine.

RSV was initially isolated in 1956 from chimpanzees with nasal inflammation (coryza), and the unknown viral pathogen was simply named chimpanzee coryza agent or CCA (Blount et al. 1956). During this study, one of the investigators working with both CCA in culture and with infected chimpanzees developed cold symptoms. This individual subsequently developed antibodies against CCA, indicating an exposure to the agent. The timing of the illness and the appearance of anti-CCA antibodies were consistent with CCA causing the disease symptoms but not definitive. It was possible that another virus caused the cold, and that the patient was coincidentally infected at roughly the same time with both the actual cold virus and CCA. Interestingly, a follow-up examination of random patients hospitalized for other reasons found that about 15% of them had antibodies that reacted with the CCA agent, suggesting that they had been infected with CCA or some closely related virus. While this still didn't prove that CCA caused any illness in humans it at least demonstrated that the agent was naturally circulating and fairly prevalent. Within a year of CCA's discovery, another group of investigators was trying to identify viruses associated with severe lower respiratory tract illness in infants (Chanock et al. 1957). An unknown viral agent was isolated from two unrelated patients, and extensive testing revealed that the two isolates were indistinguishable from each other and appeared identical to the CCA agent. This pathogen was unrelated to any known viruses and caused an unusual cytopathic effect in culture cells, the rapid formation of syncytia. As described for measles (Chap. 7), syncytia refer to fused cells. A cell infected with the CCA agent would fuse to neighboring uninfected cells resulting in giant cells that were easily discerned under the microscope. Surveying healthy populations for antibodies against CCA confirmed that infections with this virus were common among children, and the investigators suggested renaming CCA as the respiratory syncytial virus (RSV) to acknowledge its site of isolation and distinctive cytopathic feature (Chanock and Finberg 1957). But once again, showing that the virus infected humans and was in general circulation didn't prove that the virus caused disease. The dilemma was that respiratory infections are caused by many different viruses that all give similar symptoms. Unlike diseases such as measles or chickenpox which have unique and distinctive clinical features, this multitude of cold viruses can't be distinguished by clinical presentation. Even when RSV was isolated from sick children it might just be a harmless passenger with no clinical impact while another virus was producing the disease. Determining whether or not RSV was a pathogen would take the collective results of disparate types of studies.

One approach for ascribing a specific disease to a new virus is to show that the disease can be produced by infection of an experimental animal system. This was quickly tested by intranasally infecting seven common lab animal species (mice, rats, guinea pigs, minks, chinchillas, marmosets, and ferrets) with RSV (Coates and Chanock 1962). Most of the species showed evidence of infection through their development of anti-RSV antibodies along with syncytia and increased mucus in their nasal passages. Disappointingly, none of the tested species developed clinical signs of disease such as coughing, sneezing, or fevers. Without frank disease symptoms, the disease-causing ability of RSV remained uncertain, and human studies would be needed to solve this puzzle.

As virologic and cell culture techniques became more available to hospitals and clinicians in the 1950s, there was a growing application of these methods to address diseases where a specific pathogen hadn't yet been identified, including the myriad of respiratory illnesses. Several reports in the late 1950s into the early 1960s were able to isolate RSV in a significant number of children hospitalized for respiratory illness but not from patients without respiratory disease (Chanock et al. 1961; Beem et al. 1960; Hamparian et al. 1961). While adding support for RSV being the relevant pathogen, this was still not quite definitive proof that RSV was causing these children's illnesses as there could have been concurrent infection with another undetected pathogen. An important and fortuitous event that substantially strengthened the association between RSV and disease was a respiratory disease outbreak in 1960 at the

Junior Village facility in the District of Columbia (Kapikian et al. 1961). The Junior Village was an institution for homeless children that housed around 400 kids ranging from infants to 18 years of age. The institution was part of an ongoing study of infectious diseases in children, and the residents were continuously monitored for a variety of pathogens (Bell et al. 1961). During the spring of that year, 90 residents developed pneumonia, and they were immediately assessed for viruses and antibodies against both viral and bacterial organisms. The results of this assessment indicated that over 90% of the patients had evidence of RSV infection, either through direct isolation of the virus or by developing anti-RSV antibodies during the progression of the disease. The antibody data was particularly important because there was baseline data before the disease onset. Previous studies found RSV antibodies in many children with respiratory disease but couldn't determine whether those antibodies arose due to current disease or remained from some previous exposure. The Junior Village outbreak was able to establish that the anti-RSV antibodies first appeared during the disease and increased over the course of several weeks, the typical immune response to an infection.

As with the pediatric population, epidemiologic studies showed a high prevalence of RSV antibodies in adults with respiratory illnesses, consistent with RSV being a widespread infectious agent throughout life (Hamre and Procknow 1961; Johnson et al. 1962). The presence of RSV disease in adults suggested that childhood exposure did not provide lifelong immunity and that adult volunteers could be used as subjects to test RSV's pathogenic properties. In one convincing study, volunteers (males from 21 to 35 years old) were recruited from federal prisons and administered live RSV by spraying the inoculum into their noses and throats (Kravetz et al. 1961). Among the 41 infected volunteers, 20 men (49%) developed clinically recognizable colds, while no colds occurred in 26 control volunteers. Importantly, eight of the control patients had been inoculated with the RSV preparation that was first treated with RSV-neutralizing antibodies (the remaining controls were just given virus-free fluid). If some other virus in the preparation was causing the disease then the RSV antibodies should have no effect and roughly half of these 8 patients would have contracted colds. The absence of symptoms in these 8 controls confirmed that RSV was the active pathogen responsible for the cold symptoms in the 20 test patients. The lack of symptoms in the other 21 RSV-infected volunteers was likely due to preexisting immunity resulting from previous natural RSV infections. Also, in the 20 patients with colds, the symptoms and duration of the illness were less than observed in children, again likely due to these adults having partial immunity from previous exposure to RSV.

By the early 1960s, the cumulative results from these studies of children and adults conclusively established RSV as a common human pathogen causing respiratory illness. First exposures in infancy or early childhood were typically the most severe and could result in cases of pneumonia and other serious lower respiratory tract issues requiring hospitalizations. Upon recovery, immunity was either incomplete and/or waned significantly over time as individuals could contract the virus again and develop cold symptoms. Generally, these subsequent exposures tended to produce milder disease confined to the upper airway due to at least partial acquired immunity. So, RSV joined the growing cohort of respiratory viruses with "cold-causing" abilities. With the pathogen isolated and identified, and because RSV caused serious to life-threatening illness in many infants, attention quickly turned to vaccine development.

A Tragic Start

In the 1960s there were only two vaccine platforms available, live attenuated vaccines and inactivated vaccines. The methodology for producing attenuated strains of a virus by passaging it through cell culture and/or animals was well established, but there was one caveat to this approach for RSV. To assess the virulence of a potential vaccine strain there had to be a test animal where disease production could be evaluated. For RSV, no animal had been found where the wild-type virus caused measurable disease mimicking the human illness. Without an animal disease model, testing strains for attenuation was not possible, and the only alternative was to test strains directly in humans. One study in the late 1960s reported on this approach using an RSV strain that had been passaged 52 times in embryonic bovine kidney tissue, with the last 16 passages at 26 °C to select for a temperature-sensitive version of the virus (Parrott et al. 1970). To assess its safety, the final strain was administered to 45 young adults via the nose and mouth, and none developed disease symptoms. It was subsequently given to 34 children (2–13 years of age) and again none developed symptoms although most produced antibodies against RSV as expected for a vaccination. With these preliminary safety studies showing good results, the next step was to assess the vaccine in young infants. Infants and children under 2 years of age were the primary target audience for the vaccine as this was the cohort most susceptible to serious disease from RSV infection. In a small pilot study, the virus was next given to 5 infants ranging in age from 6 to 17 months, but the results were not encouraging. Two of 5 infants developed cold symptoms with fever and ear inflammations, and 1 of the 2 had more serious bronchitis. As the strain was not sufficiently attenuated for infants, further work with this strain was not pursued. Over the next decade, there were other attempts to develop an attenuated vaccine strain without success. Chimpanzees were eventually shown to be a reasonable animal model for RSV as they developed clinical cold symptoms similar to humans (Belshe et al. 1977). Multiple attenuated RSV strains were selected and tested in chimpanzees, but in humans, these strains failed to function adequately as a protective vaccine (Wright et al. 1982). A natural strain that was sufficiently attenuated so that it did not elicit disease in infants or young children yet induced protective immunity was never found, and the standard approach for attenuating viruses by passaging in cell culture was inadequate for RSV. Fast forward to the current era and precise molecular manipulation of the RSV genome has produced numerous modified RSV strains that are potential vaccine candidates (Topalidou et al. 2023). Many of these strains are in phase I or phase II trials, so a live, attenuated RSV vaccine may still be possible.

The difficulties associated with developing an attenuated strain led most investigators to focus instead on inactivated vaccine candidates. During the 1960s, four groups independently tested inactivated RSV vaccines, including once again Maurice Hilleman at Merck. Hilleman's group took an interesting approach of trying to combine individual vaccines for multiple respiratory pathogens into one combined vaccine. They hoped that a single vaccination series could protect against a panel of common childhood respiratory pathogens and reduce the overall rate of respiratory illness. In a series of studies over several years, they looked at a formalin-inactivated RSV preparation alone or combined with up to 6 other vaccines (Weibel et al. 1966, 1967; Woodhour et al. 1966; Sweet et al. 1966; Potash et al. 1966). Studies were conducted both with institutionalized children as well as nursery school and kindergarten children in the general community. In both groups of children, the different RSV vaccine preparations tested were safe and did not elicit disease symptoms or any significant adverse effects after intramuscular injection. However, while the efficacy results against some other pathogens were encouraging for inactivated vaccines, the data for RSV was less impressive. Children who already had antibodies against RSV generally showed only modest if any improvements in antibody levels after vaccination. For seronegative children (those without detectable antibodies to RSV), there was a better response, but still less than 50% of the vaccinees showed a four-fold increase in antibody titer, the usual standard for an effective immune response. More importantly, when the community vaccinated children were followed for disease during the 1964–1965 and 1965–1966 cold seasons there was no evidence that the vaccine reduced the incidence or severity of RSV infections (Weibel et al. 1966, 1967). Merck's attempt at an RSV vaccine was a failure, but more disturbing issues would arise in similar studies by other research groups.

In 1969, three groups, one from the University of Colorado (Fulginiti et al. 1969), one from the California Department of Public Health (Chin et al. 1969), and one from Washington, D.C. (Kapikian et al. 1969; Kim et al. 1969), reported the results of clinical trials with an inactivated RSV vaccine. The vaccines used in these three different studies were essentially the same and all were composed of formalin-treated RSV of the Bernett strain (a clinical sample originally isolated at NIH in 1961). The Colorado and California studies involved children of military personnel with ages ranging from 4 months to 9 years, and together these two studies administered the RSV vaccine to 683 children with a comparable number of matched control children, all volunteered by their parents. The Washington, D.C. study group was much smaller and younger, consisting of 31 vaccinated infants (ages 2–7 months) whose parents brought them to a Child Health Center in D.C. This cohort was predominantly black children from lower socioeconomic families. All three studies were devised as field trials to not only evaluate antibody responses but also to gauge the vaccine's protective efficacy against community-acquired RSV.

Like the Merck trials, all three studies found that the vaccination process was safe and tolerable. It did not induce RSV disease and invoked little in the way of localized or systemic responses to the vaccination. However, antibody responses were modest, and no protective effect was observed, with RSV case numbers essentially equal in the vaccinated and control groups. Unexpectedly, not only was the vaccine not protective, but each trial observed that RSV vaccination made the recipients more prone to serious disease when they contracted RSV infections, especially infants. Instead of helping these children, this vaccine was making things worse. RSV infection caused more hospitalizations in the vaccinated groups than the control groups and a higher incidence of severe symptoms such as bronchiolitis (inflammation of the small airways in the lungs) and pneumonia. For example, in the Colorado study, 13.7% of the vaccinated infants who contracted RSV were hospitalized while less than 1% of the RSV infections in the control children needed hospitalization (Fulginiti et al. 1969). The results were even worse in the small Washington, D.C. study. Out of the 31 vaccinated infants, 23 later developed RSV infections, 18 needed hospitalization for pneumonia and/or bronchiolitis, and tragically, 2 died (Kim et al. 1969). While this study had no specific control group, the hospitalization rate (18 out of 23 infections) was much higher than typical for RSV, consistent with the results from the Colorado study. Tragically, rather than protecting these children, the vaccine potentiated disease from subsequent infection with RSV. With a clearly established danger from this type of RSV vaccine and no scientific explanation for this phenomenon, hopes for an inactivated RSV vaccine abruptly ended. The increased virulence caused by the inactivated RSV vaccine would eventually be seen for several other vaccines and given the generic name Vaccine Associated Enhanced Disease (VAED) (Munoz et al. 2021). Scientists striving to prevent RSV disease and save lives must have been horrified and dismayed that their efforts harmed the test children instead of protecting them. Stymied by the lack of molecular tools to investigate this virus further, and unwilling to risk testing other potentially dangerous RSV vaccines, RSV vaccines were abandoned, and researchers moved on to tackle other virus vaccines. What was needed before revisiting RSV vaccines were details about the virion's physical properties, knowledge of how the virus infected cells, and an understanding of how the immune system responded to the virus. It would be 50 years before research on RSV biology finally revealed a concrete explanation for this heartbreaking vaccine failure and provided a path forward for a safe vaccine.

About F'ing Time

In the 1960s, very little was known about RSV's molecular features or its interaction with host cells during infection (Canchola et al. 1965). This was not unusual as very little was known about the specific genes or the protein components for any virus as the tools for cloning and sequencing had not yet been developed. While many viruses, including RSV, had been isolated and passaged in cell culture, their biochemical and genetic properties were largely unstudied. Viruses could be propagated in the lab, their gross pathological effects on cells or test animals could be evaluated, and their immunologic properties could be investigated, but beyond that, the science was limited. For some viruses, this was sufficient to develop effective vaccines, even if the molecular mechanisms by which they entered and disrupted cells were still mysterious. For other viruses, such as RSV, a much greater understanding of virus biology would be necessary before a rational approach to a vaccine could be developed. When the initial vaccine development work for RSV was conducted, only a few basic physical properties were known. Electron microscopy revealed that RSV virion was spherical (although filamentous and asymmetric forms would be identified later and it is now considered a filamentous virus (Sweet et al. 1966)), enveloped, and about 150 nanometers in diameter with surface projections. Crude biochemical analysis confirmed that the virion was composed of lipids, protein, and RNA. Based on these characteristics, RSV was classified as an RNA virus in the myxovirus family, a broad family that included measles, mumps, influenza, and several animal viruses. (Eventually, more detailed characterization confirmed that RSV was distinct from the myxoviruses and in 2016 it was reclassified into its own family called the pneumoviruses.) With only this limited structural knowledge, and without any understanding of how RSV functioned during an infection or how the immune system responded to this invader, the adverse response engendered by the vaccine was enigmatic and unsolvable.

As the 1970s and 1980s progressed, the explosion in molecular biology brought new techniques for deciphering viruses down to individual genes and proteins. Numerous studies utilized these powerful methods and began to dissect the RSV genome to identify each viral gene and its protein product. RSV mRNAs were cloned and sequenced, and their encoded proteins were revealed. Painstakingly the entire genome was pieced together to reveal an RNA of roughly 15,000 nucleotides that encoded 11 proteins (Collins 1991). Among these 11 proteins, the F and G proteins comprise the surface spikes protruding from the virion (Peeples and Levine 1979). As expected for surface proteins, both F and G were recognized by the immune system with the F protein being the primary target for neutralizing antibodies (Olmsted et al. 1986). The G protein functions in viral attachment to host cells and F mediates the fusion between the virion and the host membrane that allows the virus to enter the cell (McLellan et al. 2013b). F also induced fusion between infected and uninfected cells and was responsible for the syncytia observed in patients.

With the critical F protein in hand, it seemed that a subunit vaccine strategy would be feasible as was done for the hepatitis B surface protein (HBsAg—Chap. 8). Just express the cloned F gene and use the purified full-length F protein as the antigen in a vaccine. By the early 1990s, there was a flurry of studies using recombinant F protein to test this vaccine strategy. While the purified F protein produced no adverse effects and induced antibodies in both children (Tristram et al. 1993; Belshe et al. 1993) and the elderly (Paradiso et al. 1994; Falsey and Walsh 1996), ultimately there was little or no protective effect elicited by this type of RSV vaccine, so this platform failed. Once again the obvious approach was thwarted by a lack of understanding of the virus and the properties of the F protein, but at least the science was progressing and each small advance enabled additional research to tease apart the workings of this virus.

The F protein biology turned out to be surprisingly complex and it would take another 10 years to fully understand its secrets. Unlike some viral surface proteins, the presentation of F protein on the virion surface was not direct and instead involved intricate processing and assembly to generate the final surface form of the F protein (Battles and McLellan 2019). Inside an infected cell, RSV produces the mRNA for the F protein, and this mRNA is translated into the full-length protein called F_0. The F_0 form is 574 amino acids long and was the form used for the subunit vaccines. Unfortunately, this full-length protein is nonfunctional and is not the form present on the virion surface. To generate the surface form of F, several cleavage events must occur, mediated by cellular protease enzymes (Fig. 14.1a) (Zimmer et al. 2001; Gonzalez-Reyes

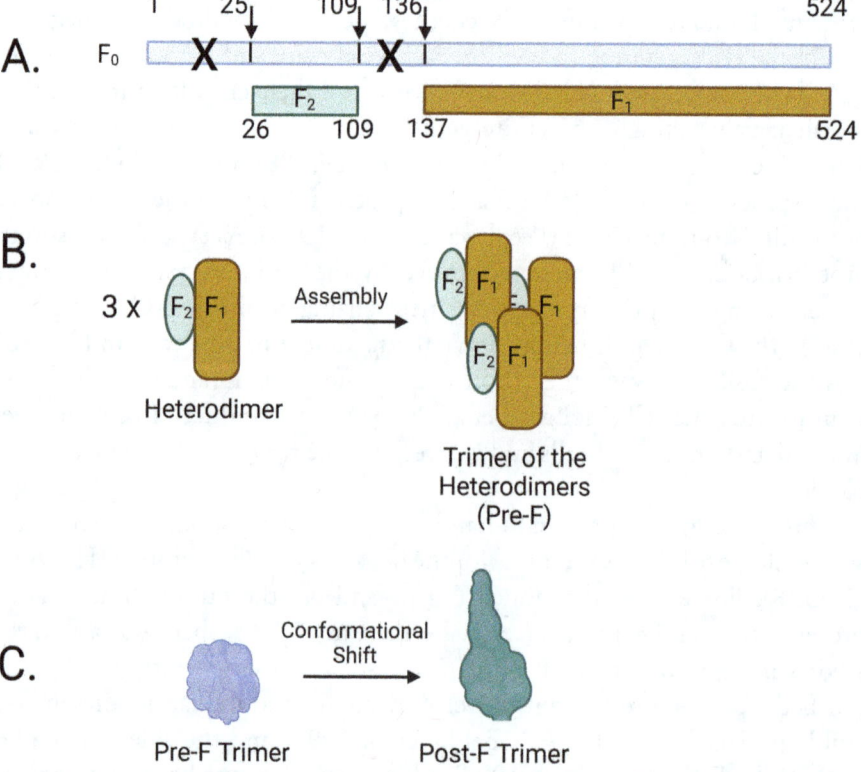

Fig. 14.1 RSV protein forms. (**a**) A linear representation of the full-length, 524 amino acid F_0 protein and the two cleavage products, F1 and F2. The numbers above F_0 indicate the protease cleavage sites and the **X**s mark the portions of F_0 that are removed by the cleavage events. The numbers on F1 and F2 indicate the beginning and end amino acids present in these two cleavage products. (**b**) F1 and F2 are covalently joined to form a heterodimer. Three heterodimers assemble to form the trimeric Pre-F protein. (**c**) The overall shape of the Pre-F trimer is roughly spherical as shown. This form is unstable and can spontaneously convert to the Post-F shape, a process that naturally occurs when the virion binds its receptor on the host cell

et al. 2001). Twenty-five amino acids are cleaved off one end of the protein (amino acids 1–25), and then more importantly, two internal cleavages remove 27 internal amino acids (amino acids 110–136), thus creating two smaller F proteins designated F_2 (amino acids 26–109) and F_1 (amino acids 137–524). Although F_0 is cleaved into separate smaller proteins, the F_1 and F_2 proteins remain chemically joined through covalent linkages known as disulfide bonds. This process converts the single F_0 protein into a dimer with two distinct subunits ($F_0 \rightarrow F_1$-F_2). In addition to this cleavage conversion process, three copies of each dimer bind together to assemble into a trimeric structure (F_1-F_2/F_1-F_2/F_1-F_2) that is displayed on the virion surface and is known as the prefusion F form (pre-F) (Fig. 14.1b). As the overall shape of the trimeric pre-F is drastically different than the shape of the monomeric F_0, antibodies generated against F_0 don't recognize pre-F effectively. Consequently, anti-F_0 antibodies aren't sufficiently neutralizing and don't prevent RSV infection, dooming the F_0 subunit vaccine to failure.

As this molecular characterization of the F protein and its subproducts clarified some aspects of how the RSV virion spike was composed, the structural biologists began to investigate the pre-F shape in precise detail. Structural biology is the field that attempts to solve the 3-dimensional shape of macromolecules (proteins, DNA, and RNA) down to the position of individual atoms. Knowing the exact shape of a large molecule at the atomic level can provide important insights into how the molecule functions and interacts with other molecules. These structural biologists use an array of methods from multiple types of electron microscopy to X-ray crystallography to solve structures in greater and greater detail. Using this tool kit, research beginning in 2000 began to reveal critical additional clues about the F protein. The first indication of another level of complexity was the determination that the trimeric F protein could fold into two drastically distinct shapes (Fig. 14.1c), one that was somewhat spherical and the other like an inverted lollipop (Calder et al. 2000). Subsequent studies from many labs resolved and explained this observation, demonstrating that there with two functional forms of the trimeric F protein complex on the virion surface, the prefusion form (pre-F) and a second form designated the post-fusion form (post-F). On the virion, the pre-F form predominates (Liljeroos et al. 2013) but during the attachment and entry of the virion into the host cell pre-F converts to the post-F form. Importantly, the pre-F is very unstable and easily transitions to the stable post-F form, and this is a permanent, irreversible change. The procedures used to isolate the F protein from the virions for biochemical and structural studies all triggered this transition, meaning that these studies were all examining only the post-F form. The key to characterizing the pre-F form was to trap it in this form using virion-neutralizing antibodies (McLellan et al. 2013a). These antibodies bound to the pre-F form and acted like a molecular scaffold that trapped and stabilized it, thus preventing its transition to the post-F form. Now the structure of the pre-F could be solved and compared to that of the post-F form.

With the pre-F form now available, two seminal advances were made that helped solve the confounding issues with RSV vaccines. First, the immune response to the pre-F and post-F proteins was compared, revealing that there were differences in the antigenic sites present on the pre-F and post-F forms (Gilman et al. 2016). Antigenic sites are the specific locations on a protein that are recognized and bound by antibodies. Pre-F contains six major antigenic sites, but two of these sites are lost when pre-F converts to the post-F form. Critically, these two lost sites turned out to be the most important binding sites for neutralizing antibodies. The absence of these two sites in post-F meant that antibodies raised against the post-F protein lacked effective neutralizing capacity against the pre-F form found on virions and would not prevent infection. Examination of the failed inactivated RSV vaccines determined that the inactivation process had converted the pre-F on the virions to the post-F form (Killikelly et al. 2016). Consequently, these vaccines were only inducing antibodies against post-F and not pre-F in the recipients. The inability of these vaccines to induce neutralizing anti-pre-F antibodies finally explained why they were ineffective and failed to protect the vaccinees from disease. (Why they caused enhanced virulence to subsequent RSV infection is still being investigated (Knudson et al. 2015).) Subsequent direct testing of a vaccine based on post-F confirmed that post-F did not generate protective immunity and was not suitable as a vaccine antigen for RSV (Falloon et al. 2017). This inadvertent conversion of all the pre-F forms to post-F during processing likely also explains the failure of the subunit vaccines consisting of purified F protein.

With pre-F being so unstable, any attempt to isolate or purify the active protein resulted in a post-F preparation that was useless for vaccines. The second major advance occurred in 2013 when researchers used structural biology to create stabilized pre-F for potential use in vaccines (McLellan et al. 2013a). Examining the atomic structure of pre-F, these scientists explored how changes in specific amino acids in the protein would make its shape more stable and prevent its transition to post-F. These predicted stabilizing changes were introduced by making mutations in the F gene. The mutant F proteins were expressed, and these newly created versions of the F protein were tested for maintenance of the pre-F form. This is akin to engineers examining a bridge design and toying with struts and supports to ensure greater strength and structural integrity. The end result was that these protein engineers produced a highly stable pre-F protein (designated DS-Cav1) that was locked into this desired conformation by the intentionally designed amino acid changes. This stable pre-F protein induced highly potent RSV-neutralizing antibodies in animals, moving the field to a threshold of a true RSV vaccine. After decades of painstaking biochemical, molecular, and structural analysis of RSV and its proteins, there was finally sufficient understanding to move forward with the rational design of an RSV vaccine.

Parenthetically, the structure-based protein stabilization methodology developed for RSV was also applicable to several other viruses, including SARS-CoV-2 (Chap. 13) (Hsieh et al. 2020). Like the F protein, the SARS-CoV-2 spike protein is an unstable, trimeric molecule on the virion surface. For use as a vaccine antigen, the spike protein needs to be stabilized in the form that elicits neutralizing antibodies. As COVID-19 erupted, the knowledge learned in engineering the RSV F protein was immediately applied to the spike protein by vaccine developers. The ability to quickly analyze and modify the spike protein structure ensured the rapid generation of a stable version of the spike protein for use in the various vaccine platforms. This was another illustration of why the scientific knowledge base was poised and able to respond so quickly to this new pathogen and produce SARS-CoV-2 vaccines in record time.

The Homestretch

With the know-how to build stabilized pre-F proteins and the proof that this form induced robust RSV-neutralizing antibodies; the vaccine industry responded by generating a large array of potential RSV vaccine candidates. By the early 2020s, there were around 25 candidates in development, most based on slightly different versions of stabilized pre-F or utilizing different delivery platforms. The platforms being investigated are wide-ranging and include live attenuated viruses, recombinant vectors, nanoparticles, mRNAs, and straightforward protein subunit vaccines (Topalidou et al. 2023). It was this last category that was first to market, with both GlaxoSmithKline (GSK) and Pfizer receiving FDA approval for their purified pre-F protein subunit vaccines in 2023. Both companies designed their version of pre-F starting with the prototype DS-Cav1 structure and introducing additional mutations to increase stability and immunogenicity. After successfully passing through phase I and phase II clinical trials, both vaccines performed extremely well in phase III trials. The GSK vaccine (RSVPreF3 OA; marketed as Arexvy) was tested in approximately 12,500 adult volunteers of 60 years of age or older with a matched control group (Papi et al. 2023). Participants were from 17 countries across Africa, Asia, Australia, Europe, and North America. Each received a single dose of the vaccine or a placebo and was followed for two or three RSV seasons. The vaccine was 83% effective in preventing RSV disease and 94% effective in preventing serious disease (requiring hospitalization) while having no evidence of significant adverse effects. Very similar results were observed with the Pfizer vaccine (RSVPreF; marketed as Abrysvo). Pfizer's version was a bivalent vaccine with stabilized pre-F protein from the two major strains of RSV, type A and type B. Tested in over 17,000 volunteers (with a similar number of controls), the Pfizer vaccine had an overall efficacy of 67% and 86% efficacy against serious disease, with no significant adverse effects detected. Based on these positive phase III trial results, both vaccines received FDA approval for use in the older adult population (60 years and older) and are now on the market. As with all new vaccines, there will be post-marketing evaluation to monitor for efficacy and safety in the broader community of recipients. For example, because of a slightly increased risk of Guillain-Barré syndrome in RSV vaccine recipients, in mid-2024 the CDC changed its recommendation for the vaccine. The new recommendation is that the vaccine should be used by people over 74 years of age or younger people with underlying conditions such as chronic respiratory disease, cardiac problems, or immune deficiencies.

The availability of vaccines to protect the elderly population from RSV is an important achievement, but a vaccine for the other vulnerable population, infants, has not been achieved. The tragic results from the 1960s precluded any further attempts to test RSV vaccines in infants for decades until the virus and its infectious properties were thoroughly understood. Even now with a solid grasp of how to manipulate the F protein for use as a vaccine antigen, there are still issues that prevent its immediate application to a pediatric vaccine. The immune response of infants is limited and different from that of older individuals in complex ways that may require the use of unique adjuvants or novel strategies to evoke a protective immune response to RSV and avoid vaccine-enhanced adverse reactions (van Haren et al. 2022). As an alternative to infant vaccination, both GSK and Pfizer explored maternal vaccination during pregnancy as a method for protecting newborns. This approach relied on the transfer of maternal antibodies to the developing fetus, especially during the last weeks of pregnancy. Neonates are born with circulating antibodies reflective of their mother's antibody repertoire. These maternal antibodies eventually degrade and are lost but offer substantial protection initially until the newborns begin to respond immunologically and manufacture their own antibodies. The proposed strategy was to vaccinate pregnant women before 36 weeks of gestation so that they would develop high titers of RSV-neutralizing antibodies. These RSV antibodies should pass to the fetus and be present at birth, thereby protecting the newborn for the critical first few months when RSV infections are most dangerous.

In parallel with their older adult studies, GSK and Pfizer tested their pre-F protein vaccines in pregnant women to evaluate the protection of the newborns. GSK stopped their study when they observed a small increase in the number of preterm births in the vaccine cohort (7% of births were preterm) compared to the control group (5% of births were preterm), with the difference occurring mostly in countries considered low or middle income (Topalidou et al. 2023). While not a proven consequence of the vaccine and possibly just a sample size anomaly, they elected to halt the study and no longer pursue this usage for Arexvy. In contrast, in a Pfizer study involving over 3000 pregnant women who received the Abrysvo vaccine, they concluded that the vaccine was safe even though there were 1% more preterm births in the vaccinated cohort (Kampmann et al. 2023). In both the GSK and Pfizer trials, the vaccines were highly effective, with the Pfizer study showing 82% efficacy in preventing serious lower respiratory tract infections in newborns. Based on the high efficacy of the vaccine and the large potential for reducing serious to life-threatening disease in infants, the FDA concluded that the benefits outweighed the possible risk of increased preterm births. In late 2023 the FDA approved Pfizer's Abrysvo for pregnant women with the recommendation that it only be used during weeks 32–36 rather than the 24–36 week period that Pfizer used in its phase III trial. Having only been in clinical use for a short while, additional data about the risks of this vaccine are limited. One subsequent study found no evidence for an increased risk of preterm births among vaccinated pregnant women (Son et al. 2024), but this prenatal usage of the vaccine will require much additional scrutiny for effects on pregnancy duration and any other possible complications.

As important as Abrysvo and Arexvy are to the fight against RSV infections, these are both just first-generation vaccines. Science has peeled open the complexities of RSV and provided the initial framework for protective vaccines, but many possibilities exist for improvement (Schaerlaekens et al. 2024). Both Abrysvo and Arexvy are protein subunit vaccines that primarily elicit a B-cell response to produce neutralizing antibodies. The duration of protection with these vaccines is not yet known but is likely limited since even natural infection with RSV doesn't provide lifelong immunity. This would necessitate revaccination at some regular interval which increases cost and decreases compliance. Other vaccine platforms such as attenuated RSV (with stabilized F protein) or recombinant virus vaccines that use innocuous viruses to deliver the RSV F protein may be better choices. These live virus type vaccines evoke both the B-cell and T-cell response and typically provide more robust and/or longer lasting protection, so there may be ways to develop a vaccine that provides permanent immunity. In addition to improved adult vaccines, finding better approaches for protecting infants is also a critical need. It will be several years before the true scope of the maternal vaccination campaign can be assessed so work on alternative approaches will continue. Hopefully, the next decade will solve the current challenges and present the public with safe, effective, and long-lasting vaccines to protect both seniors and infants from RSV disease.

References

Battles MB, McLellan JS (2019) Respiratory syncytial virus entry and how to block it. Nat Rev Microbiol 17(4):233–245. https://doi.org/10.1038/s41579-019-0149-x

Beem M, Wright FH, Hamre D, Egerer R, Oehme M (1960) Association of the chimpanzee coryza agent with acute respiratory disease in children. N Engl J Med 263:523–530. https://doi.org/10.1056/NEJM196009152631101

Bell JA, Cole RM, Chanock RM, Shvedoff RA, Huebner RJ, Mastrota FM, Rowe WP, Rosen L, Floyd TM (1961) Illness and microbial experiences of nursery children at junior village. Am J Hyg 74(3):267. https://doi.org/10.1093/oxfordjournals.aje.a120219

Belshe RB, Richardson LS, London WT, Sly DL, Lorfeld JH, Camargo E, Prevar DA, Chanock RM (1977) Experimental respiratory syncytial virus infection of four species of primates. J Med Virol 1(3):157–162. https://doi.org/10.1002/jmv.1890010302

Belshe RB, Anderson EL, Walsh EE (1993) Immunogenicity of purified F glycoprotein of respiratory syncytial virus: clinical and immune responses to subsequent natural infection in children. J Infect Dis 168(4):1024–1029. https://doi.org/10.1093/infdis/168.4.1024

Blount RE Jr, Morris JA, Savage RE (1956) Recovery of cytopathogenic agent from chimpanzees with coryza. Proc Soc Exp Biol Med 92(3):544–549. https://doi.org/10.3181/00379727-92-22538

Calder LJ, Gonzalez-Reyes L, Garcia-Barreno B, Wharton SA, Skehel JJ, Wiley DC, Melero JA (2000) Electron microscopy of the human respiratory syncytial virus fusion protein and complexes that it forms with monoclonal antibodies. Virology 271(1):122–131. https://doi.org/10.1006/viro.2000.0279

Canchola JG, Chanock RM, Jeffries BC, Christmas EE, Kim HW, Vargosko AJ, Parrott RH (1965) Recovery and identification of human myxoviruses. Bacteriol Rev 29(4):496–503. https://doi.org/10.1128/br.29.4.496-503.1965

Chanock R, Finberg L (1957) Recovery from infants with respiratory illness of a virus related to chimpanzee coryza agent (CCA). II. Epidemiologic aspects of infection in infants and young children. Am J Hyg 66(3):291–300. https://doi.org/10.1093/oxfordjournals.aje.a119902

Chanock R, Roizman B, Myers R (1957) Recovery from infants with respiratory illness of a virus related to chimpanzee coryza agent (CCA). I. Isolation, properties and characterization. Am J Hyg 66(3):281–290. https://doi.org/10.1093/oxfordjournals.aje.a119901

Chanock RM, Kim HW, Vargosko AJ, Deleva A, Johnson KM, Cumming C, Parrott RH (1961) Respiratory syncytial virus. I. Virus recovery and other observations during 1960 outbreak of bronchiolitis, pneumonia, and minor respiratory diseases in children. JAMA 176:647–653

Chin J, Magoffin RL, Shearer LA, Schieble JH, Lennette EH (1969) Field evaluation of a respiratory syncytial virus vaccine and a trivalent parainfluenza virus vaccine in a pediatric population. Am J Epidemiol 89(4):449–463. https://doi.org/10.1093/oxfordjournals.aje.a120957

Coates HV, Chanock RM (1962) Experimental infection with respiratory syncytial virus in several species of animals. Am J Hyg 76:302–312. https://doi.org/10.1093/oxfordjournals.aje.a120285

Cohen C, Zar HJ (2022) Deaths from RSV in young infants-the hidden community burden. Lancet Glob Health 10(2):e169–e170. https://doi.org/10.1016/S2214-109X(21)00558-1

Collins PL (1991) The molecular biology of human respiratory syncytial virus (RSV) of the genus pneumovirus. In: Kingsbury DW (ed) The paramyxoviruses. Springer US, Boston, pp 103–162. https://doi.org/10.1007/978-1-4615-3790-8_4

Falloon J, Yu J, Esser MT, Villafana T, Yu L, Dubovsky F, Takas T, Levin MJ, Falsey AR (2017) An adjuvanted, postfusion F protein-based vaccine did not prevent respiratory syncytial virus illness in older adults. J Infect Dis 216(11):1362–1370. https://doi.org/10.1093/infdis/jix503

Falsey AR, Walsh EE (1996) Safety and immunogenicity of a respiratory syncytial virus subunit vaccine (PFP-2) in ambulatory adults over age 60. Vaccine 14(13):1214–1218. https://doi.org/10.1016/s0264-410x(96)00030-8

Fulginiti VA, Eller JJ, Sieber OF, Joyner JW, Minamitani M, Meiklejohn G (1969) Respiratory virus immunization. I. A field trial of two inactivated respiratory virus vaccines; an aqueous trivalent parainfluenza virus vaccine and an alum-precipitated respiratory syncytial virus vaccine. Am J Epidemiol 89(4):435–448. https://doi.org/10.1093/oxfordjournals.aje.a120956

Gilman MS, Castellanos CA, Chen M, Ngwuta JO, Goodwin E, Moin SM, Mas V, Melero JA, Wright PF, Graham BS, McLellan JS, Walker LM (2016) Rapid profiling of RSV antibody repertoires from the memory B cells of naturally infected adult donors. Sci Immunol 1(6). https://doi.org/10.1126/sciimmunol.aaj1879

Gonzalez-Reyes L, Ruiz-Arguello MB, Garcia-Barreno B, Calder L, Lopez JA, Albar JP, Skehel JJ, Wiley DC, Melero JA (2001) Cleavage of the human respiratory syncytial virus fusion protein at two distinct sites is required for activation of membrane fusion. Proc Natl Acad Sci USA 98(17):9859–9864. https://doi.org/10.1073/pnas.151098198

Hamparian VV, Ketler A, Hilleman MR, Reilly CM, Mc CL, Cornfeld D, Stokes J Jr (1961) Studies of acute respiratory illnesses caused by respiratory syncytial virus. 1. Laboratory findings in 109 cases. Proc Soc Exp Biol Med 106:717–722. https://doi.org/10.3181/00379727-106-26452

Hamre D, Procknow JJ (1961) Viruses isolated from natural common colds in the U.S.A. Br Med J 2(5264):1382–1385. https://doi.org/10.1136/bmj.2.5264.1382

Hsieh CL, Goldsmith JA, Schaub JM, DiVenere AM, Kuo HC, Javanmardi K, Le KC, Wrapp D, Lee AG, Liu Y, Chou CW, Byrne PO, Hjorth CK, Johnson NV, Ludes-Meyers J, Nguyen AW, Park J, Wang N, Amengor D, Lavinder JJ, Ippolito GC, Maynard JA, Finkelstein IJ, McLellan JS (2020) Structure-based design of prefusion-stabilized SARS-CoV-2 spikes. Science 369(6510):1501–1505. https://doi.org/10.1126/science.abd0826

Johnson KM, Bloom HH, Mufson MA, Chanock RM (1962) Natural reinfection of adults by respiratory syncytial virus. Possible relation to mild upper respiratory disease. N Engl J Med 267:68–72. https://doi.org/10.1056/NEJM196207122670204

Kampmann B, Madhi SA, Munjal I, Simões EAF, Pahud BA, Llapur C, Baker J, Marc GP, Radley D, Shittu E, Glanternik J, Snaggs H, Baber J, Zachariah P, Barnabas SL, Fausett M, Adam T, Perreras N, Houten MAV, Kantele A, Huang L-M, Bont LJ, Otsuki T, Vargas SL, Gullam J, Tapiero B, Stein RT, Polack FP, Zar HJ, Staerke NB, Padilla MD, Richmond PC, Koury K, Schneider K, Kalinina EV, Cooper D, Jansen KU, Anderson AS, Swanson KA, Gruber WC, Gurtman A (2023) Bivalent prefusion F vaccine in pregnancy to prevent RSV illness in infants. N Engl J Med 388(16):1451–1464. https://doi.org/10.1056/NEJMoa2216480

Kapikian AZ, Bell JA, Mastrota FM, Johnson KM, Huebner RJ, Chanock RM (1961) An outbreak of febrile illness and pneumonia associated with respiratory syncytial virus infection. Am J Hyg 74:234–248. https://doi.org/10.1093/oxfordjournals.aje.a120216

Kapikian AZ, Mitchell RH, Chanock RM, Shvedoff RA, Stewart CE (1969) An epidemiologic study of altered clinical reactivity to respiratory syncytial (RS) virus infection in children previously vaccinated with an inactivated RS virus vaccine. Am J Epidemiol 89(4):405–421. https://doi.org/10.1093/oxfordjournals.aje.a120954

Killikelly AM, Kanekiyo M, Graham BS (2016) Pre-fusion F is absent on the surface of formalin-inactivated respiratory syncytial virus. Sci Rep 6:34108. https://doi.org/10.1038/srep34108

Kim HW, Canchola JG, Brandt CD, Pyles G, Chanock RM, Jensen K, Parrott RH (1969) Respiratory syncytial virus disease in infants despite prior administration of antigenic inactivated vaccine. Am J Epidemiol 89(4):422–434. https://doi.org/10.1093/oxfordjournals.aje.a120955

Knudson CJ, Hartwig SM, Meyerholz DK, Varga SM (2015) RSV vaccine-enhanced disease is orchestrated by the combined actions of distinct CD4 T cell subsets. PLoS Pathog 11(3):e1004757. https://doi.org/10.1371/journal.ppat.1004757

Kravetz HM, Knight V, Chanock RM, Morris JA, Johnson KM, Rifkind D, Utz JP (1961) Respiratory syncytial virus. III. Production of illness and clinical observations in adult volunteers. JAMA 176:657–663

Liljeroos L, Krzyzaniak MA, Helenius A, Butcher SJ (2013) Architecture of respiratory syncytial virus revealed by electron cryotomography. Proc Natl Acad Sci USA 110(27):11133–11138. https://doi.org/10.1073/pnas.1309070110

McLaughlin JM, Khan F, Begier E, Swerdlow DL, Jodar L, Falsey AR (2022) Rates of medically attended RSV among US adults: a systematic review and meta-analysis. Open Forum Infect Dis 9(7):ofac300. https://doi.org/10.1093/ofid/ofac300

McLellan JS, Chen M, Joyce MG, Sastry M, Stewart-Jones GB, Yang Y, Zhang B, Chen L, Srivatsan S, Zheng A, Zhou T, Graepel KW, Kumar A, Moin S, Boyington JC, Chuang GY, Soto C, Baxa U, Bakker AQ, Spits H, Beaumont T, Zheng Z, Xia N, Ko SY, Todd JP, Rao S, Graham BS, Kwong PD (2013a) Structure-based design of a fusion glycoprotein vaccine for respiratory syncytial virus. Science 342(6158):592–598. https://doi.org/10.1126/science.1243283

McLellan JS, Ray WC, Peeples ME (2013b) Structure and function of respiratory syncytial virus surface glycoproteins. Curr Top Microbiol Immunol 372:83–104. https://doi.org/10.1007/978-3-642-38919-1_4

Munoz FM, Cramer JP, Dekker CL, Dudley MZ, Graham BS, Gurwith M, Law B, Perlman S, Polack FP, Spergel JM, Van Braeckel E, Ward BJ, Didierlaurent AM, Lambert PH, Brighton Collaboration Vaccine-associated Enhanced Disease Working G (2021) Vaccine-associated enhanced disease: case definition and guidelines for data collection, analysis, and presentation of immunization safety data. Vaccine 39(22):3053–3066. https://doi.org/10.1016/j.vaccine.2021.01.055

Nam HH, Ison MG (2021) Respiratory syncytial virus. Semin Respir Crit Care Med 42(6):788–799. https://doi.org/10.1055/s-0041-1736182

Olmsted RA, Elango N, Prince GA, Murphy BR, Johnson PR, Moss B, Chanock RM, Collins PL (1986) Expression of the F-glycoprotein of respiratory syncytial virus by a recombinant vaccinia virus - comparison of the individual contributions of the F-glycoprotein and G-glycoprotein to host immunity. Proc Natl Acad Sci U S A 83(19):7462–7466. https://doi.org/10.1073/pnas.83.19.7462

Papi A, Ison MG, Langley JM, Lee DG, Leroux-Roels I, Martinon-Torres F, Schwarz TF, van Zyl-Smit RN, Campora L, Dezutter N, de Schrevel N, Fissette L, David MP, Van der Wielen M, Kostanyan L, Hulstrom V, Group AR-S (2023) Respiratory syncytial virus prefusion F protein vaccine in older adults. N Engl J Med 388(7):595–608. https://doi.org/10.1056/NEJMoa2209604

Paradiso PR, Hildreth SW, Hogerman DA, Speelman DJ, Lewin EB, Oren J, Smith DH (1994) Safety and immunogenicity of a subunit respiratory syncytial virus vaccine in children 24 to 48 months old. Pediatr Infect Dis J 13(9):792–798. https://doi.org/10.1097/00006454-199409000-00008

Parrott RH, Kim HW, Brandt CD, Chanock RM (1970) Vaccination against pediatric illness caused by parainfluenza virus, respiratory syncytial virus and certain adenoviruses. Current status and perspective. Med Ann Dist Columbia 39(11):612–614

Peeples M, Levine S (1979) Respiratory syncytial virus polypeptides: their location in the virion. Virology 95(1):137–145. https://doi.org/10.1016/0042-6822(79)90408-2

Potash L, Tytell AA, Sweet BH, Machlowitz RA, Stokes J Jr, Weibel RE, Woodhour AF, Hilleman MR (1966) Respiratory virus vaccines. I. Respiratory syncytial and parainfluenza virus vaccines. Am Rev Respir Dis 93(4):536–548. https://doi.org/10.1164/arrd.1966.93.4.536

Schaerlaekens S, Jacobs L, Stobbelaar K, Cos P, Delputte P (2024) All eyes on the prefusion-stabilized F construct, but are we missing the potential of alternative targets for respiratory syncytial virus vaccine design? Vaccines (Basel) 12(1). https://doi.org/10.3390/vaccines12010097

Shi T, McAllister DA, O'Brien KL, Simoes EAF, Madhi SA, Gessner BD, Polack FP, Balsells E, Acacio S, Aguayo C, Alassani I, Ali A, Antonio M, Awasthi S, Awori JO, Azziz-Baumgartner E, Baggett HC, Baillie VL, Balmaseda A, Barahona A, Basnet S, Bassat Q, Basualdo W, Bigogo G, Bont L, Breiman RF, Brooks WA, Broor S, Bruce N, Bruden D, Buchy P, Campbell S, Carosone-Link P, Chadha M, Chipeta J, Chou M, Clara W, Cohen C, de Cuellar E, Dang DA, Dash-Yandag B, Deloria-Knoll M, Dherani M, Eap T, Ebruke BE, Echavarria M, de Freitas Lazaro Emediato CC, Fasce RA, Feikin DR, Feng L, Gentile A, Gordon A, Goswami D, Goyet S, Groome M, Halasa N, Hirve S, Homaira N, Howie SRC, Jara J, Jroundi I, Kartasasmita CB, Khuri-Bulos N, Kotloff KL, Krishnan A, Libster R, Lopez O, Lucero MG, Lucion F, Lupisan SP, Marcone DN, McCracken JP, Mejia M, Moisi JC, Montgomery JM, Moore DP, Moraleda C, Moyes J, Munywoki P, Mutyara K, Nicol MP, Nokes DJ, Nymadawa P, da Costa Oliveira MT, Oshitani H, Pandey N, Paranhos-Baccala G, Phillips LN, Picot VS, Rahman M, Rakoto-Andrianarivelo M, Rasmussen ZA, Rath BA, Robinson A, Romero C, Russomando G, Salimi V, Sawatwong P, Scheltema N, Schweiger B, Scott JAG, Seidenberg P, Shen K, Singleton R, Sotomayor V, Strand TA, Sutanto A, Sylla M, Tapia MD, Thamthitiwat S, Thomas ED, Tokarz R, Turner C, Venter M, Waicharoen S, Wang J, Watthanaworawit W, Yoshida LM, Yu H, Zar HJ, Campbell H, Nair H, Network RSVGE (2017) Global, regional, and national disease burden estimates of acute lower respiratory infections due to respiratory syncytial virus in young children in 2015: a systematic review and modelling study. Lancet 390(10098):946–958. https://doi.org/10.1016/S0140-6736(17)30938-8

Son M, Riley LE, Staniczenko AP, Cron J, Yen S, Thomas C, Sholle E, Osborne LM, Lipkind HS (2024) Nonadjuvanted bivalent respiratory syncytial virus vaccination and perinatal outcomes. JAMA Netw Open 7(7):e2419268. https://doi.org/10.1001/jamanetworkopen.2024.19268

Suh M, Movva N, Jiang X, Bylsma LC, Reichert H, Fryzek JP, Nelson CB (2022) Respiratory syncytial virus is the leading cause of United States infant hospitalizations, 2009-2019: a study of the national (Nationwide) inpatient sample. J Infect Dis 226(Suppl 2):S154–S163. https://doi.org/10.1093/infdis/jiac120

Sweet BH, Tytell AA, Potash L, Weibel RE, Stokes J Jr, Drake ME, Woodhour AF, Hilleman MR (1966) Respiratory virus vaccines. 3. Pentavalent respiratory syncytial-parainfluenza-mycoplasma pneumoniae vaccine. Am Rev Respir Dis 94(3):340–349. https://doi.org/10.1164/arrd.1966.94.3.340

Topalidou X, Kalergis AM, Papazisis G (2023) Respiratory syncytial virus vaccines: a review of the candidates and the approved vaccines. Pathogens 12(10). https://doi.org/10.3390/pathogens12101259

Tristram DA, Welliver RC, Mohar CK, Hogerman DA, Hildreth SW, Paradiso P (1993) Immunogenicity and safety of respiratory syncytial virus subunit vaccine in seropositive children 18–36 months old. J Infect Dis 167(1):191–195. https://doi.org/10.1093/infdis/167.1.191

van Haren SD, Pedersen GK, Kumar A, Ruckwardt TJ, Moin S, Moore IN, Minai M, Liu M, Pak J, Borriello F, Doss-Gollin S, Beijnen EMS, Ahmed S, Helmel M, Andersen P, Graham BS, Steen H, Christensen D, Levy O (2022) CAF08 adjuvant enables single dose protection against respiratory syncytial virus infection in murine newborns. Nat Commun 13(1):4234. https://doi.org/10.1038/s41467-022-31709-2

Weibel RE, Stokes J Jr, Leagus MB, Mascoli CC, Hilleman MR (1966) Respiratory virus vaccines. V. Field evaluation for efficacy of heptavalent vaccine. Am Rev Respir Dis 94(3):362–379. https://doi.org/10.1164/arrd.1966.94.3.362

Weibel RE, Stokes J, Mascoli CC, Leagus MB, Woodhour AF, Tytell AA, Vella PP, Hilleman MR (1967) Respiratory virus vaccines .7. Field evaluation of respiratory syncytial parainfluenza 1 2 3 and mycoplasma pneumoniae vaccines 1965 to 1966. Am Rev Respir Dis 96(4):724-+

Woodhour AF, Sweet BH, Tytell AA, Potash L, Stokes J Jr, Weibel RE, Metzgar DP, Hilleman MR (1966) Respiratory virus vaccines. IV. Heptavalent respiratory syncytial-parainfluenza-mycoplasma-influenza vaccine in institutionalized persons. Am Rev Respir Dis 94(3):350–361. https://doi.org/10.1164/arrd.1966.94.3.350

Wright PF, Belshe RB, Kim HW, Van Voris LP, Chanock RM (1982) Administration of a highly attenuated, live respiratory syncytial virus vaccine to adults and children. Infect Immun 37(1):397–400. https://doi.org/10.1128/iai.37.1.397-400.1982

Zimmer G, Budz L, Herrler G (2001) Proteolytic activation of respiratory syncytial virus fusion protein. Cleavage at two furin consensus sequences. J Biol Chem 276(34):31642–31650. https://doi.org/10.1074/jbc.M102633200

15

The Fight Continues: Virus Without Vaccines

Keywords AIDS • Cytomegalovirus • Epstein-Barr virus • Exotic viruses • Hepatitis C virus • Herpes simplex virus • Human Immunodeficiency Virus (HIV)

Abbreviations

AIDS	Acquired Immunodeficiency Syndrome
bnAbs	Broadly neutralizing antibodies
CMV	Cytomegalovirus
EBV	Epstein-Barr virus
gB	Glycoprotein B
HAV	Hepatitis A virus
HBV	Hepatitis B virus
HCV	Hepatitis C virus
HIV	Human Immunodeficiency Virus
HSV	Herpes simplex virus
IM	Infectious mononucleosis
PC	Pentameric complex
SARS-CoV-2	Severe acute respiratory syndrome-Coronavirus −2
VZV	Varicella-zoster virus
WHO	World Health Organization

Three Centuries of Progress

We've come a long way since Jenner's primitive cowpox vaccine for smallpox. In the 1700s there was no concept of the microbial world and the pathogens that abound there. The nature of infectious diseases and how our bodies respond to protect us from these invisible marauders was a complete mystery. Yet curious men and women were beginning to make careful observations and develop rational explanations for the causes of illnesses. From these rudimentary attempts to understand disease grew the foundations of the scientific method: observe, hypothesize, test, and repeat. During the 1800s, advances in light microscopy and culture media fully revealed the microscopic world and allowed bacteria to be seen and grown by investigators. By the end of the 1800s, the brilliance of Pasteur, Koch, and their contemporaries disproved primitive ideas about infectious illnesses and established the causal relationship between specific microbes and diseases. The paradigm that each infectious disease was caused by a unique microbe was both revolutionary and transformative. No longer could sickness be ascribed to elusive and unsubstantial ill winds or bad humors. Instead, having a distinct, physical entity as the causative agent meant that pathogens could be isolated and studied to reveal their secrets. There was hope that understanding a microbe would lead to cures or preventative measures to combat its disease, and progress on this goal was being made on important illnesses such as anthrax, cholera, and rabies. It was a golden era of early microbiology as numerous bacteria were isolated, characterized, and named, giving order and structure to the previously unknown microscopic realm. But there was one final conundrum of that century, the filterable agents.

By the late 1800s, the Dutch biologist Martinus Beijerinck recognized that there were infectious agents that could not be seen by light microscopy, could not be grown on bacterial culture media, and would pass through filters that retained typical bacteria. Without knowing what they were, he named these filterable agents viruses to distinguish them from the larger and more experimentally tractable bacteria. For decades, this filterable characteristic was the only functional definition of this invisible class of pathogens until tools were finally developed in the twentieth century to reveal the structures and mechanisms of viruses. An early advance pioneered by Pasteur was the use of animals to propagate virus infectious activity. Samples from diseased patients could be injected into test animals to generate disease. Subsequent samples from an infected animal could be transferred to a second animal, and then from the second to a third, and so on. Even if it couldn't be seen directly, this sequential passaging confirmed the pathogenic properties of the tested virus and demonstrated that it must be replicating. While crude and having many practical limitations, these animal models nonetheless allowed the production of viral stocks for early vaccines. The next critical advance in the early to mid-1900s was the development of egg, tissue, and cell cultures as systems to grow and characterize viruses. Eggs were and remain an excellent and economical vehicle for growing many types of viruses and for producing massive stocks for vaccine production. Likewise, cell culture provided a scalable alternative media for virus production, especially for viruses that wouldn't reproduce in eggs. Equally, importantly, cell culture eventually allowed researchers to explore virus-cell interactions in a stable and reproducible system.

The implementation of whole animal and culture systems was a seminal advance that opened virology to experimental dissection. Before these surrogate hosts, viruses could only be studied by examining infected patients. While the history of virus research is replete with using human subjects as guinea pigs, these studies mostly helped define disease symptoms and modes of transmission. Isolating large quantities of virus from humans was often problematic, so generating enough material to study virion structure and function was difficult. Similarly, studying symptoms in humans offered little insight into the biology of viruses or the effects of viruses at the cellular level. In contrast, experiments on infected cells in culture would reveal many aspects of the viral life cycle, from attachment and entry to reproduction and exit. Furthermore, host animals and permissive cell cultures provided the propagation medium for developing attenuated viral strains for vaccines and for testing inactivated vaccines. In parallel with these advances in virology, our knowledge of human immune function expanded greatly throughout the 1900s. Our concept of antibodies went from an undefined substance in serum with the ability to block or neutralize viruses to specific immunoglobulin proteins whose shape and antigen-binding activity were well understood. By the mid-1900s the antibody-producing B cells and the cell-killing T cells were identified and the complementary roles of these two cell types in acquired immunity were gaining clarity. These advances ushered in the modern vaccine era that gave us early successes against some of the serious viral diseases of that era such as yellow fever (Chap. 4), influenza (Chap. 5), and polio (Chap. 6).

With the explosion of molecular biology in the 1960s and 1970s, there were wonderful new tools to sequence viral genomes, clone viral genes, express individual viral proteins, and interrogate viral infections at the cellular and molecular level. This molecular era began to resolve viruses down to the individual genes and the proteins they encoded. Scientists slowly pieced together the structures of whole virions along with the sequence, conformation, and function of each viral protein. This enormous assembly of knowledge is still incomplete and ongoing but has provided incredible insight into how different viruses attack and subvert our cells. With this molecular information now we can identify specific viral proteins that are bound by neutralizing antibodies and conceive novel vaccine strategies directed at these critical viral surface proteins. Viruses that were refractory to attenuation or inactivation were now amenable to these new subunit approaches that have given us vaccines against hepatitis B (Chap. 8), papillomaviruses (Chap. 12), and SARS-CoV-2 (Chap. 13).

As far as we've come in virology and immunology we have not yet achieved our goal: the complete conquest of viral disease. We've managed to control and reduce the incidence of many once-common viral diseases although the viruses are still in circulation and the diseases will rebound if vaccination levels decline. Ideally, through vaccination and herd immunity, we would eradicate a virus globally, like smallpox (Chap. 2), so that it no longer existed and would never plague mankind again. We are close to the eradication of poliovirus (Chap. 6) and the next decade may well see its final defeat. Several viruses in the previous chapters are unique to humans (e.g. measles, mumps, rubella—Chap. 7) so their extinction might be possible but that would require united global initiatives that are lacking so far. As health standards rise in countries around the planet, the possibility of attacking and eradicating some of these threats may materialize. Sadly, even though we have many good vaccines, certain viruses are much more problematic to eradicate. Viruses causing chronic infections that persist for life (e.g. hepatitis B (Chap. 8) and herpes varicella-zoster (Chap. 10)) are difficult because these persistently infected individuals are a source to reintroduce the virus. Once every man, woman, and child on the planet was vaccinated, we'd still have to wait decades until every last chronic carrier was gone before declaring the virus eradicated. Even more difficult are viruses that have animal reservoirs (influenza (Chap. 5) and SARS-CoV-2 (Chap. 13)). Viruses circulating in wild animals would be nearly impossible to eliminate as vaccinating these entire wildlife populations is a daunting task beyond our current capabilities. Regardless of the current limitations, better, cheaper, and more convenient vaccines may someday confront these difficult-to-eradicate viruses, and continuing scientific progress may ultimately provide the tools for a concerted effort to eliminate them from the world. In the meantime, maintaining high levels of vaccination across countries is the most effective strategy for keeping these diseases in check.

The Next Challenge

While our current viral vaccine arsenal is large, there are still dangerous viruses lacking an effective vaccine. Even with all the scientific advances over the last 100 years, we remain at the mercy of these uncontrolled viruses and significant challenges remain for vaccine developers. HIV, hepatitis C (HCV), and several herpesviruses are worldwide pathogens that are well-known and have been studied for years, but they have evasive properties that have impeded vaccine development. Each of these viruses can cause life-threatening illnesses, and except for HCV, there is no curative therapy, so preventative vaccines are desperately needed. Decades of failed vaccine attempts have been frustrating, although each experiment reveals new information and nuances about these pathogens that may someday provide the key to their defeat. Other viruses like Zika, Chikungunya, and dengue are mostly tropical diseases that have had little impact on the United States and Europe and have been somewhat neglected. This is shortsighted as each of these viruses is spread by mosquitoes and could potentially invade new geographical locations as did the African West Nile virus that is now endemic in the U.S. (Kilpatrick 2011). Even if these tropical viruses don't spread more widely, they continue to cause misery and death in their current locations and those affected populations deserve vaccines to protect the vulnerable. Fortunately, science never stops. Generation after generation of innovative and creative researchers continue to attack these viruses with new knowledge and new techniques. Many of these recalcitrant viruses are now being targeted by advanced vaccine methodologies and there is optimism that effective vaccines may be within reach.

Still others are viruses that have yet to be discovered. There is an enormous cadre of viruses lurking in animal populations that might someday spill over into humans to cause new diseases like COVID-19. Collectively, these zoonotic viruses present many known and unknown challenges for vaccine developers. It is probably inevitable that someday an unexpected animal virus will explode into human populations, causing another pandemic. This chapter will explore some of the known remaining viruses without vaccines and the issues they present. Hopefully, as science continues to move forward, many of the viruses in this chapter may someday be relegated to the category of "no longer a concern".

The Other Herpesviruses

Chapter 10 detailed the development and implementation of the vaccines for chickenpox and shingles. Both of these diseases are manifestations of infection with herpes varicella-zoster virus (VZV). Chickenpox is an acute disease that occurs upon first exposure to VZV while shingles (zoster) results from the reactivation of latent VZV that has been lurking in the patient since the initial infection. This property of establishing latent infections that persist for the lifetime of the infected individual is a feature of all herpesviruses. Every member of this family of viruses can cause acute disease upon first exposure and will then go into a latent state where the virus hides and remains in an inactive form. This latent form can be triggered to reactivate years to decades later to cause a second disease event. Besides VZV, the three most important human herpesviruses are herpes simplex virus (HSV), Epstein-Barr virus (EBV), and cytomegalovirus (CMV). These three herpesvirus infections can all cause significant disease and there are no preventative measures, cures, or means to remove these latent invaders. Vaccines that prevent the initial infection would be ideal if they can ever be developed, but the success with VZV hasn't transferred to the other herpesviruses.

Herpes simplex virus (HSV) is generally associated with painful, blistering skin lesions that occur orally or genitally (Zhu and Viejo-Borbolla 2021). There are two subtypes, type 1 and type 2, with type 1 predominating orally and type 2 genitally. For our purpose, we won't distinguish between the subtypes and will lump them together as HSV. The virus is spread primarily through direct skin-to-skin contact such as kissing and sexual activities. Upon exposure, skin lesions develop at the site of the initial infection and then the virus goes latent in the nerve cells (similar to VZV) that innervate that region of skin. It is estimated that 1.5 billion people are infected with HSV, roughly 19% of the world's population, and more young people are exposed each year as they enter the dating realm (James et al. 2020). Fortunately, in most individuals, the virus never reactivates and simply remains latent in the nerve cells. In other patients, viral reactivation can be triggered by different stimuli, and this may occur rarely or frequently depending on the individual. Upon reactivation, the virus travels back down the nerve cell to the original skin location and causes another round of painful blisters. Not only are these lesions exquisitely painful, but the blisters are full of viruses and are the source of infectious material that can be transmitted to other people. For individuals who suffer frequent recurrences, these lesions produce a great deal of physical and emotional suffering from the pain and the danger of passing the virus to others.

As troublesome as recurrent skin lesions can be, HSV can also cause even more dangerous infections. Transmitting the virus to one's eyes (touching a lesion then inadvertently rubbing the eyes) can lead to ocular herpes which is quite painful and can result in corneal damage leading to vision loss. An even more serious and potentially fatal manifestation of HSV infection is encephalitis. In rare cases, instead of remaining localized to the skin lesions, the virus spreads to the brain where it can cause permanent damage or death. While there are antiviral drugs to treat HSV infections, brain infections are still extremely grave and do not always respond to the available medications. Another particularly dire form of herpes infection is neonatal herpes simplex. This is typically acquired during a vaginal birth where the mother has active genital HSV, but infants can also be infected after birth if they are exposed to the virus (Pinninti and Kimberlin 2018). Neonates and young infants are particularly susceptible to the virus and often get generalized infections that spread throughout the body rather than remaining localized skin infections. This disseminated infection can cause multiorgan damage, including in the brain, often leading to death or severe injuries. Again, aggressive antiviral drug treatment can ameliorate the disease although fatalities and permanent brain damage still occur at a high frequency.

Preventing HSV infections and the subsequent poor clinical outcomes with a vaccine is a long-sought but not yet achieved goal (Krishnan and Stuart 2021). Potentially, two types of vaccines might be useful. A prophylactic vaccine would prevent the initial infection (and subsequent latency) as is typical for most vaccines while a therapeutic vaccine would prevent reactivation in individuals already infected (like the shingle vaccine). However, the properties of the virus create complications for either approach. HSV has a large and complex virion with 12 surface proteins and numerous internal proteins, so which protein or combination of proteins would make the most effective target for preventing infection or reactivation is still not resolved. Additionally, HSV has numerous mechanisms for counteracting and evading the host immune responses, so trying to develop a vaccine that creates an immune response resistant to HSV's weapons has been challenging.

Nonetheless, over the last 5 years, numerous clinical trials for HSV vaccines have been conducted or are in progress. With no obvious ideal approach for an HSV vaccine, investigators are exploring multiple options including live attenuated vaccines, protein subunit vaccines, recombinant vaccines, and mRNA vaccines. Several of these candidate vaccines have shown promise in animal models or limited human testing, so there is hope that the next decade could bring an HSV vaccine to market at last.

As prevalent as HSV is globally, the Epstein-Barr virus (EBV; named after its discoverers, Drs. Epstein and Barr) is even more pervasive with over 90% of the world's population infected (Smatti et al. 2018). It is spread via saliva and is commonly transmitted by mothers to their infants. Early-life infections are mostly asymptomatic although the virus does establish lifelong permanent residency in B cells. Acute illness from EBV results from initial infection in adolescence and beyond. At these older ages, the virus is usually transmitted by intimate behaviors such as kissing and can produce infectious mononucleosis (IM, often called the kissing disease). IM presents as fatigue, fever, sore throat, general malaise, and swollen lymph glands in the neck and/or armpits (Yu and Robertson 2023). The disease is generally self-limiting and resolves without treatment after 4–6 weeks, though some patients suffer fatigue for months afterward. As for early-life infections, infections at older ages still lead to permanent maintenance of the virus in B cells. While the results of an initial infection are relatively innocuous (ranging from no symptoms to IM) numerous post-infection sequelae are extremely dangerous. These complications include chronic fatigue syndrome, multiple types of cancer (Burkitt's lymphoma, Hodgkin's lymphoma, nasopharyngeal carcinoma, and some T cell lymphomas), and several autoimmune diseases (lupus, rheumatoid arthritis, and multiple sclerosis) (Houen and Trier 2020; Soldan and Lieberman 2023; Patel et al. 2022; Ruiz-Pablos et al. 2021). As with all members of the herpesvirus family, EBV is a large and complex virus that expresses dozens of proteins and viral RNAs that can negatively impact the infected host over decades of viral persistence. Stopping the spread of the virus to generation after generation with an effective vaccine would reduce the incidence of the many associated conditions and would have enormous health benefits (Zhong et al. 2022). Additionally, even if initial infections cannot be dramatically reduced and the latent virus cannot be eliminated, a vaccine that prevented the post-infection consequences would be a huge boon clinically.

Unlike HSV, EBV vaccine work has focused almost exclusively on subunit vaccines due to the properties of the virus. EBV is very difficult to grow in culture and achieve the large quantities needed for an inactivated vaccine, so this approach hasn't been feasible. Likewise, the growth difficulties, coupled with its inherent cancer-causing ability, have made the development of a live, attenuated vaccine problematic. Consequently, research has focused on subunit vaccines that could induce neutralizing antibodies against the virion. Unfortunately, similar to HSV, EBV has numerous surface proteins that all play a role in binding to and penetrating host cells. Identifying which protein or combination of proteins would be the best target for a vaccine and finding the right delivery system has been challenging. At least ten different viral proteins have been tested using platforms including direct injection of purified protein, protein delivery by nanoparticles or VLPs, or intracellular protein expression from viral vectors or mRNA (Zhong et al. 2022). Sadly, none of the clinical trials completed so far have shown efficacy in preventing EBV infections, and the evaluation of the protective effects of candidate vaccines against EBV sequelae is still in progress. While the results so far have been disappointing, work continues towards a more detailed and complete understanding of EBV's properties and its interplay with the host immune system. New approaches and new vaccine candidates are in development and optimism remains high that this ubiquitous virus can one day be tamed through scientific advances.

The last of the major herpesviruses lacking a vaccine is cytomegalovirus (CMV). Like EBV, CMV is incredibly ubiquitous with 70–90% of the world's population infected, and nearly 100% of adults in some countries (Shang and Li 2024). Initial infections are often asymptomatic, yet the virus inevitably establishes latency and remains with the infected individual for life. Unfortunately, the primary infection doesn't confer protective immunity, so both viral reactivation and subsequent reinfection with other CMV strains can occur. In immunocompetent people, neither reactivation nor subsequent infections are generally problematic, but there are two groups at particular risk: The immunocompromised, especially transplant patients, and infants in utero (congenital infections). For transplant recipients, CMV infections can cause serious disease and/or transplant rejection. Antiviral therapy has reduced the CMV risk and made this problem manageable although a vaccine would be useful for this group and candidate vaccines are still being investigated. For congenital infections, mothers who silently reactivate during pregnancy or who develop asymptomatic new CMV infections can unknowingly pass the virus to the developing child. These CMV infections are the most common congenital infections in the world, and in the U.S. alone there are up to 40,000 cases annually (Nishikawa and Sanchez 2024). While the majority of infected newborns will appear normal and healthy, more than half of this cohort will have clinical symptoms in their bloodwork or imaging studies, though most of these issues will resolve with no lasting effects. In contrast, 10–15% of newborns infected in utero will have more serious and permanent complications from the infection. This can include low birth weight, microencephaly, neurological impairments (cerebral palsy, cognitive deficits), eye disorders, and most commonly hearing loss; CMV accounts for roughly 25% of all cases of hearing loss in children under 4 years of age (Goderis et al.

2014). Given the prevalence of CMV and its potential for serious effects on children, what is badly needed is a vaccine that protects women and prevents them from transmitting the virus to their babies during pregnancy.

As for EBV and HSV, CMV has a large and complex virion with numerous virion surface proteins that function in attachment to and penetration of the host cell. The complexity of this process means that no simple vaccine targeting a single CMV protein is likely to be successful. Similarly, the immune response to this virus is diverse and multifaceted with the role of various humoral (antibodies) and cellular (T cell) responses still not completely understood. These challenges have stymied the field and prevented successful vaccine development for 50 years. Initial vaccine attempts in the 1970s using attenuated strains of CMV failed to show protection from community-acquired infections and these early vaccines were eventually abandoned (Plotkin et al. 1975; Elek and Stern 1974). In retrospect, this result is not surprising since if wild-type infections fail to provide lasting immunity then expecting an attenuated version of the virus to elicit better immunity may be overly optimistic, although modifications of this approach are still being tested (Das et al. 2023). Over the next 50 years, numerous different strategies and viral targets were tested as vaccine candidates with only limited success (Plotkin 2015). Some elicited decent immune responses but none met the criteria for a successful vaccine. Nonetheless, each attempt provides new information and our understanding of this virus and its interaction with humans slowly progressed. Among the positive advances was the identification of two important viral surface proteins that are targets for our immune responses, glycoprotein B (gB) and the pentameric complex (PC; a complex of five CMV proteins). PC binds receptors on epithelial and endothelial cells to attach the virus and gB is the primary viral fusion protein that triggers viral entry into the host cell. Candidate vaccines against either or both of these proteins elicit some degree of protection, but so far nothing that is sufficiently protective or long-lasting enough to be an effective vaccine for reducing community-acquired CMV infections. Still, the quest continues and new approaches continue to be developed, including a recent mRNA vaccine expressing both gB and PC that evokes a very potent immune response (Hu et al. 2024). With expanding vaccine platforms and increasing insight into CMV biology and our immune response to this virus, there is reason for optimism that a long-awaited CMV vaccine will come to the market in the not-too-distant future.

The Other Hepatitis Virus

Effective vaccines have greatly diminished the impact of hepatitis A virus (HAV; Chap. 9) and hepatitis B (HBV; Chap. 8), but the third major hepatitis virus, hepatitis C virus (HCV), has remained refractory to vaccine development. HCV is an RNA virus and a member of the Flavivirus family (as is the yellow fever virus—Chap. 4), putting it in a different family from either HAV (picornavirus family) or HBV (hepadnavirus family). However, similar to HAV and HBV, HCV targets the liver hepatocytes for infection and can cause acute hepatitis. More importantly, like HBV, HCV also can establish chronic infections, with up to 80% of HCV-infected individuals developing life-long infections, putting these carriers at high risk for cirrhosis and/or liver cancer (Fiehn et al. 2024). Transmitted via blood, bodily fluids, and sexual activity, HCV is a global problem with an estimated 58 million chronic carriers worldwide. While highly effective anti-HCV drugs with a cure rate of 90–95% now exist, reducing and keeping down the HCV disease burden would be greatly aided by an effective vaccine.

Since its discovery in 1989 (a Nobel Prize-winning achievement), HCV has confounded vaccine attempts due to several intrinsic features. First, HCV cannot be grown in cell culture like typical viruses, precluding relatively easy investigation of its biology. Second, there is no natural animal model system replicating human disease except chimpanzees, and these animals are now rarely used in disease research due to ethical concerns. Together, the inability to grow the virus or study its disease mechanisms except in humans has severely slowed our understanding of the virus and our response to infections. A third critical factor is the large genetic diversity of this virus. Like the influenza virus (Chap. 5), the genome replicative enzyme of HCV is highly error-prone, leading to considerable mutational heterogeneity in the viral population (Abbasi 2022). Over time, this led to the establishment of eight distinct genotypes (variants differing by 30% or more in the genome sequences) worldwide which are classified into roughly 100 subtypes (varying by 15–25% in genome sequence). This genome variation translates into differences in the sequences of the viral surface proteins that are the targets for neutralizing antibodies. Consequently, an effective vaccine would have to somehow induce neutralizing antibodies capable of recognizing all these protein variants. Furthermore, within any chronically infected individual, there can be additional extensive mutational variation due to the replication errors. These diverse cohorts of viruses are termed quasispecies and functionally they generate even more individual amino acid variety in the viral surface proteins (Tsukiyama-Kohara and

Kohara 2017). As the person develops antibodies that neutralize the majority of virions, these antibodies may not recognize some of the variants. These so-called escape mutants that avoid the immune system will then replicate and help maintain the viral population. Eventually, antibodies will develop against the original escape mutants, but by this time new escape mutants will have arisen. As this process is repeated continuously the virus persists and the chronic infection is not resolved. On top of the mutational heterogeneity, some of the important surface proteins have structural flexibility that makes it hard for antibodies to find and bind their target sequences on these proteins, adding another viral mechanism for evading the host immune response. Lastly, antibodies alone may not even be sufficient to prevent infections and a vaccine may need to induce a cellular immune response (T cells) as well as neutralizing antibodies.

All of the above issues have made the development of an HCV vaccine perplexing, frustrating, and so far unsuccessful, though not from lack of trying (Adugna 2023). Multiple vaccine platforms have been tested and a variety of viral proteins have been targeted. Unfortunately, most attempts have failed to establish immune responses sufficient to justify moving forward with clinical trials. While there is no clear path forward for this vaccine, research groups around the world continue to explore novel and creative approaches to HCV vaccines (Rzymski et al. 2024). Every new nuance of understanding about this virus and the host response to HCV infections brings the opportunity for the breakthrough that will finally unlock the secret to an effective vaccine and help stop the yearly toll of illness and deaths from this cryptic virus.

A Persistent STD—Human Immunodeficiency Virus (HIV)

If the HCV vaccine is a challenge, the human immunodeficiency virus (HIV) vaccine is the grand challenge. While around 30% of people infected with HCV develop a clearing immunity and become virus-free, essentially 100% of HIV-infected individuals become chronically infected. Our inability to develop a natural immunity to defeat HIV poses a significant hurdle to vaccine development, and the quest for a protective HIV vaccine that blocks infection and the subsequent acquired immunodeficiency syndrome (AIDS) is 40 years of frustration and some glimmers of hope at last. When the AIDS epidemic exploded in the early 1980s, the victims suffered from an immune system collapse that made them highly susceptible to cancers and infections, including infections by normally harmless organisms. There were multiple hypotheses proposed about the underlying cause such as drug use, lifestyle factors, and environmental toxins. When the trigger for AIDS was finally shown to be a sexually transmitted virus that precipitated the immune system decline, there was initial optimism that cures or preventative vaccines could be developed quickly. Unfortunately, HIV is a unique type of virus known as a retrovirus. While retroviruses were first identified in the early 1900s and had been studied for many years (Rubin 2011), they had proven very intractable to vaccine development.

Retroviruses have a particularly pernicious lifecycle where the viral genome inserts itself into the host chromosome and becomes a permanent part of the host genome; the integrated viral genome is called a provirus. For HIV, the process starts with the infection of the host $CD4^+$ T cell, a critical immune cell known as a helper T cell. These cells are critical for initiating and regulating the immune response to infectious agents. The incoming viral genome is a single-stranded RNA molecule, but once inside the cell, a viral enzyme carried in the capsid converts the viral RNA genome into a double-stranded DNA. This double-stranded DNA traverses to the nucleus and is inserted randomly into the host chromosomal DNA to create the provirus. The viral genes in the proviral DNA are transcribed by the host cell RNA polymerase and the resultant mRNAs are translated into viral proteins by the host ribosomes. Newly produced viral genome length RNAs and the viral proteins are assembled into progeny virions that bud from the cell without directly killing the cell. Consequently, the cell becomes a permanent viral factory as long as it lives and there is no way to remove the provirus from the cells. Over years of infection, the stress of viral infection slowly kills and depletes the $CD4^+$ cell population by multiple mechanisms leading to immunodeficiency (Cummins and Badley 2014). We now have an extensive repertoire of antiviral drugs that block the production of virions from the provirus, stopping the further spread of the virus and preventing the slow decline of the $CD4^+$ population, but none of them "cure" the patient. Even after years of therapy, if the drugs are stopped the proviruses present in long-lived $CD4^+$ begin making virions again, and the infection proceeds. Now there are even drugs that can prevent an HIV infection but they need to be taken regularly so access, cost, and compliance can be problematic which makes them less than an ideal solution for the general public (Rubin 2011). Developing a vaccine to protect uninfected individuals and stop the transmission cycle remains our best hope of controlling and possibly defeating this virus.

The first attempts at an HIV vaccine in 2005 and 2006 tried a standard approach of inducing antibodies against an HIV surface protein known as gp120 (Pitisuttithum et al. 2006; Flynn et al. 2005). While antibodies were induced, clinical trials showed no evidence of protection from HIV infection so that simplistic approach failed. Over the next 10 years, a variety of more sophisticated methodologies were developed and tested, including using combinations of multiple HIV proteins as the immunogen, inducing T-cell responses, and using different delivery systems and adjuvants to enhance the immune response (Walker 2016). None were successful with the highest reported protective efficacy at around 31%, a rate far too low to be suitable for a general vaccine against this highly devastating virus (Rerks-Ngarm et al. 2009).

One of the confounding issues with HIV is that like HCV it mutates rapidly and generates enormous genetic diversity within both the infected individual and the global population of chronic carriers. This genetic diversity translates into widely different amino acids present in the viral surface proteins that are the targets for neutralizing antibodies. By the time antibodies develop that can bind the antigenic sites on the surface proteins and neutralize the initial virions, new virions with differences in their antigenic sites have arisen. These new virions are not recognized by the initial antibodies, so they escape neutralization and continue to propagate the infection. This process is continuous, so the virus remains constantly ahead of the antibody response and is never cleared like a typical virus. The one glimmer of hope is the discovery that some individuals do develop broadly neutralizing antibodies (bnAbs) that can neutralize many of the viral variants (Spencer et al. 2021). If these bnAbs could be induced by a vaccine they might provide broad enough protection to thwart initial HIV infections and be the long-sought defense against this virus. The recent implementation of a strategy to induce bnAbs showed excellent promise in a humanized mouse model and will likely be moved into human testing soon (Xie et al. 2024; Wang et al. 2024). It is still too early to assess whether or not bnAbs will lead to a safe and effective HIV vaccine, but many other approaches are also being developed and tested. Each new study or trial gives us new information and leads us closer and closer to the goal of a protective HIV vaccine. The next 10 years should see dramatic advances and we may finally have the elusive vaccine to end the AIDS scourge.

The Tropical Diseases

The viruses covered so far in this book have a fairly global distribution. From the Americas to Asia and from Europe to Africa, these common viruses infect populations in almost every country and cause well-known diseases. Because of their ubiquitous nature, these common viruses have been well-studied and targeted throughout the twentieth century for vaccine development. In contrast, there are many tropical viruses that those of us living in temperate climate zones are only vaguely familiar with. Some have restricted geographical locations and may only be found in Africa (e.g. Lassa Fever virus and Ebola virus) or Asia (e.g. Nipah virus) while others such as Zika virus are more broadly distributed and are found throughout the tropics in South America, Africa, and Asia. Collectively these tropical viruses cause tens of thousands to millions of infections and deaths each but have often been ignored by the major vaccine-producing countries. This is short-sighted as many of these viruses are spread via mosquitoes and other biting insects. As global commerce and travel spread infected individuals or insect vectors to other countries these exotic diseases can establish themselves in new regions if the foreign vector can find a foothold or if there are local insects that can serve as a host for the virus. This is exactly what happened with West Nile Virus which was originally endemic to the African continent but is now well-established in mosquitoes and birds in North America and Europe (Kramer et al. 2019). Likewise, cases of Zika, monkeypox (renamed mpox in 2022 by the WHO), and chikungunya virus have been found in several temperate climate nations and have a least some potential to become endemic in these countries. The potential for even more widespread distribution of these viruses to high-population areas demands that more research efforts must be directed at finding effective vaccines to protect the world's population. Additionally, the tropical areas also likely contain a vast repertoire of unknown animal viruses that could jump into human populations and cause new zoonotic epidemics, so continued identification and study of novel viruses is essential preparation for future public health challenges.

While the existence of all these uncontrolled viruses is troubling, there is some positive news on the horizon. Redoubled efforts and significant scientific advances have promising vaccines in development for several of the most important exotic viruses (Vuitika et al. 2022). There are two licensed vaccines for Ebola and several more are in development as well as two vaccines for mpox. Dengue virus has a vaccine, but adverse effect issues have limited its use so alternative versions are in phase III testing. Likewise, Zika, Nipah, Lassa, and chikungunya viruses have multiple vaccine candidates in clinical trials using different platforms. Even where the current vaccines have limitations or issues, they are useful measures to help prevent disease in endemic areas until second and third-generation vaccines are created against these viral agents. Fortunately, we are no longer limited to just inactivated or attenuated vaccines. The availability of diverse platform approaches including subunits,

VLPs, heterologous viral vectors, nanoparticles, DNA, and mRNA has greatly enhanced our ability to generate better and safer vaccines. The next 10 years should see an explosion of new life-saving vaccines as novel approaches, new technologies, and a better understanding of immune responses propel vaccine science forward. Even currently intractable viruses may succumb to vaccines and be controllable, providing a boon to the millions of people worldwide who are now at risk for these horrid diseases.

References

Abbasi J (2022) RSV vaccines, finally within reach, could prevent tens of thousands of yearly deaths. JAMA-J Am Med Assoc 327(3):204–206. https://doi.org/10.1001/jama.2021.23772

Adugna A (2023) Therapeutic strategies and promising vaccine for hepatitis C virus infection. Immun Inflamm Dis 11(8):e977. https://doi.org/10.1002/iid3.977

Cummins NW, Badley AD (2014) Making sense of how HIV kills infected CD4 T cells: implications for HIV cure. Mol Cell Ther 2:20. https://doi.org/10.1186/2052-8426-2-20

Das R, Blazquez-Gamero D, Bernstein DI, Gantt S, Bautista O, Beck K, Conlon A, Rosenbloom DIS, Wang D, Ritter M, Arnold B, Annunziato P, Russell KL, group Vs (2023) Safety, efficacy, and immunogenicity of a replication-defective human cytomegalovirus vaccine, V160, in cytomegalovirus-seronegative women: a double-blind, randomised, placebo-controlled, phase 2b trial. Lancet Infect Dis 23(12):1383–1394. https://doi.org/10.1016/S1473-3099(23)00343-2

Elek SD, Stern H (1974) Development of a vaccine against mental retardation caused by cytomegalovirus infection in utero. Lancet 1(7845):1–5. https://doi.org/10.1016/s0140-6736(74)92997-3

Fiehn F, Beisel C, Binder M (2024) Hepatitis C virus and hepatocellular carcinoma: carcinogenesis in the era of direct-acting antivirals. Curr Opin Virol 67:101423. https://doi.org/10.1016/j.coviro.2024.101423

Flynn NM, Forthal DN, Harro CD, Judson FN, Mayer KH, Para MF, rgp HIVVSG (2005) Placebo-controlled phase 3 trial of a recombinant glycoprotein 120 vaccine to prevent HIV-1 infection. J Infect Dis 191(5):654–665. https://doi.org/10.1086/428404

Goderis J, De Leenheer E, Smets K, Van Hoecke H, Keymeulen A, Dhooge I (2014) Hearing loss and congenital CMV infection: a systematic review. Pediatrics 134(5):972–982. https://doi.org/10.1542/peds.2014-1173

Houen G, Trier NH (2020) Epstein-Barr virus and systemic autoimmune diseases. Front Immunol 11:587380. https://doi.org/10.3389/fimmu.2020.587380

Hu X, Karthigeyan KP, Herbek S, Valencia SM, Jenks JA, Webster H, Miller IG, Connors M, Pollara J, Andy C, Gerber LM, Walter EB, Edwards KM, Bernstein

DI, Hou J, Koch M, Panther L, Carfi A, Wu K, Permar SR (2024) Human cytomegalovirus mRNA-1647 vaccine candidate elicits potent and broad neutralization and higher antibody-dependent cellular cytotoxicity responses than the gB/MF59 vaccine. J Infect Dis 230(2):455–466. https://doi.org/10.1093/infdis/jiad593

James C, Harfouche M, Welton NJ, Turner KM, Abu-Raddad LJ, Gottlieb SL, Looker KJ (2020) Herpes simplex virus: global infection prevalence and incidence estimates, 2016. Bull World Health Organ 98(5):315–329. https://doi.org/10.2471/BLT.19.237149

Kilpatrick AM (2011) Globalization, land use, and the invasion of West Nile virus. Science 334(6054):323–327. https://doi.org/10.1126/science.1201010

Kramer LD, Ciota AT, Kilpatrick AM (2019) Introduction, spread, and establishment of west Nile virus in the Americas. J Med Entomol 56(6):1448–1455. https://doi.org/10.1093/jme/tjz151

Krishnan R, Stuart PM (2021) Developments in vaccination for herpes simplex virus. Front Microbiol 12:798927. https://doi.org/10.3389/fmicb.2021.798927

Nishikawa JK, Sanchez PJ (2024) Congenital cytomegalovirus infection and hearing loss: it's time to screen. Otol Neurotol 45(10):e702–e709. https://doi.org/10.1097/MAO.0000000000004323

Patel PD, Alghareeb R, Hussain A, Maheshwari MV, Khalid N (2022) The association of Epstein-Barr virus with cancer. Cureus 14(6):e26314. https://doi.org/10.7759/cureus.26314

Pinninti SG, Kimberlin DW (2018) Neonatal herpes simplex virus infections. Semin Perinatol 42(3):168–175. https://doi.org/10.1053/j.semperi.2018.02.004

Pitisuttithum P, Gilbert P, Gurwith M, Heyward W, Martin M, van Griensven F, Hu D, Tappero JW, Choopanya K, Bangkok Vaccine Evaluation G (2006) Randomized, double-blind, placebo-controlled efficacy trial of a bivalent recombinant glycoprotein 120 HIV-1 vaccine among injection drug users in Bangkok, Thailand. J Infect Dis 194(12):1661–1671. https://doi.org/10.1086/508748

Plotkin S (2015) The history of vaccination against cytomegalovirus. Med Microbiol Immunol 204(3):247–254. https://doi.org/10.1007/s00430-015-0388-z

Plotkin SA, Furukawa T, Zygraich N, Huygelen C (1975) Candidate cytomegalovirus strain for human vaccination. Infect Immun 12(3):521–527. https://doi.org/10.1128/iai.12.3.521-527.1975

Rerks-Ngarm S, Pitisuttithum P, Nitayaphan S, Kaewkungwal J, Chiu J, Paris R, Premsri N, Namwat C, de Souza M, Adams E, Benenson M, Gurunathan S, Tartaglia J, McNeil JG, Francis DP, Stablein D, Birx DL, Chunsuttiwat S, Khamboonruang C, Thongcharoen P, Robb ML, Michael NL, Kunasol P, Kim JH, Investigators M-T (2009) Vaccination with ALVAC and AIDSVAX to prevent HIV-1 infection in Thailand. N Engl J Med 361(23):2209–2220. https://doi.org/10.1056/NEJMoa0908492

Rubin H (2011) The early history of tumor virology: Rous, RIF, and RAV. Proc Natl Acad Sci USA 108(35):14389–14396. https://doi.org/10.1073/pnas.1108655108

Ruiz-Pablos M, Paiva B, Montero-Mateo R, Garcia N, Zabaleta A (2021) Epstein-Barr virus and the origin of myalgic encephalomyelitis or chronic fatigue syndrome. Front Immunol 12:656797. https://doi.org/10.3389/fimmu.2021.656797

Rzymski P, Jibril AT, Rahmah L, Abarikwu SO, Hashem F, Lawati AA, Morrison FMM, Marquez LP, Mohamed K, Khan A, Mushtaq S, Minakova K, Poniedzialek B, Zarebska-Michaluk D, Flisiak R (2024) Is there still hope for the prophylactic hepatitis C vaccine? A review of different approaches. J Med Virol 96(9):e29900. https://doi.org/10.1002/jmv.29900

Shang Z, Li X (2024) Human cytomegalovirus: pathogenesis, prevention, and treatment. Mol Biomed 5:20241125. https://doi.org/10.1186/s43556-024-00226-7

Smatti MK, Al-Sadeq DW, Ali NH, Pintus G, Abou-Saleh H, Nasrallah GK (2018) Epstein-Barr virus epidemiology, serology, and genetic variability of LMP-1 oncogene among healthy population: an update. Front Oncol 8:211. https://doi.org/10.3389/fonc.2018.00211

Soldan SS, Lieberman PM (2023) Epstein-Barr virus and multiple sclerosis. Nat Rev Microbiol 21(1):51–64. https://doi.org/10.1038/s41579-022-00770-5

Spencer DA, Shapiro MB, Haigwood NL, Hessell AJ (2021) Advancing HIV broadly neutralizing antibodies: from discovery to the clinic. Front Public Health 9:690017. https://doi.org/10.3389/fpubh.2021.690017

Tsukiyama-Kohara K, Kohara M (2017) Hepatitis C virus: viral quasispecies and genotypes. Int J Mol Sci 19(1). https://doi.org/10.3390/ijms19010023

Vuitika L, Prates-Syed WA, Silva JDQ, Crema KP, Cortes N, Lira A, Lima JBM, Camara NOS, Schimke LF, Cabral-Marques O, Sadraeian M, Chaves LCS, Cabral-Miranda G (2022) Vaccines against emerging and neglected infectious diseases: an overview. Vaccines (Basel) 10(9). https://doi.org/10.3390/vaccines10091385

Walker BD (2016) AIDS vaccines. In: Vaccine book, 2nd edn, pp 401–422. https://doi.org/10.1016/B978-0-12-802174-3.00020-5

Wang X, Cottrell CA, Hu X, Ray R, Bottermann M, Villavicencio PM, Yan Y, Xie Z, Warner JE, Ellis-Pugh JR, Kalyuzhniy O, Liguori A, Willis JR, Menis S, Ramisch S, Eskandarzadeh S, Kubitz M, Tingle R, Phelps N, Groschel B, Himansu S, Carfi A, Kirsch KH, Weldon SR, Nair U, Schief WR, Batista FD (2024) mRNA-LNP prime boost evolves precursors toward VRC01-like broadly neutralizing antibodies in preclinical humanized mouse models. Sci Immunol 9(95):eadn0622. https://doi.org/10.1126/sciimmunol.adn0622

Xie Z, Lin YC, Steichen JM, Ozorowski G, Kratochvil S, Ray R, Torres JL, Liguori A, Kalyuzhniy O, Wang X, Warner JE, Weldon SR, Dale GA, Kirsch KH, Nair U, Baboo S, Georgeson E, Adachi Y, Kubitz M, Jackson AM, Richey ST, Volk RM, Lee JH, Diedrich JK, Prum T, Falcone S, Himansu S, Carfi A, Yates JR 3rd, Paulson JC, Sok D, Ward AB, Schief WR, Batista FD (2024) mRNA-LNP HIV-1 trimer boosters elicit precursors to broad neutralizing antibodies. Science 384(6697):eadk0582. https://doi.org/10.1126/science.adk0582

Yu H, Robertson ES (2023) Epstein-Barr virus history and pathogenesis. Viruses 15(3). https://doi.org/10.3390/v15030714

Zhong L, Krummenacher C, Zhang W, Hong J, Feng Q, Chen Y, Zhao Q, Zeng MS, Zeng YX, Xu M, Zhang X (2022) Urgency and necessity of Epstein-Barr virus prophylactic vaccines. NPJ Vaccines 7(1):159. https://doi.org/10.1038/s41541-022-00587-6

Zhu S, Viejo-Borbolla A (2021) Pathogenesis and virulence of herpes simplex virus. Virulence 12(1):2670–2702. https://doi.org/10.1080/21505594.2021.1982373

Epilogue

While the story of vaccines has evolved from the primitive beginnings in the 1700s to the sophisticated molecular approaches of the 2000s, the goal has been the same since Jenner's first cowpox vaccine: prevent disease and death from viral infections. Tens of thousands of scientists and physicians have dedicated their research to improving human health through prophylactic vaccines. From small academic labs to giant pharmaceutical companies, these dedicated researchers have crafted marvelous, life-saving vaccines against many of our most feared viral agents. Coupled with massive public health campaigns to promote widespread vaccine coverage, once fearsome diseases like smallpox and polio have disappeared and many others like measles, mumps, rubella, and chickenpox have become rare in most developed nations. Multiple generations have now escaped these diseases that commonly plagued their forebears. As these diseases fade into historical curiosities the current population has mostly forgotten about them. Modern parents often lack any sense of how prevalent and devastating these diseases were to children or about the crippling and fatal outcomes associated with several of these illnesses. This lack of familiarity can breed complacency and a reduced sense of urgency about the need for continued vaccine compliance. In the absence of an immediate threat, multiple vaccinations may seem a low priority or even unnecessary. Additionally, even the very low risks associated with a vaccine may evoke more concern than the seemingly nonexistent chance of contracting the actual disease. Unfortunately, except for smallpox, all the viral diseases that we vaccinate against still exist and are circulating on the planet. Visitors from endemic countries can reintroduce any of these viral agents into countries or regions that are currently disease-free. Only continuation of high vaccination levels to maintain individual and herd immunity keeps us all safe and healthy, so constant efforts to ensure public acceptance of routine vaccination are essential. Coupled with these public health efforts, science must be open and transparent about vaccine issues and continue to improve existing vaccines to reduce adverse effects, increase their safety, and enhance their protective efficacy.

Sadly, as science is making excellent progress with new vaccines and safer versions of older vaccines, public resistance to vaccines continues to grow. Skepticism is healthy and important, but it must be based on facts and data, not misinformation, distortions, and outright lies. In science, skepticism is one of the core tenets. We are taught to be critical thinkers who examine the evidence of each study and debate the conclusions. We try to find any flaws, errors, assumptions, or oversights that might invalidate the results and lead to wrong inferences about the data. Our work is publically shared at meetings and conferences where scientific colleagues can scrutinize and critique the results. Many manuscripts are now posted on preprint servers where anyone, scientist or nonscientist, can examine the work and post comments, suggestions, and evaluations of the quality and correctness of the study. Ultimately, studies are submitted to scientific journals where they are formally peer-reviewed by subject and content experts prior to publication. At each step, the studies are viewed, analyzed, and reviewed by qualified experts who try to ensure that only the best and most accurate studies make it into the scientific literature. The process is not perfect and flawed studies do get published (e.g. the Wakefield measles study—Chap. 7), but science tends to be self-correcting. Published work is accessible worldwide on the internet and will be read and studied by many researchers. If the study is important and impactful, its results will be tested, repeated, and extended to try to verify or disprove the validity of the results. Errors, flaws, and mistakes will eventually be exposed and corrected in subsequent published work. The process can be slow and cumbersome, but over time there is a slow accretion of valid information and a rejection of inaccurate and flawed data. As the sentences and paragraphs of the "book" of scientific knowledge are written, it is critical that we rely on this tome for the best and most reliable information available about vaccines. Our public and individual health is too important to be misled by falsehoods and inaccurate portrayals of the risks and benefits of these vital medical interventions. Anecdotes and hearsay are not science and do not belong in legitimate discussions about vaccine safety and necessity. There is plenty of room for honest dissent and debate about any vaccine and the tension between public good and personal autonomy, but the substance of the debate must be grounded in verified science. Hopefully, this book has provided some relevant background and accurate information about how vaccines are created, how they function, and how effective they have been in thwarting viral diseases. Let us all work towards a future free from pathogenic viruses where no one has to suffer, be crippled, or die needlessly.

Index

A
ACE2 receptor, 328
Acquired immunity, 112, 168, 355, 375
Acquired immunodeficiency syndrome (AIDS), 172, 205, 324, 384–386
Anthrax, 42–46, 54, 374
Attenuated vaccines, 11, 13, 33, 57, 82, 128, 141, 145, 147, 227–231, 249, 251, 252, 254, 259, 283, 355, 356, 379, 380, 387
Australia antigen, 195, 209, 216

B
BioNTech, 338, 339
Blumberg, Baruch (Barry), 195–197, 204

C
Cell cultures, 57, 58, 110, 133–135, 138, 141, 162, 171–173, 180, 181, 203, 219, 226, 227, 240, 267, 271, 274, 297, 329, 341, 353, 355, 356, 358, 375, 383

Cervical cancers, 294, 300–307, 309, 312–316
Chimpanzee coryza agent (CCA), 352, 353
Chronic infections, 172, 188, 189, 194, 197, 200, 202, 205, 207, 209, 223, 309, 376, 383, 384
Congenital rubella syndrome (CRS), 177–179, 182
Congenital varicella syndrome (CVS), 241
Cowpox, 6, 11, 12, 27–35, 39, 41, 45, 54, 238, 274, 374, 393
Cytomegalovirus (CMV), 239, 338, 378, 381, 382
Cytopathic effect (CPE), 136, 179, 180, 353

D
Dane particles, 197

E
Edmonston strain, 162, 164, 165
Enders, J.F., 134–136, 160–165, 171, 179, 180, 182, 268

Epstein-Barr virus (EBV), 239, 250, 302, 303, 378, 380–382
Exotic viruses, 387

F

Filterable agents, 74, 94, 121, 299, 374
Flexner, S., 123–126, 128–130
F proteins, 359–364, 366

G

Gardasil, 313–316
Genetic drifts, 106, 107
Genetic shift, 107
Germ theory, 41–43, 67, 70
German measles, 10, 136, 156, 175–178
G proteins, 359

H

Hepatitis B surface antigen (HBsAg), 197–199, 201, 203, 204, 206, 207, 231, 311, 359
Hepatitis C virus (HCV), 192, 209, 230, 377, 383–386
Hepatocellular carcinoma (HCC), 194, 199–202, 205, 207, 209
Herpes simplex virus (HSV), 245, 302, 378–380, 382
Hilleman, M.R., 163–165, 171–173, 181, 182, 204, 206, 225–227, 312, 356
HN types, 106
Human herpesvirus-3 (HHV-3), 239
Human immunodeficiency virus (HIV), 172, 324, 377, 384–386

I

IgM antibodies, 225
Immunoelectron microscopy (IEM), 220, 221

Inactivated vaccines, 11, 13, 14, 77, 82, 99, 111, 133, 141–143, 145, 227–230, 284, 310, 355–358, 362, 375, 380, 387
Innate immunity, 51, 336
Intussusception, 275–282, 284

J

Jenner, E., 27, 29–36, 39, 41, 46, 54, 56, 274, 374

K

Koch, R., 41–44, 67, 70, 71, 92, 119, 121, 126, 196, 197, 374

L

Latent infections, 242, 248, 378
Lipid nanoparticle (LNP), 336, 338, 339, 341, 342
Lynn, J. strain, 173

M

Memory cells, 168
Middle East respiratory syndrome (MERS), 325, 329, 331, 332, 339
Moderna, 339
Mosquito vectors, 73
mRNA vaccines, 14, 79, 82, 284, 333, 334, 336–343, 379, 382

N

NSP4, 271
1918 flu, 87–89, 91–95, 97, 106

O

Oka strain, 248–250, 253, 255, 258
Oncoproteins, 308, 309

P

Paralytic polio, 119, 121, 128, 131, 140, 142, 147, 238
Pasteur, L., 26, 32, 33, 41–47, 49, 52–56, 58, 67, 70, 75, 78, 93, 203, 374
Persistent viruses, 188
Plaque assay, 136–138, 144

R

Reassortment, 104–106, 268, 275
Reed, W., 67, 70–75, 97, 181
Restriction endonuclease mapping, 246, 247, 305
Reverse transcription, 189
Reversion, 13, 111, 146, 147, 250, 254
Rotarix, 281–283
RotaShield, 273–280, 340
RotaTeq, 279–283
Rubeola, 159, 161, 162, 167, 168, 175

S

Sabin, A., 125, 126, 129, 130, 133, 138, 144–148
Salk, J.E., 99, 129, 130, 133, 141–146
Serotypes, 133, 275, 280
17D strain, 79, 81, 96
Severe acute respiratory syndrome (SARS), 325, 328–332
Sexually transmitted diseases (STDs), 300–302, 307–309, 311, 312, 384–386
Shingles, 242, 243, 246, 250, 253, 259, 378
Shope, R.E., 94, 95, 299
Simian virus 40 (SV40), 143, 181, 250
Spike proteins, 327–329, 334, 338, 339, 363
Subunit vaccines, 11, 14, 15, 110, 188, 231, 232, 259, 284, 334, 338, 340, 359–363, 365, 379, 380

Syncytia, 162, 353, 359

T

Takahashi, M., 246–253
Theiler, M., 77–81, 99

V

Vaccinations, 2, 6, 7, 10, 11, 13, 14, 18, 20, 27–36, 39, 54, 55, 58, 79, 99, 103, 111–113, 118, 123, 127, 142, 143, 145–150, 157, 158, 160, 161, 165–168, 173–175, 182, 191, 202–204, 206–209, 225, 226, 230, 231, 250, 252–258, 277, 299, 309, 313–316, 324, 340, 343, 355–357, 364, 366, 376, 393, 394
Vaccine Adverse Events Reporting System (VAERS), 277, 278
Vaccine Associated Enhanced Disease (VAED), 358
Vaccine-associated paralytic polio (VAPP), 147–149
Vaccinia, 12, 33–35, 96, 142
Variola virus (VARV), 19, 21–25, 27–29, 34, 35
Variolation, 24–28, 30–33, 45, 54
Viremia, 19, 73, 74, 132, 170, 240, 242
Virion, 3, 50, 80, 101, 139, 193, 219, 240, 268, 296, 327, 358, 375
Virus-like particles (VLPs), 206, 284, 309–315, 381, 388
VP proteins, 4, 5, 9, 19, 50, 51, 80, 101, 104, 108, 222, 231, 232, 259, 268, 269, 272, 284, 285, 296, 308, 310, 375, 376, 381, 384, 385

W

Wakefield, A.F., 277, 394
Wakefield, A.J., 156–158
Warts, 294, 298–301, 303–305, 307, 312, 314, 315

Y

Yellow berets, 217–221

Z

Zoonotic diseases, 49, 52, 101, 331
zur Hausen, H., 301–305, 311, 316

GPSR Compliance

The European Union's (EU) General Product Safety Regulation (GPSR) is a set of rules that requires consumer products to be safe and our obligations to ensure this.

If you have any concerns about our products, you can contact us on

ProductSafety@springernature.com

In case Publisher is established outside the EU, the EU authorized representative is:

Springer Nature Customer Service Center GmbH
Europaplatz 3
69115 Heidelberg, Germany

www.ingramcontent.com/pod-product-compliance
Lightning Source LLC
LaVergne TN
LVHW010959250326
834688LV00003B/25